Chamomile
Industrial Profiles

Medicinal and Aromatic Plants — Industrial Profiles

Individual volumes in this series provide both industry and academia with in-depth coverage of one major genus of industrial importance.

Series Edited by Dr. Roland Hardman

Chamomile
Industrial Profiles

Edited by Rolf Franke and Heinz Schilcher

Medicinal and Aromatic Plants — Industrial Profiles

CRC Press
Taylor & Francis Group
Boca Raton London New York

CRC Press is an imprint of the
Taylor & Francis Group, an **informa** business
A TAYLOR & FRANCIS BOOK

CRC Press
Taylor & Francis Group
6000 Broken Sound Parkway NW, Suite 300
Boca Raton, FL 33487-2742

First issued in paperback 2019

ISBN-13: 978-0-415-33463-1 (hbk)
ISBN-13: 978-0-367-39280-2 (pbk)

Library of Congress Card Number 2004061667

Library of Congress Cataloging-in-Publication Data

Chamomile : industrial profiles / edited by Rolf Franke and Heinz Schilcher
     p. cm. -- (Medicinal and aromatic plants--industrial profiles ; v. 42)
  Includes bibliographical references and index.
  ISBN 0-415-33463-2 (alk. paper)
  1. German chamomile--Therapeutic use. I. Franke, Rolf. II. Series.

RS165.C24C48 2004
615'.32399--dc22
                                                      2004061667

Visit the Taylor & Francis Web site at
http://www.taylorandfrancis.com

and the CRC Press Web site at
http://www.crcpress.com

Preface to the Series

There is increasing interest in industry, academia, and the health sciences in medicinal and aromatic plants. In passing from plant production to the eventual product used by the public, many sciences are involved. This series brings together information that is currently scattered through an ever-increasing number of journals. Each volume gives an in-depth look at one plant genus, about which an area specialist has assembled information ranging from the production of the plant to market trends and quality control.

Many industries are involved, such as forestry, agriculture, chemicals, food, flavor, beverage, pharmaceutical, cosmetics, and fragrance. The plant raw materials are roots, rhizomes, bulbs, leaves, stems, barks, wood, flowers, fruits, and seeds. These yield gums, resins, essential (volatile) oils, fixed oils, waxes, juices, extracts, and spices for medicinal and aromatic purposes. All these commodities are traded worldwide. A dealer's market report for an item may say "drought in the country of origin has forced up prices."

Natural products do not mean safe products, and account of this has to be taken by the above industries, which are subject to regulation. For example, a number of plants that are approved for use in medicine must not be used in cosmetic products.

The assessment of "safe to use" starts with the harvested plant material, which has to comply with an official monograph. This may require absence of, or prescribed limits of, radioactive material, heavy metals, aflatoxin, pesticide residue, as well as the required level of active principle. This analytical control is costly and tends to exclude small batches of plant material. Large-scale, contracted, mechanized cultivation with designated seed or plantlets is now preferable.

Today, plant selection is not only for the yield of active principle, but for the plant's ability to overcome disease, climatic stress, and the hazards caused by mankind. Such methods as *in vitro* fertilization, meristem cultures, and somatic embryogenesis are used. The transfer of sections of DNA is giving rise to controversy in the case of some end uses of the plant material.

Some suppliers of plant raw material are now able to certify that they are supplying organically farmed medicinal plants, herbs, and spices. The Economic Union directive CVO/EU No. 2092/91 details the specifications for the *obligatory* quality controls to be carried out at all stages of production and processing of organic products.

Fascinating plant folklore and ethnopharmacology lead to medicinal potential. Examples are the muscle relaxants based on the arrow poison, curare, from species of *Chondrodendron*, and the antimalarials derived from species of *Cinchona* and *Artemisia*. The methods of detection of pharmacological activity have become increasingly reliable and specific, frequently involving enzymes in bioassays and avoiding the use of laboratory animals. By using bioassay-linked fractionation of crude plant juices or extracts, compounds can be specifically targeted which, for example, inhibit blood platelet aggregation, or have antitumor, or antiviral, or any other required activity. With the assistance of robotic devices, all the members of a genus may be readily screened. However, the plant material must be fully authenticated by a specialist.

The medicinal traditions of ancient civilizations such as those of China and India have a large armamentarium of plants in their pharmacopoeias that are used throughout Southeast Asia. A similar situation exists in Africa and South America. Thus, a very high percentage of the world's population relies on medicinal and aromatic plants for their medicine. Western medicine is also responding. Already in Germany all medical practitioners have to pass an examination in phytotherapy before being allowed to practice. It is noticeable that medical, pharmacy, and health-related schools throughout Europe and the United States are increasingly offering training in phytotherapy.

Multinational pharmaceutical companies have become less enamored of the single compound, magic-bullet cure. The high costs of such ventures and the endless competition from "me-too" compounds from rival companies often discourage the attempt. Independent phytomedicine companies have been very strong in Germany. However, by the end of 1995, 11 (almost all) had been acquired by the multinational pharmaceutical firms, acknowledging the lay public's growing demand for phytomedicines in the Western world.

The business of dietary supplements in the Western world has expanded from the health store to the pharmacy. Alternative medicine includes plant-based products. Appropriate measures to ensure their quality, safety, and efficacy either already exist or are being answered by greater legislative control by such bodies as the U.S. Food and Drug Administration and the recently created European Agency for the Evaluation of Medicinal Products based in London.

In the United States, the Dietary Supplement and Health Education Act of 1994 recognized the class of phytotherapeutic agents derived from medicinal and aromatic plants. Furthermore, under public pressure, the U.S. Congress set up an Office of Alternative Medicine, which in 1994 assisted the filing of several Investigational New Drug (IND) applications, required for clinical trials of some Chinese herbal preparations. The significance of these applications was that each Chinese preparation involved several plants and yet was handled as a *single* IND. A demonstration of the contribution to efficacy of *each* ingredient of *each* plant was not required. This was a major step toward more sensible regulations in regard to phytomedicines.

My thanks are due to the staff of CRC Press who have made this series possible and especially to the volume editors and their chapter contributors for the authoritative information.

Dr. Roland Hardman

Preface

For more than 2000 years, preparations of chamomile flowers count among the medicinal treasures of various cultural groups. Since ancient times, the chamomile has survived the storms of time as well as different trends in the art of healing throughout the world.

It is certainly one of the most fascinating medicinal plants of our globe, although the "True chamomile," often called "German chamomile," is native only to Europe and the Near East, but naturalized in many other regions too. There are only a few medicinal plants with a millennium-lasting successful therapeutic use that can claim to be part of a wide interdisciplinary scientific research.

Over 100 years ago, the first attempts in cultivation and breeding were made and these are still up to date. In 1921, Chemiewerke Homburg received the patent for the first chamomile extract. Today, university and industrial research groups work on the optimal extraction and optimal stability of the ingredients that have an influence on the efficacy in different galenic preparations. The relatively good findings on hydrophilic and lipophilic ingredients are not finalized yet, as the recent analytic results show.

As other modern pharmacopoeae, the *European Pharmacopoeia* in its 5th edition (published in February 2004) determines minimum and maximum values for three ingredients that have an influence on the efficacy of the essential chamomile oil and require a test for the so-called "chromatographic profile" by gas chromatography (GCP).

Chamomile flowers are also interesting objects for the research of the biosynthesis of mono- and sesquiterpenes.

Of great interest are naturally the numerous tests on efficacy and safety studies of chamomile flower preparations. They shall serve as scientifically well-founded confirmations of therapeutic reports from prescientific times and also for precision of application fields and correct dosage. Here, too, chamomile flower preparations and essential chamomile flower oil have a special position.

In the 1930s, pharmacological studies with chamazulen and chamomile flower extracts were realized and were confirmed later on with modified testing methods of a more recent date. Only very few medicinal plants exist, of which a similarly high number of qualified pharmacological or experimental studies are available. There are also a surprising number of clinical studies; most of which, however, do not correspond to the GCP test design, as they were conducted before the GCP test design was developed in 1994. Whether these clinical studies, realized between 1960 and 1992, are less meaningful is questionable because they are of high scientific level. A similar case is the reported allergenic potential of chamomile flowers, particularly if the myriad uses of chamomile flower preparations are taken into account.

When Roland Hardman suggested that we edit a book for his series *Medicinal and Aromatic Plants — Industrial Profiles*, we accepted with pleasure, although we are aware that "chamomile" has many faces. Therefore, it was our aim to involve various competent persons from different specialized fields and countries.

The present compendium, *Chamomile: Industrial Profiles,* provides an interdisciplinary inventory of the scientific level of knowledge about "True chamomile" as well as Roman chamomile and shall discuss controversial questions too.

We would like to thank the co-authors for their factual, well-founded articles, the project coordinators for their friendly support, and CRC Press for the printing of this compendium. Finally, we would like to thank all the people who contributed to the success of the cultivation and use of this fascinating plant.

Rolf Franke and Heinz Schilcher
Munich, April 2005

Contributors

Dr. sc. Rolf Franke, Salus Haus Dr. med. Otto Greither Nachf. GmbH & Co. KG, Bahnhofstraße 24, D-83052 Bruckmühl/Obb., Germany
e-mail: rolf.franke@salus.de

Prof. Dr. rer. nat. Dr. med h.c. Heinz Schilcher, Zaumberg 25, D-87509 Immenstadt/Allgäu, Germany
e-mail: schilcher_h@hotmail.com

Prof. Dr. Jenö Bernáth, Corvinus University, Faculty of Horticultural Sciences, Department of Medicinal and Aromatic Plants, Villányi str., 29, H-1118 Budapest, Hungary
e-mail: jbernath@omega.kee.hu

Prof. Dr. Horst Böttcher, Institut für Ernährungswissenschaften der Martin-Luther-Universität Halle-Wittenberg, EmilAbderhaldenStr. 25 b, D-06108 Halle (Saale), Germany
e-mail: Guenther@landw.uni-halle.de

Prof. Dr. Reinhold Carle, Hohenheim University, Institute of Food Technology, Department of Plant Foodstuff Technology, Garbenstraße 25, D 70599 Stuttgart, Germany
e-mail: carle@uni-hohenheim.de

Tamer Fahmi, 32 Abdallah Ben Taher St., Nasr City, Cairo, Egypt
e-mail: tfahmi@hotmail.com

Norberto R. Fogola, Av. Gral. Chenaut 1757 Piso 9 B, RA-1426 Buenos Aires, Argentina
e-mail: nrfogola@vianw.com.ar

Dr. Susanne Goeters, c/o Boehringer Ingelheim Pharma GmbH & Co. KG, Birkendorfer Straße 65, D-88397 Biberach/Riß, Germany
e-mail: Susanne.Goeters@bc.boehringer-ingelheim.com

Prof. Dr. Peter Imming, Institut für Pharmazeutische Chemie, Martin-Luther-Universität, Halle-Wittenberg, Wolfgang-Langenbeck-Str. 4, D-06120 Halle (Saale), Germany
e-mail: peter.imming@pharmazie.uni-halle.de

Dušan Jedínak, Mierova 20, SK-06401 Stará L'ubovňa, Slovakia

Ingeborg Günther, Institut für Ernährungswissenschaften der Martin-Luther-Universität Halle-Wittenberg, Emil-Abderhalden-Str. 25 b, D-06108 Halle (Saale), Germany
e-mail: Guenther@landw.uni-halle.de

Dr. Hans-Jürgen Hannig, Martin Bauer GmbH & Co. KG, Dutendorfer Straße 5-7, D-91487 Vestenbergsgreuth, Germany
e-mail: hans-juergen.hannig@martin-bauer.de

Dr.-Ing. Albert Heindl, Marktplatz 5, D-84048 Mainburg, Germany
e-mail: heindl-gmbh@t-online.de

Jozef Holubář, Ústřední kontrolní a zkušební ústav zemědělský, Hroznova 3, CZ-65606 Brno, Czech Republic

Prof. Dr. Éva Németh, Corvinus University, Department of Medicinal and Aromatic Plants; Villányi str., 29, H-1118 Budapest, Hungary
e-mail: eva.nemeth@uni-corvinus.hu

Dr. Viliam Oravec, 17. Novembra 32, SK-064 01 Stará L'ubovňa, Slovakia
e-mail: vilora_sk@hotmail.com

Viliam Oravec, Jr., Helsinska 5, SK-04000 Košice, Slovakia
e-mail: gaia_sk@hotmail.com

Dr. Andreas Plescher, Pharmaplant GmbH, Straße am Westbahnhof, D-06556 Artern, Germany
e-mail: info@pharmaplant.de

Asst. Prof. Dr. Miroslav Repčák, Department of Experimental Botany and Genetics, Prírodovedecká fakulta, Universitat P.J. Šafarik, Mánesova 23, SK-04154 Košice, Slovakia
e-mail: repcak@kosice.upjs.sk

L'ubomír Šebo, 17. Novembra 7, SK-06401 Stará L'ubovňa, Slovakia

Ivan Varga, Veterna 11, SK-92027 Hlohovec, Slovakia

Eduardo Weldt S., Puelche S.A., P.O. Box 902, Los Angeles, Chile
e-mail: e.weldt@puelche.cl

Table of Contents

1 Introduction

Rolf Franke

The medicinal plant of chamomile and the species related to it are of manifold interest as both the drug *Matricariae flos* (*Ph. Eur.* 4.6, 2004) [1] and the native plants. Cultivated and wild plants offer a number of subjects that belong to the range of basic and applied research; on the other hand, they are of great economic interest as well.

Considering the fact that the traded chamomile flowers show a very heterogeneous spectrum of content [2, 3], a drug bound to a pharmacopoeia can — if it is to be recognized by the medical sciences in the rational sense — only reach a high standard relating to natural science provided there is a comprehensive knowledge about the biology of the medicinal plant and the species related to it, as well as the drug obtained from it.

Chamomile flowers belong to those drugs that experienced a wide medical application in ancient times [2, 3]. The curative effect of chamomile has been known by physicians for about 2500 years. Hippocrates gives a description of the drug in the 5th century B.C., and chamomile appears as a medicinal plant in the work *De Materia Medica* written by Dioscorides (1st century A.D.). Galen and Asclepios describe the application of a chamomile infusion at some length. In Palladius' writings dating back to the 4th or 5th century, notes about chamomile are to be found as well. Medical applications continued in the Middle Ages. Bock, for example, gives the following report about chamomile in his *Kreutterbuch* of 1539 (described as "gantz gemein Chamill": "quite ordinary/common chamomile"): "Es ist bei allen menschen kein breuchlicher kraut in der artznei als eben Chamillenblumen/dann sie werden beinahe zu allen presten gebraucht." (In Old High German, "There is no herb in medicine for people being more usual than chamomile flowers because they are used against nearly all kinds of ailments.")

From the *New Kreuterbuch* written by Mathiolus (1626) it may be gathered that "...das Kamillenöl dienet sonderlich wol wider den krampf" ("chamomile oil serves as a remedy against convulsion"). Tabernaemontanus mentions chamomile in 1664 [4]: "...zu mancherley gebrauch in der Artzency/als in Pflaster/Salben/Behung/Säcklein/Bäder und dergleichen nützlich gebraucht wird/und vielerley Artzeneyen darauß bereiten" ("for manifold use in medicine such as in plasters/ointments/pouches, [medicinal] baths and for other useful purposes, many different medicines can be prepared out of it"). In his *Medizinisches Lexikon* (*Medical Encyclopaedia*, 1755) von Haller makes honorable mention of the pain-relieving and spasmolytic effect of chamomile flowers, and in his *Praktische Arzneimittellehre* (1814, Vol. 1) (*Practical Pharmacology*), Hecker as well as Hufeland and colleague Collenbuch report the successful use of chamomile preparations in cases of ulcerations (in *Hufeland's Journal*) [5]. Saladin von Asculum mentioned the blue volatile oil of chamomile in 1488 and Hieronimus Brunschwig described the distillation of the volatile chamomile oil in 1500. High appreciation is also given to chamomile preparations in the various works written by Sebastian Kneipp (1821–1897) [6]. In the first edition of the *Lehrbuch der Phytotherapie* (*Textbook of Phytotherapy*) (1942) Weiss refers to chamomile as being one of the most significant medicinal plants, giving wider space to it in the review [7].

The summary of a choice of internationally known pharmacopoeias (see Chapter 2, Table 2.1) shows that since the publication of the *German Pharmacopoeia* in 1882, chamomile flowers and preparations from chamomile flowers have been included in a number of pharmacopoeias. In the current pharmacopoeia for the Federal Republic of Germany, the drug called *Matricariae flos* is to be found in the *European Pharmacopoeia* 4.6, edition 2004 [1].

Roman chamomile has also been significant as a drug from ancient times. Tabernaemontanus had already known filled and unfilled forms in 1664 [4]. This Roman chamomile is included in a number of pharmacopoeias and also in the *European Pharmacopoeia*, Suppl. 4.3 2003 (see Chapter 2, Table 2.2).

In 1956 Heeger [8] reported that true chamomile has been worked on to a small extent [only] and that it is "on its way to becoming a culture plant." In the past 40 years chamomile has developed into a real culture plant that is cultivated on a wide scale.

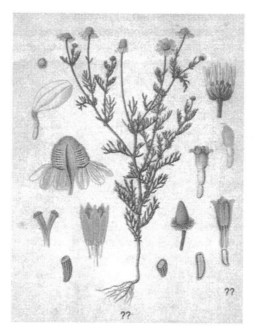

FIGURE 1.1 True chamomile (*Matricaria recutita* L.). (Fuchs, L. [1543] *New Kreüterbuch*. Basel. With permission of Botanical Garden and Botanical Museum, Berlin-Dahlem.)

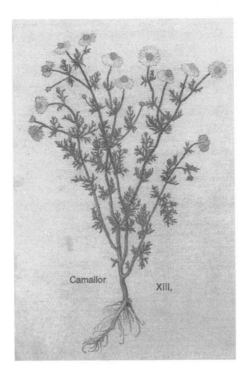

FIGURE 1.2 The capitulum of the True chamomile has a cone-shaped, hollow bottom. (Köhler [1987] *Medizinal-Pflanzen in naturgetreuen Abbildungen mit kurz erläuterndem Text.* Gera. Vol. With permission of Botanical Garden and Botanical Museum, Berlin-Dahlem.)

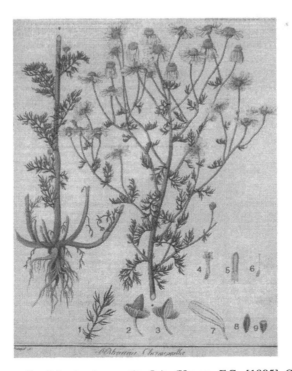

FIGURE 1.3 True chamomile (*Matricaria recutita* L.). (Hayne, F.G. [1805] *Getreue Darstellung und Beschreibung der in der Arzneykunde gebräuchlichen Gewächse.* Berlin. Vol. 1. With permission of Botanical Garden and Botanical Museum, Berlin-Dahlem.)

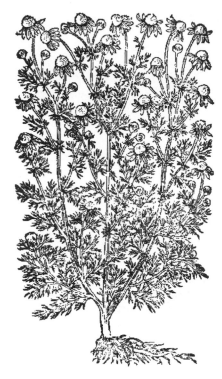

FIGURE 1.4 True chamomile (*Matricaria recutita* L.) [4].

FIGURE 1.5 Roman chamomile (*Chamaemelum nobile* [L.] All.). (Chaumeton, F.P. [1815] *Flore médicinale.* Paris. Vol. 2. With permission of Botanical Garden and Botanical Museum, Berlin-Dahlem.)

FIGURE 1.6 Roman chamomile (*Chamaemelum nobile* [L.] All.) (*Tabernaemontanus*, 1664 p. 58) [4].

FIGURE 1.7 Roman chamomile (Chamaemelum nobile [L.] All.) (*Tabernaemontanus*, 1664 p. 59) [4].

REFERENCES

1. *European Pharmacopoeia* 4.6 (edition 2004) Deutscher Apotheker Verlag, Stuttgart, Germany.
2. Isaac, O., Schimpke, H. (1965) *Mitt. Dtsch. Pharmazeut. Ges.*, 35, 133.
3. Schilcher, H. (1973) Neuere Erkenntnisse bei der Qualitätsbeurteilung von Kamillenblüten bzw. Kamillenöl — Einteilung der Handelskamille in vier chemische Typen. *Planta Medica*, 23, 132.
4. Tabernaemontanus, J.T. (1664) *New vollkommenlich Kräuter-Buch,* Jacob Werenfels, Basel, 1529 pp.
5. *Hufelands Journal*, ref. in Madaus, G. (1979) *Lehrbuch der biologischen Heilmittel.* Band I.G. Olms, Hildesheim, New York.
6. Kneipp, S. (1882) *Öffentliche Vorträge*, p. 232 and (1985) p. 46; (1866) *Meine Wasserkur,* p. 137; (1893) *Kneippblätter,* p. 52; (1889) *Ratgeber für Gesunde und Kranke*, p. 283; (1894) *Mein Testament*, p. 122.
7. Weiss, R.F. (1942) *Lehrbuch der Phytotherapie,* Hippokrates, Stuttgart, Germany.
8. Heeger, E.F. (1956) *Handbuch des Arznei- und Gewürzpflanzenanbaus. Drogengewinnung.* Deutscher Bauernverlag, Berlin, pp. 186, 494–495.

2 Legal Situation of German Chamomile: Monographs

Heinz Schilcher

CONTENTS

2.1 OVERVIEW OF THE PHARMACOPOEIAS: MONOGRAPHS

TABLE 2.1
Pharmacopoeias Containing Monographs on Chamomile Flowers, Chamomile Oil, or Chamomile Preparations (e.g., Extracts)

Year	Pharmacopoeia	Part of the Plant
1882	*DAB 2 – German Pharmacopoeia* 2nd edition	Flowers
1893	*Pharmacopoeia Helvetica* 3rd edition	Flowers and volatile oil
1894	*DAB 3 – German Pharmacopoeia* 3rd edition	Flowers
1897	Supplement for *DAB 3*, drugs not being included in *DAB 3*, 2nd edition	Volatile oil, extract with fatty oil, volatile oil combined with lemon oil, spissum extract
1900	*DAB 4 – German Pharmacopoeia* 4th edition	Flowers
1901	*Svenska Farmakopen* 8	Flowers and spissum extract
1905	*Pharmacopoeia of the USA,* 1900 edition	Flowers under the name *"Matricaria"*
1905	*Pharmacopoea Nederlandica* 4 (Dutch Ph)	Flowers
1905	*Farmacopea Española* 7	Flowers under the name *"Manzanilla"* and extract with fatty oil
1906	3rd supplementary volume for *DAB 4*	Volatile oil, extract with fatty oil, volatile oil combined with lemon oil, spissum extract
1906	*Pharmacopée Belgicae* 3	Flowers under *Flores Chamomillae* that originate from *Anthemis nobilis* (!)
1906	*Pharmacopoea Austria* 8	Flowers
1907	*Pharmacopoeia of Japan,* English edition	Flowers
1908	Pharmacopée Française	Flowers of *Arthemis nobilis* (!) extract from fatty oil, extract with camphor oil from *A. nobilis*
1910	*DAB 5* German Pharmacopoeia 5th edition	Flowers
1916	*DAB 5* – 4th supplementary volume	Volatile oil, extract with fatty oil, volatile oil combined with lemon oil, spissum extract
1925	*Svenska Farmacopen* X	Flowers
1926	*DAB 6 – German Pharmacopoeia* 6th edition	Flowers with a volatile oil content of 0.4%
1934	*HAB* – Dr. Willmar Schwabe	Fresh plant collected at flowering time
1939	*Den Norske Farmakopo* V (Norwegian PhP)	Flowers
1940	*Pharmacopoea Nederlandica* 5 (2nd print)	Flowers
1941	*DAB 6* – Supplementary volume	Volatile oil, extract with fatty oil, fluid extract, chamomile tincture, chamomile syrup, chamomile water
1941	*Pharmacopoea Helvetica* 5	Flowers and volatile oil
1941	Farmacopea Chilena	Flowers
1946	*Svenska Farmakopen* XI (Sweden)	Flowers
1946	*Farmacopea Portuguesa* 46	Flowers
1948	Pharmacopoea Danica	Flowers
1953	*Pharmacopée Belge* V	Flowers, volatile oil, chamomile water and (ethyl) alcohol
1953	Indian Pharmaceutical Codex	Flowers
1954	*Farmacopea oficial Española* IX	Flowers with an volatile oil content of 0.4%
1958	*Nederlandse Pharmacopee* 6	Flowers with an volatile oil content of 0.6%
1959	Farmacopeia dos Estados Unidos do Brasil	Flowers
1960	*Österreichisches Arzneibuch 9 – Austrian Pharmacopoeia* 9	Flowers with an volatile oil content of 0.4%
1962	*Pharmacopée Belge* VI	Flowers
1965	*Farmakopea Polska* IV (Polish Ph.)	Flowers

TABLE 2.1
**Pharmacopoeias Containing Monographs on Chamomile Flowers, Chamomile Oil, or
Chamomile Preparations (e.g., Extracts) (continued)**

Year	Pharmacopoeia	Part of the Plant
1966	*Farmacopea Argenina* 6	Flowers
1967	*Pharmacopoea Hungarica* VI	Flowers with a volatile oil content of 0.4%
1967	*Martindale* 25 (*The Extra Pharmacopoeia*, London)	Flowers with a volatile oil content of 0.4%
1968	*DAB 7* – BRD – *German Pharmacopoeia of the FRG*, 7th edition	Flowers with a volatile oil content of 0.4%
1971	*Pharmacopoea Helvetica* 6	Flowers with a volatile oil content of 0.5% and fluid extract
1971	*Nederlandse Pharmacopee* 8	Flowers
1972	*Farmacopea ufficiale della Repubblica Italiana* VIII	Flowers
1973	*Pharmacopée Française* 9	Flowers with a volatile oil content of 0.5%
1974	Estra farmakope Indonesia	Flowers
1975	*DAB 7* – DDR – *German Pharmacopoeia of the GDR*, 7th edition (2nd issue)	Flowers and fluid extract
1975	*Europ. Pharmacopoeia*, 1st edition, volume III	Flowers with a volatile oil content of 0.4%
1976	*Pharmacopoea Bohemeslovenica* III (addendum) (Czechoslovakia)	Flowers
1976	*Farmacopea Romana* IX (Rumania)	Flowers
1978	*DAB 8* – BRD – *German Pharmacopoeia of the FRG*, 8th edition	Flowers with a volatile oil content of 0.4%
1978	1st official edition of the *Homeopathic Pharmacopoeia* with 4th addendum of 1985	Fresh whole plants of *Chamomilla recutita* (L.) Rauschert collected at flowering time
1980	British Pharmacopoeia	Flowers of *Arthemis nobilis* (!) under "Chamomile Flowers" with a volatile oil content of 0.7%
1981	*Österreichisches Arzneibuch 1981* – *Austrian Pharmacopoeia* 1981	Flowers, volatile oil, chamomile fluid extract, and tincture
1982	*Standard Admission* § 36 AMG 76	Flowers with a volatile oil content of 0.4%
1982	*Martindale* 28	Flowers with a volatile oil content of 0.4%
1983	British Herbal Pharmacopoeia	Flowers
1984	Egyptian Pharmacopoeia	Flowers
1984	*Farmacopea Jugoslavica* IV	Flowers
1985	*African Pharmacopoeia* 1st ed.	Flowers
1986	*DAB 9* – BRD – *German Pharmacopoeia of the FRG*, 9th edition	Flowers with a volatile oil content of 0.4%
1986	*Pharmacopea Hungarica* VII	Flowers and volatile oil
1987	*Pharmacopea Helvetica* VII	Flowers with a volatile oil content of 0.4%, standardized chamomile fluid extract with a volatile oil content of 0.12–0.18%
1988	*Pharmacopée Française* X	Fresh flowers for homeopathic preparation
1989	*Martindale* (London)	Flowers and volatile oil
1990	Österreichisches Arzneibuch – Austrian Pharmacopoeia	Flowers with a volatile oil content of 0.4%, volatile oil, chamomile fluid extract (with a volatile oil content of 0.3%), and tincture
1990	*Commission E Monograph* (revised version) (Germany)	Flowers
1990	*British Herbal Pharmacopoeia*, (Volume 1)	Flowers
1992	*DAB 10* – *German Pharmacopoeia*, 10th edition, 1st addendum	*Chamaemelum nobile* (L.) All. with not less than 0.7% of volatile oil

TABLE 2.1
Pharmacopoeias Containing Monographs on Chamomile Flowers, Chamomile Oil, or Chamomile Preparations (e.g., Extracts) (continued)

Year	Pharmacopoeia	Part of the Plant
1992	*British Herbal Compendium* (Volume 1)	Flowers
1993	*DAB 10 – German Pharmacopoeia*, 10th edition, 2nd addendum	Flowers with a minimum volatile oil content of 0.4%
1993	*Martindale* (London)	Flowers
1993	Pharmacopée Française	Flowers
1993	*Pharmacopoeia Helvetica* VII (Addendum)	Flowers
1994	Austrian Pharmacopoeia	Fluid extract/tincture
1995	Polish Pharmacopoeia	Flowers
1996	*British Herbal Pharmacopoeia* 1996	Flowers
1996	Pharmacopée Française	Flowers
1996	*Martindale* (London), 31st edition	Flowers and volatile oil
1997	*European Pharmacopoeia*, 3rd edition	Flowers with a minimum volatile oil content of 0.4%
1997	*DAB 1997* – German Pharmacopoeia 1997	Volatile oil, chamomile liquid extract (with a volatile oil content of min. 0.3%)
1997	*Real Farmacopea Española*, 1st edition	Flowers with a minimum volatile oil content of 4 ml/kg
1997	*European Pharmacopoeia* (valid until 2001)	Flowers
1999	ESCOP Monograph	Flowers
1999	WHO Monograph	Flowers
1999	*Martindale* (London), 32nd edition	Flowers and volatile oil
1999	*DAB 1999* – German Pharmacopeia	Liquid extract/volatile oil
1999	USP 24/NF 19	Flowers
2000	British Pharmacopoeia	Flowers of *Chamaemelum nobile* (L.) All. with not less than 7 ml/kg of volatile oil
2001	*European Pharmacopoeia* (Addendum 2001)	Liquid extract with 0.3% blue volatile oil
2001	USP 24/NF 19 (Addendum 3)	Flowers (changes to read)
2002	European Pharmacopeia	Flowers/liquid extract
2003	USP 26/NF 21	Flowers
2003	*European Pharmacopoeia* 4.5	Liquid extract/volatile oil
2004	*European Pharmacopoeia* 4.6 (since 01/2004)	Flowers

TABLE 2.2
Some Pharmacopoeias Containing Monographs on Roman Chamomile Flowers, Roman Chamomile Oil, or Roman Chamomile Preparations (e.g., Extracts)

Year	Pharmacopoeia	Part of the plant
1906	*Pharmacopée Belgicae* 3	Monograph title: *Flores Chamomillae* originating from *Anthemis nobilis*
1908	Pharmacopée Française	Flowers of *Arthemis nobilis*
1934	*HAB* – Dr. Willmar Schwabe	Fresh total plant from *Anthemis nobilis*
1940	*Nederlandsche Pharmacopee,* 5th edition	Flowers
1941	*Pharmacopoeia Helvetica,* 5th edition	Flowers
1953	*Pharmacopée Belge* V	Flowers under *Chamomillae flos* that originate from *Anthemis nobilis*
1954	British Pharmaceutical Codex	Flowers of *Anthemis nobilis* L. under the name Chamomile, content of volatile oil not less than 0.4% (v/w)
1954	*Farmacopea oficial Española* IX (Hisp IX)	Flowers
1954	*Farmakopea Polska* III (Polish Ph.)	Flowers as *Anthodium anthemidis*
1959	Farmacopeia dos Estados Unidos do Brasil	Flowers
1960	*Österreichisches Arzneibuch* (ÖAB 9) – *Austrian Pharmacopoeia*	Flowers with a content of volatile oil not less than 0.7% (v/w)
1972	*Farmacopea ufficiale della Repubblica Italiana* VIII	Flowers under *Camomilla romana* (*Anthemidis flos*)
1976	British Herbal Pharmacopoeia	Flowers
1976	*Pharmacopée Française* IX	Flowers
1979	*Pharmacopoeia Helvetica* VI	Flowers
1980	British Pharmacopoeia	Flowers of *Anthemis nobilis* L. under the name Chamomile flowers, content of volatile oil not less than 0.7% (v/w)
1981	*Österreichisches Arzneibuch* 1981 – *Austrian Pharmacopoeia* 1981	Flowers
1983	British Herbal Pharmacopoeia	Flowers
07/87–01/93	*Pharmacopée Française* X	Flowers
1987	*Pharmacopea Helvetica* VII	Flowers
1988	British Pharmacopoeia	Flowers of *Anthemis nobilis* L. under the title Chamomile flowers, content of volatile oil not less than 0.7% (v/w)
1992	*DAB 10* – *German Pharmacopoeia,* 10th edition	Flowers
1993	*Pharmacopoeia Helvetica* VII	Flowers
1993	British Pharmacopoeia	Flowers of *Anthemis nobilis* L. under the title Chamomile flowers, content of volatile oil not less than 0.7% (v/w)
1997	*European Pharmacopoeia,* 3rd edition	Flowers with a content of volatile oil not less than 0.7% (v/w)
1999	British Pharmacopoeia	Flowers of *Chamaemelum nobile* (L.) All. under the title Chamomile flowers, content of essential oil not less than 7 mg/kg
2003	*European Pharmacopoeia* 4.3 (German version)	Flowers

2.2 OVERVIEW OF THE TEST REGULATIONS IN PHARMACOPOEIAS

An overview of the test regulations for chamomile flowers and chamomile preparations in pharmacopoeias since 1882 and on some pharmacopoeias containing monographs on Roman chamomile flowers, Roman chamomile oil, or Roman chamomile preparations (e.g., extracts) is given in Chapter 10, "Chemical Analysis of the Active Principles of Chamomile" (see Tables 10.1 and 10.2).

2.3 OVERVIEW OF THE QUANTITATIVE DETERMINATION OF ESSENTIAL OIL IN PHARMACOPOEIAS

An overview of the quantitative determination of the essential oil in Chamomile flowers by means of steam distillation is given in Chapter 10, "Chemical Analysis of the Active Principles of Chamomile" (see Table 10.3).

2.4 MONOGRAPHS OF THE EUROPEAN PHARMACOPOEIA

2.4.1 MATRICARIA FLOWER — MATRICARIAE FLOS

European Pharmacopoeia 4.6 [16]
Matricaria flower
Matricariae flos

2.4.1.1 Definition

Dried capitula of *Matricaria recutita* L. (*Chamomilla recutita* [L.] Rauschert)

2.4.1.2 Content

- Blue essential oil: minimum 4 ml/kg (dried drug)
- Total apigenin-7-glucoside ($C_{21}H_{20}O_{10}$): minimum 0.25% (dried drug)

2.4.1.3 Characteristics

Macroscopic and microscopic characters described under identification tests A and B

2.4.1.4 Identification

A. Capitula, when spread out, consists of an involucre made up of many bracts arranged in one to three rows; an elongated-conical receptacle, occasionally hemispherical (young capitula); 12 to 20 marginal ligulate florets with a white ligule; several dozen yellow central tubular florets. The involucre bracts are ovate to lanceolate, with a brownish-grey scarious margin. The receptacle is hollow, without paleae. The corolla of the ligulate florets has a brownish-yellow tube at the base extending to form a white, elongated-oval ligule. The inferior ovary is dark brown, ovoid to spherical, and has a long style and bifid stigma. The tubular florets are yellow and have a five-toothed corolla tube; five syngenesious, epipetalous stamens; and a gynoecium similar to that of the ligulate florets.
B. Separate the capitulum into its different parts. Examine under a microscope using chloral hydrate solution R. The bracts have thin-walled cells and a central region composed of elongated sclereids with occasional stomata. The inner epidermis of the corolla of the ligulate florets, in surface view, consists of thin-walled, polygonal cells, slightly papillose;

those of the outer epidermis are markedly sinuous and strongly striated; corolla of the tubular florets with longitudinally elongated epidermal cells; and with small groups of papillae near the apex of the lobes. Glandular trichomes each consists of a short stalk and a head of two to three tiers of two cells each occur on the outer surfaces of the bracts and on the corollas of both types of florets. The ovaries have a sclerous ring at the base and the wall is composed of vertical bands of thin-walled, longitudinally elongated cells with numerous glandular trichomes, alternating with fusiform groups of small, radially elongated cells containing mucilage. The cells at the apex of the stigmas are extended to form rounded papillae. Numerous small, cluster crystals of calcium oxalate occur in the inner tissues of the ovaries and the anther lobes. Pollen grains are spherical to triangular, about 30 μm in diameter with three pores and a spiny exine.

C. Thin-layer chromatography (TLC) (2.2.27)

Test solution: Dilute 50 μl of essential oil obtained in the assay of essential oil in 1 ml of *xylene R*.
Reference solution: Dissolve 2 μl of chamazulene R, 5 μl of levomenol R, and 10 mg of bornyl acetate R in 5 ml of toluene R.
Plate: TLC silica gel plate R
Mobile phase: Ethyl acetate R, toluene R (5:95 V/V)
Application: 10 μl, as bands
Development: Over a path of 10 cm
Drying: In air
Detection: Spray with anisaldehyde solution R and heat at 100–105°C for 5–10 min. Examine immediately in daylight.
Results: See below the sequence of the zones present in the chromatograms obtained with the reference solution and the test solution. Furthermore, other zones are present in the chromatogram obtained with the test solution.

Top of the TLC Plate

Reference solution	Test solution
	One or two blue to bluish-violet zones
Chamazulene: a red to reddish-violet zone	A red to reddish-violet zone (chamazulene)
Bornyl acetate: a yellowish-brown zone	
	A brown zone (en-yne dicycloether)
Levomenol: a reddish-violet to bluish-violet zone	A reddish-violet to bluish-violet zone (levomenol)

2.4.1.5 Tests

Broken drug: Maximum 25%, determined on 20.0 g, passes through a sieve (710).
Foreign matter (2.8.2): Maximum 2% m/m
Loss on drying (2.2.32): Maximum 12.0% is determined on 1000 g of the powdered drug (355) by drying in an oven at 100–105°C for 2 h.
Total ash (2.4.16): Maximum 13.0%

2.4.1.6 Assay

Essential oil (2.8.12): Use 30 g of whole drug, a 1000-ml flask, 300 ml of water R as distillation liquid, and 0.50 ml of xylene R in the graduated tube. Distill at a rate of 3–4 ml/min for 4 h. Toward the end of this period, stop the flow of water to the condenser assembly but continue distilling until the blue, steam-volatile components have reached the lower end of the condenser. Immediately restart the flow of water to the condenser assembly to avoid warming the separation space. Stop the distillation after a further 10 min.

Total apigenin-7-glucoside: Liquid chromatography (2.2.29)

2.4.1.7 Test Solution

Reduce 40 g of the drug to a powder (sieve 500). Place 2.00 g of the powdered drug in a 500-ml round-bottomed flask. Add 200 ml of alcohol R. Heat the mixture under a reflux condenser on a water bath for 15 min. Cool and filter. Rinse the filter and the residue with a few milliliters of alcohol R. To the filtrate add 10 ml of freshly prepared dilute sodium hydroxide solution R and heat the mixture under a reflux condenser on a water bath for about 1 h. Cool. Dilute to 250.0 ml with alcohol R. To 50.0 ml of the solution add 0.5 g of citric acid R. Shake for 5 min and filter. Dilute 5.0–10.0 ml with the mobile phase (initial mixture).

2.4.1.8 Reference Solution

A. Dissolve 10.0 mg of apigenin 7-glucoside R in 100.0 ml of methanol R. Dilute 25.0 ml of this solution to 200 ml with the mobile phase (initial mixture).

B. Dissolve 10.0 mg of 5,7-dihydroxy-4-methylcoumarin R in 100.0 ml of methanol R. Dilute 25.0 ml of this solution to 100 ml of the mobile phase (initial mixture). To 4.0 ml of this solution add 4.0 ml of reference solution (a) and dilute to 10.0 ml with the mobile phase (initial mixture).

The following chromatogram is published for information.

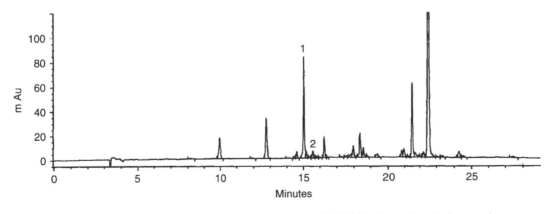

1. apigenin-7-glucoside   2. 5,7-dihydroxy-4-methylcoumarin

FIGURE 2.1 Chromatogram for the assay of total apigenin-glucoside in matricaria flower.

2.4.1.9   Precolumn

- *Size*: l = 8 mm, Ø = 4.6 mm
- *Stationary phase*: octadecylsilyl silica gel for chromatography R (5 µm)

2.4.1.10   Column

- *Size*: l = 0.25 m, Ø = 4.6 mm
- *Stationary phase*: octadecylsilyl silica gel for chromatography R (5 µm)

2.4.1.11   Mobile Phase

- *Mobile phase A*: phosphoric acid R, water R (0.5:99.5 V/V)
- *Mobile phase B*: phosphoric acid R, acetonitrile R (0.5:99.5 V/V)

Time (min)	Mobile Phase A (% V/V)	Mobile Phase B (% V/V)
0–9	75	25
9–19	75 $\Rightarrow$ 25	25 $\Rightarrow$ 75
19–24	25	75
24–29	25 $\Rightarrow$ 75	75 $\Rightarrow$ 25
29–30	75 $\Rightarrow$ 90	25 $\Rightarrow$ 10

Flow rate: 1 ml/min
Detection: Spectrophotometer at 340 nm
Injection: 20 µl
System suitability: Reference solution (b):
  Resolution: minimum 1.8 between the peaks due to apigenin-7-glucoside and 5,7-dihy-droxy-4-methylcoumarin

Calculate the percentage content of total apigenin-7-glucoside from the expression:

$$\frac{A_1 \times m_2}{A_2 \times m_1} \times P \times 0,625$$

A_1 = Area of the peak due to apigenin-7-glucoside in the chromatogram obtained with the test solution

A_2 = Area of the peak due to apigenin-7-glucoside in the chromatogram obtained with the reference solution

m_1 = Mass of the drug in the test solution, in grams

m_2 = Mass of *apigenin-7-glucoside R* in reference solution (a), in grams

P = Percentage content of apigenin-7-glucoside in the reagent

2.4.2 MATRICARIA OIL — MATRICARIAE AETHEROLEUM

European Pharmacopoeia 4.5 [15]
Matricaria oil
Matricariae aetheroleum

2.4.2.1 Definition

Blue essential oil is obtained by steam distillation from the fresh or dried flower heads or flowering tops of *Matricaria recutita* L. (*Chamomilla recutita* L. Rauschert). There are two types of matricaria oil that are characterized as rich in bisabolol oxides, or rich in levomenol.

2.4.2.2 Characteristics

Appearance: Clear, intensely blue, viscous liquid. It has an intense characteristic odor.

2.4.2.3 Identification

First identification: B
Second identification: A

A. Thin-layer chromatography (*2.2.27*)

Test solution: Dissolve 20 µl of the substance to be examined in 1.0 ml of toluene R.
Reference solution: Dissolve 2 mg of guaiazulene R, 5 µl of levomenol R, and 10 mg of bornyl acetate R in 5.0 ml of toluene R.
Plate: TLC silica gel plate R
Mobile phase: Ethyl acetate R, toluene R (5:95 V/V)
Application: 10 µl, as bands
Development: Over a path of 10 cm
Drying: In air
Detection A: Examine in daylight.
Results A: See below for the sequence of the zones present in the chromatograms obtained with the reference solution and the test solution.

Top of the TLC Plate	
Guaiazulene: a blue zone	A blue zone (chamazulene)
———	———
———	———
Reference solution	**Test solution**

Detection B: Spray with anisaldehyde solution R and heat at 100–105°C for 5–10 min. Examine immediately in daylight.

Results B: See below for the sequence of the zones present in the chromatograms obtained with the reference solution and the test solution. Furthermore, yellowish-brown to greenish-yellow zones (lower third), violet zones (lower third), and further weak zones may be present in the chromatogram obtained with the test solution.

Top of the TLC Plate	
	One or two blue to bluish-violet zones
Guaiazulene: a red to reddish-violet zone	A red to reddish-violet zone (chamazulene)
———	———
Bornyl acetate: a yellowish-brown to grayish-green zone	A brown zone (en-yne-dicycloether)
———	———
Levomenol: a reddish-violet to bluish-violet zone	A reddish-violet to bluish-violet zone (levomenol) A brownish zone
Reference solution	**Test solution**

B. Examine the chromatograms obtained in the test for chromatographic profile.

Results: The characteristic peaks corresponding to levomenol and to chamazulene in the chromatogram obtained with the test solution are similar in retention time to those in the chromatogram obtained with the reference solution.

2.4.2.4   Tests

Chromatographic profile: Gas chromatography (2.2.28): use the normalization procedure.
Test solution: Dissolve 20 μl of the oil to be examined in cyclohexane R and dilute to 5.0 ml with the same solvent.
Reference solution: Dissolve 20 μl of levomenol R, 5 mg of chamazulene R, and 6 mg of guaiazulene R in cyclohexane R and dilute to 5.0 ml with the same solvent.

Column:

- *Material*: Fused silica
- *Size*: l = 30 m (a film thickness of 1 μm may be used) to 60 m (a film thickness of 0.2 μm may be used), Ø = 0.25–0.53 mm; when using a column longer than 30 m, an adjustment of the temperature program may be necessary.
- *Stationary phase*: Macrogol 20,000 R

Carrier gas: Helium for chromatography R
Flow rate: 1–2 ml/min
Split ratio: 1:100

Temperature:

	Time (min)	Temperature ($^{\circ}$C)
Column	0–40	70 $\Rightarrow$ 230
	40–50	230
Injection port		250
Detector		250

Detection: Flame ionization

Injection: 1.0 µl

Elution order: Order indicated in the composition of the reference solution. Record the retention times of these substances.

Relative retention: With reference to chamazulene (retention time = about 34.4 min): β-farnesene = about 0.5; bisabolol oxide B = about 0.8; bisabolone = about 0.87; levomenol = about 0.9; bisabolol oxide A = about 1.02.

System suitability: Reference solution:

Resolution: Minimum of 1.5 between the peaks due to chamazulene and to guaiazulene.

Using the retention times determined from the chromatogram obtained with the reference solution, locate levomenol and chamazulene in the chromatogram obtained with the test solution; locate bisabolol oxides (bisabolol oxide B, bisabolone, and bisabolol oxide A) using Figures 1836.-1 and 1836.-2 (disregard the peak due to cyclohexane). The chromatogram obtained with the test solution does not show a peak with the retention time of guaiazulene.

Determine the percentage content of the components. The limits are within the following ranges:

	Matricaria Oil Rich in Bisabolol Oxides (%)	Matricaria Oil Rich in Levomenol (%)
Bisabolol oxides	29–81	
Levomenol		10–65
Chamazulene	$\geq$1.0	$\geq$1.0
Total of bisabolol oxides and levomenol		$\geq$20

2.4.2.5 Storage

Store in a well-filled, airtight container, protected from light at a temperature not exceeding 25°C.

2.4.2.6 Labeling

The label states the type of matricaria oil (rich in bisabolol oxides or rich in levomenol).

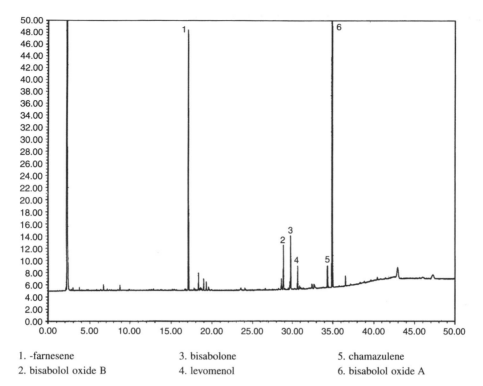

1. -farnesene   3. bisabolone   5. chamazulene
2. bisabolol oxide B   4. levomenol   6. bisabolol oxide A

FIGURE 2.2 1836.-1 — Chromatogram of matricaria oil rich in bisabolol oxides.

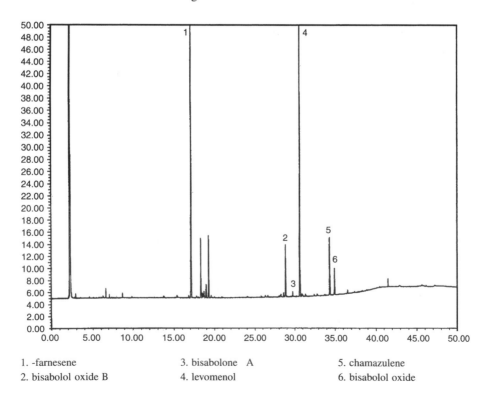

1. -farnesene   3. bisabolone   A   5. chamazulene
2. bisabolol oxide B   4. levomenol   6. bisabolol oxide

FIGURE 2.3 1836.-2 — Chromatogram of matricaria oil rich in levomenol.

2.4.3 MATRICARIA LIQUID EXTRACT — MATRICARIAE EXTRACTUM FLUIDUM

European Pharmacopoeia 4.5 [14]
Matricaria liquid extract
Matricariae extractum fluidum

2.4.3.1 Definition

Matricaria liquid extract is produced from Matricaria flower (0404). It contains not less than 0.30% of blue residual oil.

2.4.3.2 Production

The extract is produced from the drug and a mixture of 2.5 parts of a 10% m/m solution of ammonia (NH_3), 47.5 parts of water, and 50 parts of alcohol with an appropriate procedure for liquid extracts.

2.4.3.3 Characteristics

The extract is a brownish, clear liquid with an intense characteristic odor and characteristic bitter taste; miscible with water and with alcohol with development of turbidity; soluble in alcohol (50% V/V).

2.4.3.4 Identification

A. Examine by thin-layer chromatography (2.2.27), using a TLC silica gel F_{254} plate R.

> *Test solution*: Place 10 ml of the extract in a separating funnel and shake with two quantities, each of 10 ml, of pentane R. Combine the pentane layers, dry over 2 g of anhydrous sodium sulphate R, and filter. Evaporate the filtrate to dryness on a water bath and dissolve the residue in 0.5 ml of toluene R.
>
> *Reference solution*: Dissolve 4 mg of guaiazulene R, 20 mg of levomenol R, and 20 mg of bornyl acetate R in 10 ml of toluene R.

Apply 10 µl of each solution to the plate as bands. Develop over a path of 10 cm using a mixture of 5 volumes of ethyl acetate R and 95 volumes of toluene R. Allow the plate to dry in air and examine in ultraviolet light at 254 nm. The chromatogram obtained with the test solution shows several quenching zones, of which two main zones are in the middle third (en-yne-dicyclo-ether). Examine in ultraviolet light at 365 nm. The chromatogram obtained with the test solution shows an intense blue fluorescent zone (herniarin) in the middle part. Spray the plate with anisal-dehyde solution R. Examine in daylight while heating at 100–105°C for 5–10 min. The chromato-gram obtained with the reference solution shows a reddish-violet to bluish-violet zone (levomenol) in the lower third, a yellowish-brown to grayish-green zone (bornyl acetate) in the middle third, and a red to reddish-violet zone (guaiazulene) in the upper third. The chromatogram obtained with the test solution shows in the lower third yellowish-brown to greenish-yellow and violet zones and a reddish-violet to bluish-violet zone, corresponding to levomenol in the chromatogram obtained with the reference solution; a brownish zone (en-yne-dicycloether) similar in position to bornyl acetate in the chromatogram obtained with the reference solution; a red or reddish-violet zone (chamazulene) corresponding to guaiazulene in the chromatogram obtained with the reference solution and immediately above it one or two blue to bluish-violet zones; further weak zones may be present in the chromatogram obtained with the test solution.

B. Examine by thin-layer chromatography (2.2.27), using a TLC silica gel plate R.

Test solution: Use the extract.
Reference solution: Dissolve 1.0 mg of chlorogenic acid R, 2.5 mg of hyperoside R, and
2.5 mg of rutin R in 10 ml of methanol R.

Apply to the plate as bands 10 µl of each solution. Develop over a path of 15 cm using a mixture of 7.5 volumes of anhydrous formic acid R, 7.5 volumes of glacial acetic acid R, 18 volumes of water R, and 67 volumes of ethyl acetate R. Dry the plate at 100–105°C and spray the warm plate with a 10-g/l solution of diphenylboric acid aminoethyl ester R in methanol R. Subsequently spray the plate with a 50-g/l solution of macrogol 400 R in methanol R. Allow the plate to dry in air for about 30 min and examine in ultraviolet light at 365 nm. The chromatogram obtained with the reference solution shows in the middle part a light blue fluorescent zone (chlorogenic acid), below it a yellowish-brown fluorescent zone (rutin), and above it a yellowish-brown fluorescent zone (hyperoside). The chromatogram obtained with the test solution shows a yellowish-brown fluorescent zone corresponding to the zone of rutin in the chromatogram obtained with the reference solution, a light blue fluorescent zone corresponding to the zone of chlorogenic acid in the chromatogram obtained with the reference solution, a yellowish-brown fluorescent zone similar in position to the zone of hyperoside in the chromatogram obtained with the reference solution; it also shows above the yellowish-brown fluorescent zone a green fluorescent zone, then several bluish or greenish fluorescent zones and near the solvent front a yellowish fluorescent zone.

2.4.3.5 Tests

Ethanol (2.9.10): 38–53% V/V; dry residue (2.8.16): minimum 12.0%

2.4.3.6 Assay

Place 20.0 g in a 1000-ml round-bottomed flask, add 300 ml of *water R*, and distill until 200 ml has been collected in a flask. Transfer the distillate into a separating funnel. Dissolve 65 g of sodium chloride R in the distillate and shake with three quantities, each of 30 ml, of pentane R previously used to rinse the reflux condenser and the flask. Combine the pentane layers, dry over 2 g of anhydrous sodium sulphate R, and filter into a tared 100-ml round-bottomed flask that has been dried in a desiccator for 3 h. Rinse the anhydrous sodium sulphate and the filter with two quantities, each 20 ml, of pentane R. Evaporate the pentane in a water bath at 45°C. The residue of pentane is eliminated in a current of air for 3 min. Dry the flask in a desiccator for 3 h and weigh. The residual oil is blue (chamazulene).

2.5 ESCOP: MATRICARIAE FLOS — MATRICARIA FLOWER

2.5.1 ESCOP* MONOGRAPH MATRICARIAE FLOS — MATRICARIA FLOWER ([12]; SEE ALSO CHAPTERS 11 AND 12)

Matricariae flos
Matricaria Flower

2.5.1.1 Definition

Matricaria flower consists of the dried flower heads of *Matricaria recutita* L. (*Chamomilla recutita* [L.] Rauschert). It contains not less than 4 ml/kg of blue essential oil.

---

* European Scientific Cooperative on Phytotherapy.

The material complies with the *European Pharmacopoeia* [12].
Fresh material may also be used provided that when dried it complies with the *European Pharmacopoeia*.

2.5.1.2 Constituents

The main characteristic constituents of matricaria flower are the essential oil and flavone derivatives [21, 32, 60, 98, 116] such as apigenin-7-glucoside (approximately 0.5%) [32].

The essential oil contains approximately 50% of the sesquiterpenes (−)-α-bisabolol and its oxides A and B [98], bisabolonoxide, up to 25% of *cis*- and *trans*-en-yn-dicycloethers (or spiroethers) [32], and matricin, which is converted to chamazulene on distillation (up to 15%) [32].

Other constituents of matricaria flower include coumarins (herniarin and umbelliferone) [21, 32, 98, 116], phenolic acids [21, 32, 98], and polysaccharides (up to 10%) [21, 32, 52, 98].

2.5.1.3 Clinical Particulars

2.5.1.3.1 Therapeutic indications

2.5.1.3.2 Internal use
Symptomatic treatment of gastrointestinal complaints such as minor spasms, epigastric distension, flatulence, and belching [24, 27, 32, 45, 60, 65, 83, 93, 112]

2.5.1.3.3 External use
Minor inflammation and irritations of skin and mucosa, including the oral cavity and the gums (mouthwashes), the respiratory tract (inhalations), and the anal and genital areas (baths, ointments) [24, 27, 28, 35, 50, 54, 69, 74, 78, 83, 85, 86, 94, 96, 99, 105, 111, 112]

2.5.1.3.4 Posology and method of administration

2.5.1.3.5 Dosage

2.5.1.3.6 Internal use
Adults: As a tea infusion: 3 g of the drug to 150 ml of hot water, three to four times daily
Fluid extract (ethanol 45–60%): Single dose 1–4 ml [4, 32]
Dry extract: 50–300 mg three times daily [64]
Elderly: Dose as for adults
Children: Proportion of adult dose according to age or body weight

2.5.1.3.7 External use
For compresses, rinses, or gargles: 3–10% m/V infusion or 1% V/V fluid extract or 5% V/V tincture [32, 64]
For baths: 5 g of the drug, or 0.8 g of alcoholic extract, per liter of water [32]
For solid and semisolid preparations: Hydroalcoholic extracts corresponding to 3–10% m/m of the drug [60, 116]
For vapor inhalation: 10–20 ml of alcoholic extract per liter of hot water [96]

2.5.1.3.8 Method of administration
For oral administration, local application, and inhalation
Duration of administration
No restriction

2.5.1.3.9 Contraindications
Sensitivity to *Matricaria* or other *Compositae*

2.5.1.3.10   Special warnings and special precautions for use
None required

2.5.1.3.11   Interaction with other medicaments and other forms of interaction
None reported

2.5.1.3.12   Pregnancy and lactation
No harmful effects reported

2.5.1.3.13   Effects on ability to drive and use machines
None known

2.5.1.3.14   Undesirable effects
Rare cases of contact allergy have been reported in persons with known allergy to *Artemisia* species [98]. Matricaria flower of the bisabolol oxide B-type can contain traces of the contact allergen anthecotulide [27, 62, 98]. *Matricaria recutita* L. possesses a much lower allergenic potential than other chamomile species, and therefore allergic reactions to *Matricaria* must be considered as extremely rare. Most of the described allergic reactions to *Matricaria* were due to contamination with *Anthemis cotula* or related *Anthemis* species, which contain high amounts of anthecotulide. However, in cases where *Matricaria* contact allergy has been acquired, cross-reactions to other sesquiterpene lactone-containing plants are common [62].

2.5.1.3.15   Overdose
No toxic effects reported

2.5.2   PHARMACOLOGICAL PROPERTIES

2.5.2.1   Pharmacodynamic Properties

2.5.2.2   Anti-Inflammatory Effects

2.5.2.1.1   In vitro studies
Ethanolic (48% V/V) and isopropanolic (48% V/V) extracts of matricaria flower inhibited 5-lipoxygenase, cyclooxygenase, and the oxidation of arachidonic acid with IC_{50} values of 0.05–0.3%, while a supercritical carbon dioxide extract had an IC_{50} of 6–25 µg/ml for these activities [22]. Investigation of individual constituents revealed that apigenin inhibited 5- and 12-lipoxygenase (IC_{50}: 8 and 90 µM, respectively); chamazulene and (−)-α-bisabolol inhibited only 5-lipoxygenase (IC_{50}: 13 and 40 µM, respectively); apigenin, *cis*-en-yn-spiroether and (−)-α-bisabolol inhibited cyclooxygenase (IC_{50}: 70–80 µM); only chamazulene had antioxidative activity ($IC5_0$: 2 µM) [22].

2.5.2.1.2   In vitro studies
The anti-inflammatory effects of orally administered (−)-α-bisabolol have been demonstrated in carrageenan-induced rat paw edema, adjuvant arthritis of the rat, ultraviolet-induced erythema of the guinea pig, and yeast-induced fever of the rat [72]. In the carrageenan-induced rat paw edema test the following ED_{50} values (mmol/kg) were obtained after oral administration: (−)-α-bisabolol 2.69, chamazulene 4.48, guaiazulene 4.59, matricin 2.69, and salicylamide 1.53 [71].

An extract prepared from infusion of 20 g of matricaria flower in 100 ml of ethanol 42%, topically applied at a dose of 750 µg dry extract per ear, inhibited croton oil-induced edema of mouse ear by 23.4% compared to controls; benzydamine at 450 µg/ear showed comparable inhibition of 26.6% [66]. In the same test system, two polysaccharides from matricaria flower at a dose of 300 µg/ear inhibited edema by 14 and 22%, respectively [53].

2.5.2.3 Antispasmodic Effects

2.5.2.3.1 In vitro studies

A hydroethanolic extract of matricaria flower showed antispasmodic activity on isolated guinea pig ileum stimulated by various spasmogens. The ED_{50} (mg/ml) and the strength of activity relative to papaverine (= 1.0), respectively, were 1.22 and 0.0011 with barium chloride, 1.15 and 0.0019 with histamine dihydrochloride, 2.24 and 0.00074 with bradykinin, and 2.54 and ca. 0.00062 with serotonin. Pure constituents were also investigated: with barium chloride, (−)-α−bisabolol (ED_{50} = 136 μg/ml) showed activity comparable to papaverine, while apigenin (ED_{50} = 0.8 μg/ml) was more than three times as active [17].

2.5.2.4 Antiulcerogenic Effects

2.5.2.4.1 In vivo studies

The development of ulcers induced in rats by indomethacin stress or ethanol was inhibited by an orally administered extract of matricaria flower with an ED_{50} of 1 ml per rat and by (−)-α-bisabolol with an ED_{50} of 3.4 mg/kg body weight. These substances also reduced healing times for ulcers induced in rats by chemical stress (acetic acid) or heat coagulation [103].

2.5.2.5 Wound Healing Effect

2.5.2.5.1 In vivo studies

The wound healing activity of azulene has been demonstrated in studies on the thermally damaged rat tail [40] and of matricaria flower constituents in accelerated healing of experimental injuries [115].

2.5.2.6 Sedative Effects

2.5.2.6.1 In vitro studies

Apigenin competitively inhibited the binding of flunitrazepam to the central benzodiazepine receptor (K_i = 4 μM) but had no effect on muscarinic receptors, $α_1$-adreno-receptors, or the binding of muscimol to $GABA_A$ receptors [109].

HPLC fractions of a methanolic extract of matricaria flower were able to displace flunitrazepam from its receptors in rat cerebellar membranes, the ligand RO 5-4864 from "peripheral" benzodiazepine receptors in rat adrenal gland membranes, and muscimol from GABA receptors in rat cortical membranes. This last activity is mainly due to GABA present in the fractions [23].

2.5.2.6.2 In vitro studies

A sedative effect of matricaria flower was demonstrated through prolongation of hexobarbital-induced sleep, reduction of spontaneous mobility, and reduction of explorative activity in mice [43, 44].

Restricted stress-induced increases in plasma ACTH levels in normal and ovariectomized rats were decreased by administration of diazepam and inhalation of matricaria flower oil vapor. Inhaling the vapor induced greater decreases in plasma ACTH levels in ovariectomized rats than treatment with diazepam; this difference was not observed in normal rats. Furthermore, the inhalation of matricaria flower oil vapor induced a decrease in plasma ACTH level that was blocked by pretreatment with flumazenil, a potent and specific benzodiazepine receptor antagonist [113].

2.5.2.7 Antimicrobial Effects

Matricaria flower essential oil exerted a bactericidal effect against Gram-positive bacteria and a fungicidal effect against *Candida albicans* at a concentration of 0.7% V/V [98]. The essential oil was not active against Gram-negative bacteria, even in concentrations as high as 8% V/V [19].

An infusion of matricaria flower, a hydroethanolic extract, and pure herniarin showed anti-microbial activity against various bacteria and fungi in the presence of near UV light [37, 82].

2.5.2.8 Clinical Studies

2.5.2.8.1 Anti-inflammatory effects

In a bilateral comparative study, 161 patients with inflammatory dermatoses, who had been treated initially with 0.1% diflucortolone valerate, were treated during maintenance therapy with a cream containing matricara flower extract or one of three alternatives: 0.25% hydrocortisone, 0.75% fluocortin butyl ester, or 5% bufexamac. The therapeutic results with the extract were equivalent to those of hydrocortisone and superior to those of fluocortin butyl ester and bufexamac [18].

In a comparative study on 20 healthy volunteers with chemically induced toxic dermatitis, the smoothing effect on the skin of an ointment containing matricara flower extract was significantly superior ($p < 0.01$) to that of 0.1% hydrocortisone acetate or the ointment base [87].

In a study on 12 healthy subjects, a cream containing matricaria flower extract (20 mg/g) did not suppress UV-induced erythema but it reduced visual scores of skin redness in the adhesive tape stripping test ($p = 0.0625$) [76]. In an analogous study, the cream produced 69% of the effect of a hydrocortisone-27-acetate ointment [20].

In a randomized, double-blind study, 25 healthy volunteers with UVB light-induced erythema were treated with various matricara flower preparations, hydrocortisone cream, or the respective vehicle. Ranking the preparations according to visual assessment scores and mean values from chromametry, a cream containing a hydroalcoholic extract of matricaria flower gave the best result [75].

In an open study on 98 cancer patients, a matricaria flower extract preparation containing 50 mg of α-bisabolol and 150–300 mg of apigenin-7-glucoside per 100 g, applied three times daily, reduced oral mucositis caused by localized irradiation or systemic chemotherapy [30].

In a phase III double-blind, placebo-controlled clinical trial involving 164 patients, a mouthwash containing matricaria flower extract did not decrease 5-fluorouracil-induced stomatitis [49].

2.5.2.8.2 Anti-inflammatory and antispasmodic effects

In an open multicentric study, 104 patients with gastrointestinal complaints such as gastritis, flatulence, or minor spasms of the stomach were treated orally for 6 weeks with a matricaria flower extract preparation (standardized to 50 mg α-bisabolol and 150–300 mg apigenin-7-glucoside per 100 g) at a daily dose of 5 ml. Subjectively evaluated symptoms improved in all patients and disappeared in 44.2% of patients [98, 100].

2.5.2.8.3 Wound healing effects

In an open study, 147 female patients episiotomized during childbirth were treated for 6 days with either an ointment containing matricaria flower extract or a 5% dexpanthenol cream. The healing effect of the two preparations was comparable [73].

In a double-blind study on 14 patients, weeping dermatoses following dermabrasion of tattoos were treated with a matricaria flower extract preparation (standardized to 50 mg of α-bisabolol and 3 mg of chamazulene per l00 g). The decrease in the weeping wound area and the improvement in drying tendency were statistically significant ($p < 0.05$) [54].

In a randomized, open, placebo-controlled study, 120 patients with second-degree hemorrhoids were treated with rubber band ligature alone, rubber band ligature with anal dilator and vaseline, or rubber band ligature with anal dilator and an ointment containing matricaria flower extract. The last group showed the best results in amelioration of hemorrhage, itching, burning, and oozing [50].

2.5.2.9 Pharmacokinetic Properties

After cutaneous administration of $[^{14}C](-)$-α-bisabolol on mice, 82% of the radioactivity was found in the urine [59, 66].

Apigenin and luteolin were also readily absorbed by the skin. *In vivo* skin penetration studies of hydroethanolic solutions of apigenin and luteolin with nine healthy female volunteers gave a steady-state flux of 10.31 and 6.11 ng/min.cm^2, respectively [84].

After oral administration of apigenin-7-glucoside to rats free apigenin was detected in the urine [57].

After oral administration of 40 ml of a hydroethanolic matricaria flower extract (containing 225.52 mg% apigenin-7-glucoside, 22.51 mg% apigenin, 15.14 mg% herniarin) to a female volunteer, no flavones could be detected in blood plasma nor in 24-h urine, while herniarin was found in both (max. plasma concentration: ±35 mg/ml and 0.324 mg in 24-h urine) [106].

In germ-free rats no hydrolysis of flavone glycosides could be observed; obviously, intestinal microflora can affect the cleavage of the glycosidic bonds [56, 57]. Furthermore, orally administered apigenin was detected in the blood serum of animals [91].

2.5.2.10 Preclinical Safety Data

The acute oral LD$_{50}$ of matricaria flower essential oil in rats and the acute dermal LD$_{50}$ in rabbits exceeded 5 g/kg. No irritant effects of the oil were observed after application to the skin of nude mice [88].

In a 48-h patch test in volunteers, matricaria oil neither caused skin irritation nor were there any discernible sensitization reactions or phototoxic effects. Matricaria oil was granted "generally recognized as safe" (GRAS) status by FEMA and has been approved by the FDA for use in food and cosmetics [88].

The acute oral toxicity of $(-)$-α-bisabolol in mice and rats was very low with an LD$_{50}$ of about 15 ml/kg. A 6-week subacute toxicity study showed that the lowest toxic oral dose of $(-)$-α-bisabolol in rats and dogs was between 1 and 2 ml/kg. Oral doses up to 1 ml/kg of $(-)$-α-bisabolol produced no discernible effects on the prenatal development of rats or rabbits. No malformations were found at any of the tested dose levels [58].

The acute intraperitoneal LD$_{50}$ of *cis*- and *trans*-en yn-dicycloether is 670 mg/kg [17]. In the Ames test, apigenin and an aqueous matricaria flower extract showed no mutagenic or toxic activity [26, 95].

2.5.2.10.1 *Allergenicity*

Based on the fact that "German"-matricaria flower generally contains no, or only traces of, the sesquiterpene lactone anthecotulide and that millions of people come into contact with matricaria flower daily, allergic reactions due to matricaria flower can be considered to be extremely rare [64]. However, cross-reactions with other sesquiterpene lactone-containing plants are common [64]: a patient allergic to *Artemisia vulgaris* underwent severe anaphylactic reactions following ingestion of matricaria flower infusions and after eye washing with similar infusions [97]; 18 of 24 patients with *Compositae* allergy were also allergic to an ether extract of matricaria flower [89]; 96 patients of 4800 showed contact hypersensitivity to an ethanolic matricaria flower extract [39].

In a study on contact allergy performed with 540 type IV allergic patients, some of whom gave positive reactions to standard phytogenic allergens, an anthecotulide-free matricaria flower extract gave a positive reaction in none. This demonstrates the importance of using anthecotulide-free matricaria flower [70, 98].

2.6 COMMISSION E MONOGRAPH *MATRICARIAE FLOS* (CHAMOMILE FLOWERS)

Commission E is an expert committee on herbal remedies of the German Federal Health Agency, composed of physicians, pharmacists, pharmacologists, toxicologists, and statisticians.

2.6.1 *MATRICARIAE FLOS* (Chamomile flowers)

2.6.1.1 Official Name

Matricariae flos, chamomile flowers

2.6.1.2 Description

Chamomile flowers consist of fresh or dried heads of *Matricaria recutita* L. (syn. *Chamomillla recutita* [L.] Rauschert) and preparations of these in effective doses. The flowers contain not less than 0.4% (v/w) of volatile oil. The main constituents of the oil are $(-)$-α-bisabolol or bisabolol oxides A and B.

2.6.1.3 Indications

> *External use*: Skin and mucosal inflammation, bacterial skin conditions including oral cavity and parodontium
> Inflammatory conditions and irritation of respiratory tract (inhalation)
> Conditions affecting anal region and genitalia (bath and irrigation)
> *Internal use*: Gastrointestinal spasms and inflammatory diseases of the gastrointestinal tract

2.6.1.4 Contraindications

None reported

2.6.1.5 Side Effects

None reported

2.6.1.6 Dosage

Pour hot water (c. 150 ml) onto 1 heaping tablespoonful of chamomile flowers (= c. 3 g), cover, and put through a tea strainer after 5–10 min.

Unless otherwise prescribed, gastrointestinal conditions are treated by taking a cup of the freshly made tea three or four times daily between meals. For inflammation of oral and pharygeal mucosa, rinse or gargle several times daily with the freshly made tea.

> *External use*: 3–10% infusions for compresses and lavage; for baths, 50 g of the drug to 10 l of water; semisolid formulations equivalent to 3–10% of the plant drug

2.6.1.7 Method of Application

Liquid and solid formulations are for external and internal use.

2.6.1.8 Medicinal Actions

Inflammatory and antipyretic, musculotropic spasmolytic, vulnerary, deodorant, antibacterial and bacterial toxin inhibitor, stimulates skin metabolism.

2.7 WHO*: *FLOS CHAMOMILLAE*

2.7.1 WHO: Monographs on Selected Medicinal Plants [13]

Flos Chamomillae

2.7.1.1 Definition

Flos Chamomillae consists of the dried flowering heads of *Chamomilla recutita* (L.) Rauschert (*Asteraceae*) [2, 3, 8, 9].

2.7.1.2 Synonyms

Matricaria chamomilla L., *M. recutita* L., *M. suaveolens* L. [3, 98]

In most formularies and reference books, *Matricaria chamomilla* L. is regarded as the correct species name. However, according to the International Rules of Botanical Nomenclature, *Chamomilla recutita* (L.) Rauschert is the legitimate name for this species [90]. *Asteraceae* are also known as *Compositae*.

2.7.1.3 Selected Vernacular Names

Baboonig, babuna, babunah camornile, babunj, bunga kamil, camamilla, camomile, chamomile, camomilla, chamomille allemande, campomilla, chamomille commune, camomille sauvage, fleurs de petite camomille, flos chamomillae, German chamomile, Hungarian chamomile, Kamille, Kamillen, kamitsure, kamiture, manzanilla, manzanilla chiquita, manzanilla comun, manzanilla dulce, matricaire, matricaria flowers, pin heads, sweet false chamomile, sweet feverfew, wild chamomile [1, 3, 48, 80, 114]

2.7.1.4 Description

Herbaceous annual; 10–30 cm in height, with erect, branching stems and alternate, tripinnately divided leaves below and bipinnately divided leaves above, both types having almost filiform lobes; the capitulum (to 1.5 cm in diameter) comprises 12–20 white ligulate florets surrounding a conical hollow receptacle on which numerous yellow tubular (disk) florets are inserted; the inflorescence is surrounded by a flattened imbricated involucre; fruit small, smooth, yellowish [3, 29, 114].

2.7.1.5 Plant Material of Interest: Flower Heads

2.7.1.5.1 General appearance

Flos Chamomillae consists of conical flower heads, each bearing a few white ligulate florets and numerous yellowish orange to pale yellow tubular or disk florets on conical, narrow hollow receptacles with a short pedunde; disk florets are perfect and without a pappus; ray florets are pistillate, white, 3-toothed and 4-veined; involucre hemispherical, composed of 20–30 imbricate, oblanceolate, and pubescent scales; peduncles are weak brown to dusky greenish yellow, longitudinally furrowed, more or less twisted, and up to 2.5 cm long; achenes are more or less obovoid and faintly 3- to 5-ribbed; pappus none, or slightly membranous crown [5, 114].

2.7.1.5.2 Organoleptic properties

Odor is pleasant and aromatic; taste is aromatic and slightly bitter [3, 8, 9].

2.7.1.5.3 Microscopic characteristics

Receptacle and bracteoles have schizogenous secretory ducts; vascular bundles have phloem fibers; spiral, annular, and reticulate but pitted vessels; lignified cells at the bases of the ovaries are absent;

---

* World Health Organization.

nearly all parts of florets bear composite-type glandular hairs with short, biseriate stalk and enlarged head, formed of several tiers, each of two cells; ovary with longitudinal bands of small mucilage cells; stigma with elongated papillae at the apex; pollen grains, spherical or triangular, have numerous short spines [3].

2.7.1.5.4 Powdered plant material

Powdered *Flos Chamomillae* is greenish-yellow to yellowish-brown; spiny pollen grains numerous, 18–25 μm in diameter; fragments of yellow or white corolla, with polygonal, small epidermal cells having straight or slightly wavy walls, sometimes papillosed, and sometimes bearing glandular hairs of composite type; fragments of the fibrous layer of anther; fragments from ovary, with glandular hairs and rows of small mucilage cells; green fragments of parenchyma of involucre; stigma with papillae; cells of the achenes with sclariform perforations in walls; fragments of fibrovascular bundles with spiral, annular, and reticulate vessels and sclerenchyma fibers; fragments of involucral bracts with epidermis having elliptical stomata up to 30 μm in length; also vessels and fibers; occasional fiber from the stems; minute cluster crystals of calcium oxalate, up to 10 μm in diameter; fragments of lignified parenchyma of the filaments and occasional fragments of vessels [3, 29, 114].

2.7.1.6 Geographical Distribution

The plant is indigenous to northern Europe and grows wild in central European countries; it is especially abundant in eastern Europe and is also found in western Asia, the Mediterranean region of northern Africa, and the United States. It is cultivated in many countries [1, 3, 5, 6, 29, 80, 98, 108, 114].

2.7.1.7 General Identity Tests

The drug is identified by its macroscopic and microscopic characteristics, and by thin-layer chromatography [3, 8, 9].

2.7.1.8 Purity Tests

2.7.1.8.1 Microbiology

The test for *Salmonella* spp. in *Flos Chamomillae* products should be negative. The maximum acceptable limits of other microorganisms are as follows [7, 8, 11]: for preparation of decoction: aerobic bacteria, not more than 10^7/g; fungi, not more than 10^5/g; *Escherichia coli*, not more than 10^2/g. Preparations for internal use: aerobic bacteria, not more than 10^5/g or ml; fungi, not more than 10^4/g or ml; enterobacteria and certain Gram-negative bacteria, not more than 10^3/g or ml; *E. coli*, 0/g or ml. Preparations for external use: aerobic bacteria, not more than 10^2/g or ml; fungi, not more than 10^2/g or ml; enterobacteria and certain Gram-negative bacteria, not more than 10^1/g or ml.

2.7.1.8.2 Foreign organic matter

Not more than 10% stems and not more than 2% foreign organic matter [3]; no flowering heads of *Anthemis cotula* L. or *A. nobilis* L. [114]

2.7.1.8.3 Total ash

Not more than 13% [9]

2.7.1.8.4 Acid-insoluble ash

Not more than 4% [5]

2.7.1.8.5 Moisture

Not more than 12% [6]

2.7.1.8.6 Pesticide residues

To be established in accordance with national requirements. Normally, the maximum residue limit of aldrin and dieldrin for *Flos Chamomillae* is not more than 0.05 mg/kg [8]. For other pesticides, see WHO guidelines on quality control methods for medicinal plants [11] and guidelines for predicting dietary intake of pesticide residues [10].

2.7.1.8.7 Heavy metals

Recommended lead and cadmium levels are no more than 10 and 0.3 mg/kg, respectively, in the final dosage form of the plant material [11].

2.7.1.8.8 Radioactive residues

For analysis of strontium-90, iodine-131, caesium-134, caesium-137, and plutonium-239, see WHO guidelines on quality control methods for medicinal plants [11].

2.7.1.8.9 Other tests

Chemical, dilute ethanol-soluble extractive, and water-soluble extractive tests are to be established in accordance with national requirements.

2.7.1.9 Chemical Assays

Contain not less than 0.4% v/w of essential oil [3, 8, 9]. Total volatile oil content is determined by pharmacopoeial methods [3, 8, 9].

Thin-layer [8, 9] and gas-liquid [33] chromatography is for volatile oil constituents, and high-performance liquid chromatography is for flavonoids [46, 92].

2.7.1.10 Major Chemical Constituents

Flos Chamomillae contains an essential oil (0.4–1.5%), which has an intense blue color owing to its chamazulene content (1–15%). Other major constituents include α-bisabolol and related sesquiterpenes (up to 50% of the oil). Apigenin and related flavonoid glycosides constitute up to 8% (dry weight) of the drug [29, 46].

chamazulene   (–)-α-bisabolol

apigenin

2.7.1.11 Dosage Forms

Dried flower heads, liquid extract (1:1 in 45% alcohol), tinctures, and other galenicals [5]. Store in well-closed containers, protected from light [3, 8, 9].

2.7.1.12   Medicinal Uses

2.7.1.12.1   Uses supported by clinical data

2.7.1.13   Internal Use

Use for symptomatic treatment of digestive ailments such as dyspepsia, epigastric bloating, impaired digestion, and flatulence [1, 3, 5, 29, 34, 35, 114]. Infusions of camomile flowers have been used in the treatment of restlessness and in mild cases of insomnia due to nervous disorders [34, 55].

2.7.1.14   External Use

Use for inflammation and irritations of the skin and mucosa (skin cracks, bruises, frostbite, and insect bites) [29, 67], including irritations and infections of the mouth and gums, and hemorrhoids [5, 29, 34, 35, 67].

2.7.1.15   Inhalation

Symptomatic relief of irritations of the respiratory tract due to the common cold [112].

*2.7.1.15.1   Uses described in pharmacopoeias and in traditional systems of
medicine*

Adjuvant in the treatment of minor inflammatory conditions of the gastrointestinal tract [112].

*2.7.1.15.2   Uses described in folk medicine, not supported by experimental
or clinical data*

As an antibacterial and antiviral agent, an emetic, and an emmenagogue. It is also used to relieve eye strain, and to treat urinary infections and diarrhea [108].

2.7.1.16   Pharmacology

2.7.1.16.1   Experimental pharmacology

Both camomile extract and (−)-α-bisabolol demonstrated antipeptic activity *in vitro* [68, 98, 104]. A hydroalcoholic extract of camomile inhibited the growth of *Staphylococcus aureus, Streptococcus mutans,* group B *Streptococcus,* and *Streptococcus salivarius,* and it had a bactericidal effect *in vitro* on *Bacillus megatherium* and *Leptospira icterohaemorrhagiae* [38]. *In vitro,* the volatile oil of camomile also inhibited *Staphylococcus aureus* and *Bacillus subtilis* [19]. *In vitro,* camomile extracts inhibited both cyclooxygenase and lipoxygenase [110], and thus the production of prostaglandins and leukotrienes, known inducers of inflammation. Both bisabolol and bisabolol oxide have been shown to inhibit 5-lipoxygenase, but bisabolol was the more active of the two compounds [21]. Numerous *in vivo* studies have demonstrated the anti-inflammatory effects of the drug. The anti-inflammatory effects of camomile extract, the essential oil, and the isolated constituents have been evaluated in yeast-induced fever in rats and against ultraviolet radiation-induced erythema in guinea-pig models [72]. The principal anti-inflammatory and antispasmodic constituents of camomile appear to be the terpene compounds matricin, chamazulene, (−)-α-bisabololoxides A and B, and (−)-α-bisabolol [20, 41, 42, 51, 71, 77, 81, 107]. While matricin and (−)-α-bisabolol have been isolated from the plant, chamazulene is actually an artefact formed during the heating of the flowers when an infusion or the essential oil is prepared [29]. The anti-inflammatory effects of these compounds in various animal models, such as inhibition of carrageenin-induced rat paw edema, have been demonstrated [21], although their activity was somewhat less than that of salicylamide [20]. In the mouse model for croton oil-induced dermatitis, topical application of either the total camomile extract, or the flavonoid fraction only, was very effective in reducing inflammation [41]. Apigenin and luteolin were more active than indometacin and phenylbutazone [41]. Activity decreased in the following order: apigenin > luteolin > quercetin > myricetin > apigenin-7-glucoside > rutin [41]. The spasmolytic activity of camomile has been attributed to apigenin, apigenin-7-O-glucoside [29, 77], and (−)-α-bisabolol, which have activity similar to papaverine [29, 42].

Intradermal application of liposomal apigenin-7-glucoside inhibited, in a dose-dependent manner, skin inflammations induced in rats by xanthine oxidase and cumene hydroperoxide [51].

Intraperitoneal administration to mice of a lyophilized infusion of camomile decreased basal motility, exploratory and motor activities, and potentiated hexobarbital-induced sleep [43]. These results demonstrated that in mice camomile depresses the central nervous system [43].

2.7.1.17   Clinical Pharmacology

A double-blind study of the therapeutic effects of a camomile extract on re-epithelialization and drying of wound weeping after dermabrasion demonstrated a statistically significant decrease in the wound size and drying tendency [54].

In clinical trials, topical application of a camomile extract in a cream base was found to be superior to hydrocortisone 0.25% for reducing skin inflammation [18]. In an international multi-center trial camomile cream was compared with hydrocortisone 0.25%, fluocortin butyl ester 0.75%, and bufexamac 5% in the treatment of eczema of the extremities [18]. The camomile cream was shown to be as effective as hydrocortisone and superior to the other two treatments, but no statistical analysis was performed. Camomile preparations have also been found to be beneficial in the treatment of radiation mucositis owing to head and neck radiation and systemic chemotherapy [31].

2.7.1.18   Contraindications

Camomile is contraindicated in patients with a known sensitivity or allergy to plants of the *Asteraceae* (*Compositae*) such as ragweed, asters, and chrysanthemums [34].

2.7.1.19   Warnings

No information available.

2.7.1.20   Precautions

2.7.1.20.1   Carcinogenesis, mutagenesis, impairment of fertility

No mutagenic effects were found in *Salmonella typhimurium* strains TA 97a, TA 98, TA 100, and TA 104, with or without metabolic activation [95].

2.7.1.20.2   Pregnancy: teratogenic effects

No adverse effects reported *in vivo* [79].

2.7.1.20.3   Other precautions

No information available concerning general precautions, drug interactions, drug and laboratory test interactions, nonteratogenic effects on pregnancy, nursing mothers, or pediatric use.

2.7.1.21   Adverse Reactions

The presence of lactones in *Flos Chamomillae*-based preparations may cause allergic reactions in sensitive individuals, and there have been reports of contact dermatitis due to camomile preparations [47, 89, 102]. It should be noted that very few cases of allergy were specifically attributed to

German camomile [63]. A few cases of anaphylactic reactions to the ingestion of *Flos Chamomillae* have also been reported [25, 36, 101].

2.7.1.22 Posology

2.7.1.22.1   Internal use

Adult dose of flower head: average daily dose 2–8 g, three times a day [1, 5, 114]; of fluid extract 1:1 in 45% ethanol: dose 1–4 ml, three times a day [5, 48]. Child's dose of flower head: 2 g, three times daily; of fluid extract (ethanol 45–60%): single dose 0.6–2 ml [5]. Should not be used by children under 3 years old.

2.7.1.22.2   External use

For compresses, rinses, or gargles: 3–10% (30–100 g/1) infusion or 1% fluid extract or 5% tincture [5]. For baths: 5 g/l of water or 0.8 g/l of alcoholic extract. For semisolid preparations: hydroalcoholic extracts corresponding to 3–10% (30–100 g/kg) of the drug. For vapor inhalation: 6 g of the drug or 0.8 g of alcoholic extract per liter of hot water [5].

2.8 LEGAL CLASSIFICATION OF THE USE OF GERMAN CHAMOMILE IN GERMANY AND IN THE EUROPEAN COMMUNITY

Chamomile flowers and chamomile herb with flowers can be distributed both as a foodstuff as well as a drug with different fields of application.

2.8.1 FOODSTUFF

There are two substantial differences between chamomile tea as a foodstuff and as "medicinal chamomile":

1. Chamomile teas of the food category are not allowed to carry any indications but they can only be sold as so-called "domestic teas."
2. The quality of chamomile tea as a foodstuff does not have to correspond to the regulations of the national pharmacopoeias, especially the European one; a minimum content of essential oil is, for instance, not required and it is also permitted to use chamomile herb with a certain percentage of chamomile flowers (5–10% only). About 2–3 weeks after the last harvest of chamomile flowers chamomile herb is, as a rule, obtained from those aerial parts of the plants still carrying flowers of the afterbloom, and mainly after crushing it is packed into teabags. In the Federal Republic of Germany the main quality points are laid down in § 8 of the Food Regulations (prohibitions for health protection) and in § 10 of the Food Regulations (authorization for hygienic regulations).

2.8.2 DRUG

1. Chamomile flowers as a drug have to correspond to the different pharmacopoeias (see Table 1.1) and normally they must have a minimum essential oil content.
2. Depending on the indications "chamomile flowers for medicinal purposes" as such can be traded on their own or as preparations obtained thereof either as freely saleable drugs (i.e., outside the pharmacy) or as drugs being subject to sale by pharmacists only.

Chamomile flowers according to § 44.2 of AMG 76 (Second Drug Law of the Federal Republic of Germany) may — with the exception of certain limitations by the so-called list of diseases — also be prescribed outside the pharmacy to cure and alleviate diseases. A limitation by the list of diseases would, for example, be an application in case of hemorrhoids or phlebitis.

The AMG 76 of the FRG provides another possibility of free sale; viz., as a so-called "traditionally applied drug" according to § 109 a. In this case the range of application comprises: "For [health] improvement with gastric complaints and assistance for the function of the stomach." The ranges of application according to the monography of Commission E are not allowed to be used with drugs of § 109 a, because these drugs need only a minimum of about 10% of the Commission E dosage.

Chamomile preparations (e.g., tinctures, fluid extracts, dry extracts, etc.) being subject to sale by pharmacists only may carry all indications of the monography of Commission E; they are available by prescription and as a rule until April 2004 they were also reimbursed by the health insurance companies. Chamomile flower extracts, especially the pure essential chamomile oil, are parts of medical prescriptions, particularly of preparations used in dermatology.

REFERENCES

1. (1953) *The Indian Pharmaceutical Codex. Vol. I. Indigenous Drugs.* New Delhi: Council of Scientific & Industrial Research.
2. (1974) *Estra farmakope Indonesia.* Jakarta, Cetakan Kedua, Departemen Kesehatan, Republik Indonesia.
3. (1985) *African Pharmacopoeia*, 1st ed. Lagos: Organization of African Unity, Scientific, Technical & Research Commission.
5. (1983) Matricaria. *British Herbal Pharmacopoeia.* Bournemouth: British Herbal Medicine Association, 139–140.
5. (1990) *British Herbal Pharmacopoeia.* London: British Herbal Medicine Association.
6. (1995) *Polish Pharmacopoeia.* Warsaw.
7. (1996) *Deutsches Arzneibuch 1996. Vol. 2. Methoden der Biologie.* Stuttgart: Deutscher Apotheker Verlag.
8. (1996) *European Pharmacopoeia.* 3rd ed. Strasbourg: Council of Europe.
9. (1996) *Pharmacopée Francaise.* Paris: Adrapharm.
10. (1997) *Guidelines for Predicting Dietary Intake of Pesticide Residues*, 2nd rev. ed. Geneva: World Health Organization (unpublished document WHO/FSF/FOS/97.7; available from Food Safety, WHO, Geneva, Switzerland).
11. (1998) *Quality Control Methods for Medicinal Plant Materials.* Geneva: World Health Organization.
12. (1999) European Scientific Cooperative on Phytotherapy (ESCOP), Monograph, *Matricaria Flower.*
13. (1999) Flos Chamomillae. *WHO Monographs on Selected Medicinal Plants.* Vol. 1. Geneva: World Health Organization, 86–94.
14. (2003) Matricaria liquid extract. *European Pharmacopoeia.* Strasbourg: Council of Europe, 4th ed. Suppl. 4.5, 3730–3731.
15. (2003) Matricaria oil. *European Pharmacopoeia.* Strasbourg: Council of Europe, 4th ed. Suppl. 4.5, 3731–3734.
16. (2004) Matricaria flower. *European Pharmacopoeia.* Strasbourg: Council of Europe, 4th ed. Suppl. 4.6, 4064–4065.
17. Achterrath-Tuckermann, U., Kunde, R., Flaskamp, E., Isaac, O., Thiemer, K. (1980) *Planta Med.*, 39, 38–50.
18. Aertgeerts, P., Albring, M., Klaschka, F., Nasemann, T., Patzelt-Wenczler, R., Rauhut, K., Weigl, B. (1985) Vergleichende Prüfung von Kamillosan® Creme gegenüber steroidalen (0.25% Hydrocortison, 0.75% Fluocortinbutylester) und nichtsteroidalen (5% Bufexamac) Externa in der Erhaltungstherapie von Ekzemerkrankheiten. *Z. Hautkrankheiten*, 60, 270–277.
19. Aggag, M.E., Yousef, R.T. (1972) Study of antimicrobial activity of chamomile oil. *Planta Med.*, 22, 140–144.
20. Albring, M., Albrecht, H., Alcorn, G., Lücker, P.W. (1983) The measuring of the anti-inflammatory effect of a compound on the skin of volunteers. *Methods and Findings in Exp. and Clin. Pharmacol.*, 5, 75–77.

21. Ammon, H.P.T., Kaul, R. (1992) Pharmakologie der Kamille und ihrer Inhaltsstoffe. *Dtsch. Apoth. Ztg.*, 132 (Suppl. 27) 1–26.
22. Ammon, H.P.T., Sabieraj, J. (1996) *Dtsch. Apoth. Ztg.*, 136, 1821–1833.
23. Avallone, R., Zanoli, P., Corsi, L., Cannazza, 0., Baraldi, M. (1996) *Phytotherapy Res.*, 10 (Suppl. 1) 177–179.
24. Benetti, C., Manganelli, F. (1985) *Minerva Ginecol.*, 37, 799–801.
25. Benner, M.H., Lee, H.J. (1973) Anaphylactic reaction to chamomile tea. *J. Allergy and Clin. Immunol.*, 52, 307–308.
26. Birt, D.F., Walker, B., Tibbels, M.G., Bresnick, E. (1986) *Carcinogenesis*, 7, 959–963.
27. Bisset, N.G. (1994) in M. Wichtl *Herbal 2nd ed. Drugs and Phytopharmaceuticals: A Handbook for Practice on a Scientific Basis.* Boca Raton, FL: CRC Press; Stuttgart, Germany: Medpharm, 322–325.
28. Borgatti, E. (1985) *Clinica Terapeutica*, 112, 225–231.
29. Bruneton, J. (1995) *Pharmacognosy, Phytochemistry, Medicinal Plants.* Paris: Lavoisier.
30. Carl, W. (1994) in R.S. Rao, M.G. Deo, L.D. Sanghvi (Eds.) *Proceedings of the International Cancer Congress*, New Delhi, 981–986.
31. Carl, W., Emrich, L.S. (1991) Management of oral mucositis during local radiation and systemic chemotherapy: a study of 98 patients. *J. Prosthetic Dentistry*, 66, 361–369.
32. Carle, R. (1992) in R. Hänsel, K. Keller, H. Rimpler, Schneider, G. (Eds.) *Hagers Handbuch der Pharmazeutischen Praxis*, 5th ed. Vol. 4: Drogen A.-D. Berlin: Springer-Verlag, 817–831.
33. Carle, R., Fleischhauer, I., Fehr, D. (1987) Qualitätsbeurteilung von Kamillenölen. *Dtsch. Apoth. Ztg.*, 127, 2451–2457.
34. Carle, R. Gomaa, K. (1992) Chamomile: a pharmacological and clinical profile. *Drugs of Today*, 28, 559–565.
35. Carle, R. Isaac, 0. (1987) Die Kamille — Wirkung und Wirksamkeit. *Z. Phytotherapie*, 8, 67–77.
36. Casterline, C.L. (1980) Allergy to chamomile tea. *J.A.M.A.*, 244, 330–331.
37. Ceska, 0., Chaudhary, S.K., Warrington, P.J., Ashwood-Smith, M.J. (1992) *Fitoterapia*, 63, 387–394.
38. Cinco, M. et al. (1983) A microbiological survey on the activity of a hydroalcoholic extract of chamomile. *Int. J. of Crude Drug Res.*, 21, 145–151.
39. Dastychová, E., Záhejsky, J. (1992) *Ceskosl. Dermatol.*, 67, 14–18.
40. Deininger, R. (1956) *Arzneim.-Forsch./Drug. Res.*, 6, 394–395.
41. Della Loggia, R. (1985) Lokale antiphiogistische Wirkung der Kamillen-Flavone. *Dtsch. Apoth. Zt.*, 125 (Suppl. I) 9–11.
42. Della Loggia, R. et al. (1990) Evaluation of the anti-inflammatory activity of chamomile preparations. *Planta Med.*, 56, 657–658.
43. Della Loggia, R., Traversa, U., Scarica, V., Tubaro, A. (1982) Depressive effects of *Chamomilla recutita* (L.) Rausch. tubular flowers, on central nervous system in mice. *Pharrnacol. Res. Commun.*, 14, 153–162.
44. Della Loggia, R., Tubaro, A., Redaelli, C. (1981) *Riv. Neurol.*, 51, 297–310.
45. Demling, L. (1975) in L. Demling, T. Nasemann, W. Rösch (Eds.) *Erfahrungstherapie — späte Rechtfertigung.* Karlsruhe, Germany: G. Braun, 1–8.
46. Dölle, B., Carle, R., Müller, W. (1985) Flavonoidbestirrnung in Kamillcncxtraktpräparatcn. *Dtsch. Apoth. Ztg.*, 125 (Suppl. I) 14–19.
47. Dstychova, E., Zahejsky, J. (1992) Contact hypersensitivity to camomile. *Ceskoslovenska Dermatol.*, 67, 14–18.
48. Farnsworth, N.R. (Ed.) (1995) *NAPRALERT database.* Chicago, University of Illinois at Chicago, August 8, 1995 production (an online database available directly through the University of Illinois at Chicago or through the Scientific and Technical Network [STN] of Chemical Abstracts Services).
49. Fidler, P., Loprinzi, C.L., O'Fallon, J.R., Leitch, J.M., Lee, J.K., Hayes, D.L., et al. (1996) *Cancer*, 77, 522–525.
50. Förster, C.F., Süssmann, H.-E., Patzelt-Wenczler, R. (1996) *Schweiz. Rundschau Med.*, 85, 1476–1481.
51. Fuchs, J., Milbradt, R. (1993) Skin anti-inflammatory activity of apigenin-7-glucoside in rats. *D97*, 43, 370–372.
52. Füller, E., Franz, G. (1993) *Dtsch. Apoth. Ztg.*, 133, 4224–4227.
53. Füller, E., Sosa, S., Tubaro, A., Franz, 0., Della Loggia, R. (1993) *Planta Med.*, 59, A 666–667.

54. Glowania, H.J., Raulin, C., Swoboda, M. (1987) The effect of chamomile on wound healing — a controlled clinical-experimental double-blind study. *Z. Hautkr.*, 62, 1262–1271.
55. Gould, L., Reddy, C.V.R., Gomprecht, R.F. (1973) Cardiac effect of chamomile tea. *J. Clin.Pharma-col.*, 13, 475–479.
56. Griffiths, L.A., Barrow, A. (1972) *Biochem. J.*, 130, 1161–1162.
57. Griffiths, L.A., Smith, G.E. (1972) *Biochem. J.*, 128, 901–911.
58. Habersang, S., Leuschner, F., Isaac, 0., Thiemer, K. (1979) *Planta Med.*, 37, 115–123.
59. Hahn, B., Hölzl, J. (1987) *Arzneim.-Forsch./Drug Res.*, 37, 7, 16–20.
60. Hänsel, R., Haas, H. (1984) in *Therapie mit Phytopharmaka*. Berlin: Springer-Verlag, 146–150, 270–271.
61. Hatzky, K. (1930) *Med. Klin.*, 26, 819–820.
62. Hausen, B.M. (1992) in P.Ä.G.M. De Smet, K. Keller, R. Hänsel, R.F. Chandler (Eds.) *Adverse Effects of Herbal Drugs*, Vol. l. Berlin: Springer-Verlag, 243–248.
63. Hausen, B.M., Busker, E., Carle, R. (1984) Über das Sensibilisierungsvermögen von Compositenarten. VII. Experimentelle Untersuchungen mit Auszügen und Inhaltsstoffen von *Chamomilla recutita* (L.) Rauschert und *Anthemis cotula* L. *Planta Med.*, 229–234.
64. Hellemont, J. van (1988) *Fytotherapeutisch Compendium*. Bohn, Scheltema & Holkema, Utrecht, the Netherlands, 369–373.
65. Hoffmann, H.A. (1926) *Fortschritte der Therapie*, 5, 156–157.
66. Hölzl, E., Hahn, B. (1985) *Dtsch. Apoth. Ztg.*, 125 (Suppl. I) 32–38.
67. Hormann, H.P., Korting, H.C. (1994) Evidence for the efficacy and safety of topical herbal derma-tology. I. Anti-inflammatory agents. *Phytomedicine*, 1, 161–171.
68. Isaac, 0., Thiemer, K. (1975) Biochemische Untersuchungen von Kamilleninhaltsstoffen. *Arzneimittel-Forschung*, 25, 1086–1087.
69. Isar, H.J. (1930) *Dermatol. Wschr.*, 21, 712–715.
70. Jablonska, S., Rudzki, E. (1996) *Z. Hautkr.*, 71, 542–546.
71. Jakovlev, V., Isaac, 0., Flaskamp, E. (1983) Pharmakologische Untersuchungen von Kamilleninhaltsst-offen. VI. Untersuchungen zur antiphlogistischen Wirkung von Chamazulen und Matricin. *Planta Med.*, 49, 67–73.
72. Jakovlev, V., Isaac, 0., Thiemer, K., Kunde, R. (1979) Pharmacological investigations with compounds of chamomile. II. New investigations on the antiphlogistic effects of (–)α -bisabolol and bisabolol oxides. *Planta Med.*, 35, 125–140.
73. Kaltenbach, F.-J. (1991) in T. Nasemann, R. Patzelt-Wenczler (Eds.) *Kamillosan® im Spiegel der Literatur.* Frankfurt, Germany: pmi Verlag, 85–86.
74. Katz, R. (1928) *Fortschr. Med.*, 46, 388–391.
75. Kerscher, M.J. (1992) in O. Braun-Falco, H.C. Korting, H.I. Maibach (Eds.) *Liposome Dermatics*, Berlin: Springer-Verlag, 329–337.
76. Korting, H.C., Schäfer-Korting, M., Hart, H., Laux, P., Schmid, M. (1993) *Eur. J. Clin. Pharmacol.*, 44, 315–318.
77. Lang, W., Schwandt, K. (1957) Untersuchung über die glykosidischen Bestandteile der Kamille. *Dtsch. Apoth. Ztg.*, 97, 149–151.
78. Latz, B. (1927) *Fortschritte der Therapie*, 22, 796–799.
79. Leslie, G.B., Salmon, G. (1979) Repeated dose toxicity studies and reproductive studies on nine Bio-Strath herbal remedies. *Swiss Med.*, 1, 1–3.
80. Leung, A., Foster, S. (1996) *Encyclopedia of Common Natural Ingredients Used in Food, Drugs, and Cosmetics*, 2nd ed. New York: John Wiley & Sons.
81. Mann, C., Staba, J. (1986) The chemistry, pharmacology, and commercial formulations of chamomile, in L.E. Craker, J.E. Simon (Eds.), *Herbs, Spices, and Medicinal Plants: Recent Advances in Botany, Horticulture and Pharmacology,* Vol. I. Phoenix, AZ: Oryx Press, 233–280.
82. Mares, D., Romagnoli, C., Bruni, A. (1993) *Plantes Méd. Phytothér.*, 26, 91–100.
83. Matzker, J. (1975) in L. Demling, T. Nasemann, W. Rösch (Eds.) *Erfahrungstherapie — späte Rechtfertigung,* Karlsruhe, Germany: G. Graun, 77–89.
84. Merfort, I., Heilmann, J., Hagedorn-Leweke, U., Lippold, B.C. (1994) *Pharmazie*, 49, 509–511.
85. Münzel, M. (1975) *Selecta*, 24, 2258–2260.

86. Nasemann, T. (1975) in L. Demling, T. Nasemann, W. Rösch (Eds.) *Erfahrungstherapie — späte Rechtfertigung,* Karlsruhe, Germany: G. Braun, 49–53.
87. Nissen, H.P., Biltz, H., Kreysel, H.W. (1988) *Z. Hautkr.,* 63, 184–190.
88. Opdyke, D.L.J. (1974) *Fd. Cosmet. Toxicol.,* 12, 851.
89. Paulsen, E., Andersen, K.E., Hausen, B.M. (1993) Compositae dermatitis in a Danish dermatology department in one year. *Contact Dermatitis,* 29, 6–10.
90. Rauschert, S. (1990) Nomenklatorische Probleme in der Gattung *Matricaria* L., *Folia geobotanica phytotaxonomica,* 9, 249–260.
91. Redaelli, C., Formentini, L., Santaniello, E. (1980) *HPLC-determination of apigenin in natural samples, chamomile extracts and blood serum.* Poster Int. Res. Congress on Natural Products and Medicinal Agents, Strasbourg, France.
92. Redaelli, C., Formentini, L., Santaniello, E. (1981) Reversed-phase high-performance liquid chromatography analysis of apigenin and its glucosides in flowers of *Matricaria chamomilla* and chamomille extracts. *Planta Med.,* 42, 288–292.
93. Reicher, K. (1925) *Münch. Med. Wschr.,* 7, 261–262.
94. Riepelmeier, F. (1933) *Monatsschr. Ohrenheilkunde Laryngo-Rhinol.,* 67, 483–487.
95. Rivera, I.G., Martins, M.T., Sanchez, P.S., Sato, M.I.Z., Coelho, M.C.L., Akisue, M., Akisue, G. (1994) Genotoxicity assessment through the Ames test of medicinal plants commonly used in Brazil. *Environ. Toxicol. Water Quality,* 9, 87–93.
96. Saller, R., Beschorner, M., Hellenbrecht, D., Bühring, M. (1990) *Eur. J. Pharmacol.,* 183, 728–729.
97. Sánchez Palacios, A. (1992) *Rev. Esp. Alergol. Immunol. Clin.,* 7, 37–39.
98. Schilcher, H. (1987) *Die Kamille — Handbuch für Ärzte, Apotheker und andere Naturwissenschaftler,* Stuttgart, Germany: Wissenschaftliche Verlagsgesellschaft, 152 pp.
99. Schmid, F. (1975) in L. Demling, T. Nasemann, W. Rösch (Eds.) *Erfahrungstherapie — späte Rechtfertigung,* Karlsruhe, Germany: G. Braun, 43–46.
100. Stiegelmeyer, H. (1978) *Kassenarzt,* 18, 3605–366.
101. Subiza, J. et al. (1989) Anaphylactic reaction after the ingestion of chamomile tea: a study of cross-reactivity with other composite pollens. *J. Allergy and Clin. Immunol.,* 84, 353–358.
102. Subiza, J. et al. (1990) Allergic conjunctivitis to chamomile tea. *Ann. Allergy,* 65, 127–132.
103. Szelenyi, I., Isaac, O., Thiemer, K. (1979) *Planta Med.,* 35, 218–227.
104. Thiemer, V.K., Stadler, R., Isaac, O. (1972) Biochemische Untersuchungen von Kamilleninhaltsstoffen. *Arzneimittel-Forschung/Drug Res.,* 22, 1086–1087.
105. Tissot, H.C. (1929) *Schweiz. Med. Wschr.,* 39, 992–993.
106. Tschiersch, K., Hölzl, J. (1993) *Pharmazie,* 48, 554–555.
107. Tubaro, A., Zilli, C., Redaelli, C., Della Loggia, R. (1984) Evaluation of anti-inflammatory activity of chamomile extract after topical application. *Planta Med.,* 51, 359.
108. Tyler, V.E., Brady, L.R., Robbers, J.E. (Eds.) (1988) *Pharmacognosy,* 9th ed. Philadelphia: Lea & Febiger.
109. Viola, H., Wasowski, C., Levi de Stein, M., Wolfman, C., Silveira, R., Dajas, F. (1995) *Planta Med.,* 61, 213–216.
110. Wagner, H., Wierer, M., Bauer, R. (1986) In vitro inhibition of prostaglandin biosynthesis by essential oils and phenolic compounds. *Planta Med.,* 184–187.
111. Weiβ, R.F. (1982) *Moderne Pflanzenheilkunde. Neues über Heilpflanzen und ihre Anwendung in der Medizin.* 7th ed. Bad Wörishofen: Kneipp-Verlag, 15–21.
112. Weiβ, R.F. (1987) Kamille — Heilpflanze 1987. *Kneipp-Blätter,* 1, 4–8.
113. Yamada, K., Miura, T., Mimaki, Y., Sashida, Y. (1996) *Biol. Pharm. Bull.,* 19, 1244–1246.
114. Youngken, H.W. (1950) *Textbook of Pharmacognosy,* 6th ed. Philadelphia: Blakiston.
115. Zita, C. (1955) *Cas. Lek. Cesk.,* 8, 203–208.
116. Zwaving, J.H. (1982) *Pharm. Weekbl.,* 117, 157–165.

3 Plant Sources

Rolf Franke

CONTENTS

3.1 SYSTEMATICS

In 1664, Tabernaemontanus [19] distinguished six different varieties of chamomile (contrary to Dioscorides, who distinguished three):

1. "Unsere gemeine Chamillenblum" ("our common chamomile flower") is identified as the plant named *Anthemis* or *Leucanthemum* by Dioscoride and Galeno. "Dann wann wir die Beschreibung DIOSKORIDIS mit fleiß übersehen/und das Capitel von dem wolriechenden Kräutlein Anthemidis oder Leucanthemi vorhanden nehmen/die liebliche Gestalt und Abconterfeyung dieser wolriechenden Chamillenblumen dagegen-halten/darneben auch ihre Krafft und Wirckung beyderseits erwegen/so beyde diesem Kräutlein oder Blumen von DIOSCORIDE und GALENO zugeschrieben/und auch durch langwürige tägliche Erfahrung gewiß erfunden worden/können wir mit der Warheit nicht anders urtheilen/dann daß unser wolriechend gemeine Feld- Chamillen/das recht Anthe-mis und Leucanthemum der Alten seye." (Old High German: "If we duly disregard Dioscorides' description/and go through the chapter about the sweet-scented herb Anthe-midis or Leucanthemi/comparing it with the lovely appearance of these fragrant cham-omile flowers/considering also their power as well as their effect/both ascribed to this herb by Dioscoride and Galeno/and also found by long daily experience/we can only say that the sweet-scented ordinary/common field chamomile/is the true Anthemis and Leucanthemum of the ancients.")
2. "Römisch Chamillen" ("Roman chamomile"). "Das andere Geschlecht der Chamillen Leucanthemi ist den Alten unbekannt gewesen/und nicht von ihnen beschrieben worden: Das ist erstlich auß Hispanien/Engelland und andern frembden Orten zu uns gebracht worden und ist heutigen Tags in Teutschland sehr gemein …." ("The other variety of Chamomile Leucanthemi was unknown to the ancients/and it was not described by them. It was first brought to us from Spain/England and other foreign places, nowadays it is very common in Germany …")
   In this case the modern variety of *Chamaemelum nobile* (L.) All. is obviously meant.
3. "Gefüllt Römisch Chamillen" ("Filled Roman chamomile"). Here the modern variety of *Chamaemelum nobile* (L.) All. is meant as well.
4. "Gefüllt Römisch Chamillen anderer Gattung" ("Filled Roman chamomile of another genus"). In this case the determination is not clear. Among others, the designation indicated by Tabernaemontanus (1664) [19] is "Chamaemelum Anglicum flore multiplici … dieweil solche auß Engelland erstlich zu uns in Teutschland gebracht worden seynd" ("… because they were brought from England to Germany for the first time").
5. "Geel Chamillen" ("yellow chamomile"), probably meaning *Anthemis tinctoria* L.
6. "Rothe Chamillen" ("red chamomile"). This is probably *Adonis aestivalis* L. or *Adonis flammea* Jacq. being indicated by the following description: "Gegen dem Brachmonat bringt es an den Gipffeln der Stengel und Nebenästlein über die maß schöne/rothe/Men-nigfarbe/oder feuerrothe Blümlein/inwendig mit einem schwarzen Bützlin/hat ein jede Blum sieben Blättlein/die seynd am end ein wenig hintersich zurück gebogen. Nach der Blüth folgewn kleine stachlichte Kölblein/darinnen der Saamen verschlossen ist" ("towards June most beautiful/miniaceous/or small fire red flowers/are produced at the end of the stems and their small side branches/inside with a black center/every flower has seven small leaves/which are somewhat bent down at the end. After flowering small thorny spadices are following/where the seeds are to be found"). The name also points to this classification: "… Oculus daemonis. Teutsch/roth Chamillen/und in Thürin-gen/Teufels Aug/von wegen der rothen Fewerfarben Blumen" ("Oculus daemonis. Ger-man/red chamomile/and in Thuringia/devil's eye/because of the fire red flowers").

Although the systematic status is quite clear nowadays, there are a number of inaccuracies concerning the names. Apart from misdeterminations and confusion, the synonymous use of the names of *Anthemis, Chamomilla,* and *Matricaria* leads to uncertainty with regard to the botanical identification, particularly in the area of English speech. So, in the BelgV *Anthemis nobilis* L. is called "Chamomillae flos." "Marokkanische Kamille" ("Moroccan chamomile") often consists of *Ormenis multicaulis* Braun-Blanq. et Maire [4].

Moreover, the nomenclatural situation is complicated by the fact that Linneaus made mistakes in the first edition of his *Species Plantarum* that he corrected later on. According to this description "*Matricaria chamomilla* L. 1753" is definitely not the name for *True chamomile* medically used but for *Scentless chamomile*. The name applicable to *True chamomile* is the one of the species published at the same time, i.e., *Matricaria recutita* L. [12,18].

The name of the genus of "*Matricaria*" used by Linneaus is derived from matrix (womb). The popular name of "Mutterkraut" ("Mother's Herb") also points to the application for various female complaints, being surely derived from this range of applications. As, however, the Parthenion mentioned by Dioscoride does not stand for *True chamomile* but for *Tanacetum parthenium*, the popular German name of "Mutterkraut" ("Mother's Herb") also used by Zander [21] should not be used for chamomile.

The genus of *Matricaria* in a wider sense is often divided into: *Tripleurospermum* Schultz-Bip. among others with *Matricaria maritima* and *Matricaria perforata* (syn. *M. inodora*) and *Matricaria* sensu stricto among others with *Matricaria recutita* L. (syn. *Matricaria chamomilla* L.) and *Matricaria suaveolens* (syn. *M. matricarioides*).

This division needs a revision. Linneaus does not seem to have separated *Matricaria chamomilla* and *Matricaria maritima*, including *M. inodora*. In 1974 Rauschert pointed out that Linneaus used *M. chamomilla* rather than *M. maritima* for *M. chamomilla* in a modern sense [12].

It seems to be quite obvious that Linné used the name of *Matricaria recutita* for our medically applied chamomile (*Matricariae flos, Ph. Eur.* 4.6). Later Linné named this tribe called *Matricaria suaveolens* "suavius olens."

In his systematic reinvestigation Rauschert [12] also mentions that when dividing the genus Linné called *Matricaria*, the separated part with our medically used chamomile should bear the name of *Chamomilla* S.F. Gray, and that furthermore the name of *Matricaria* L. sensu stricto would be correct for the genus of *Tripleurospermum* Schultz-Bip. In volume IV (pp. 58 60) of *Flora Europea* the medically used chamomile was called *Chamomilla recutita* (L.) Rauschert, whereas the scentless chamomile was given the names *Matricaria maritima* L. and *Matricaria perforata* Mérat (syn. *M. inodora* L.).

Considering a new botanical classification there is a mere nomenclatural problem on the one hand (viz., whether the name of "*Chamomilla*," "*Matricaria*," or "*Tripleurospermum*" should be used), and on the other hand there is a systematic problem (viz., whether the genus fixed by Linné is to be segregated).

When segregating Linné's genus of *Matricaria* L. into the genera of *Chamomilla* S.F. Gray and *Matricaria* L. sensu stricto another difficulty arises because *Matricaria chamomilla* was chosen as a species type of the genus of *Matricaria* [9]; so — according to Rauschert [12] — True chamomile would be added to the genus of *Chamomilla*. What speaks against it is the fact that "*Matricaria chamomilla* 1753" does not correspond to the diagnosis of the genus, as the achenes have a coronule. Thus, the choice of the species type was made by mistake, and *Matricaria recutita* has to be regarded as a type of the genus [18]. The differences to be found in botanical literature still continue today. In connection with the aim of this book the name of *Matricaria recutita* L. (syn. *M. chamomilla* L., *Chamomilla recutita* L. Rauschert) should, however, be used; this is particularly recommended as even in the latest literature — with exception of *Flora Europea* — and in all pharmacopoeias at present the name of *Matricaria recutita* L. is used.

For a considerable length of time it was not quite clear as to whether chemical races exist within the species of *Matricaria recutita* L. Personal investigations as well as tests of other teams point to the existence of genetically conditioned chemical variations in local populations. Instead of using the word "race," the more neutral term of *"dem"* should be applied with *Matricaria recutita* L. [8,14]; even more precise are the terms "chemodem," "ecodem," and "topodem." Considering the fact that chamomile is a widely common plant the existence of chemodems, topodems, and ecodems is not only very likely but the research results of several working groups confirm that such "dems" exist. The following chamomile varieties still traded a few years ago could be called topodems, because their names give knowledge of the regions from which they originate: Holsteiner Marschkamille (Holstein Marsh chamomile), Ostfriesische Kamille (East Frisian chamomile), Fränkische Kamille (Franconian chamomile), Niederbayrische Kamille (Lower Bavarian chamomile), Quedlinburger großblütige Kamille (Quedlinburg large-flowered chamomile), Erfurter kleinblütige Kamille (Erfurt small-flowered chamomile), Böhmische Kamille (Bohemian chamomile). However, a number of goods still handled today, coming from wild collections and not from selective cultivation of breeding lines, may also be put into this category (for example, Egyptian chamomile, Hungarian Puszta chamomile, Argentinian chamomile, Spanish chamomile).

Moreover, there are chamomile chemocultivars (cultivar = cv), thus "chemical races" produced by breeding and maintained constant by maintenance breeding. Depending on the spectrum of active principles, the culture varieties could, for example, be specified as *Matricaria recutita* L. cv. "rich in bisabolol" or as *Matricaria recutita* L. cv. "rich in bisabololoxide" (see Section 5.3).

Nonradial chamomile, Scentless chamomile, Roman chamomile, as well as the species of Anthemis have to be distinguished from "True chamomile."

Different species can appear as confusion or falsification of the collected drug. Here are some distinctive features of some of these other *Anthemis* and *Matricaria* species and how they can be used [6]:

1. Without scale-like palets between the flowers of the capitulum
   - Capitulum bottom cone-shaped long, hollow
        Plant with white ligulate flowers, smells pleasantly of chamomile (typical chamomile smell), annual → True chamomile (*Matricaria recutita* L.)
        Plant without white ligulate flowers, smells like chamomile (typical chamomile scent), annual → Rayless chamomile (*Matricaria matricarioides* [Less.] Porter, syn. *Matricaria discoidea* DC.)
   - Capitulum bottom-arched or only short cone-shaped, marrowy. Plant without typical chamomile scent, but smells not unpleasant. Flower capitulum big (diameter approx. 3 cm), annual, biennial (or perennial) → Odorless, false chamomile (*Matricaria maritima* L.)
2. With, at least in the middle part of the flower capitulum, small setiform paleae between the flowers of the flower heads
   - Palets longish acuminate or with pointed stings
        Plant with revolting smell, annual → Stinking chamomile (*Anthemis cotula* L.)
        Plant without revolting smell, mostly annual → Field chamomile (*Anthemis arvensis* L.)

Palets blunt, with dry tips. Plant smells pleasantly, perennial, many-headed rootstock → Roman chamomile (*Chamaemelum nobile* [L.] All. syn. *Anthemis nobilis* L.)

3.2 LATIN BOTANICAL NAMES, SYNONYMS

In the various florae and books for the identification of plants the names used differ considerably. The synonyms already mentioned by Tabernaemontanus (1664) [19] are most interesting where they are specified in many different languages. This currently shows an amazing constancy as far as the use of the popular names in central Europe, but also in the Mediterranean area, in the Balkans, and in Arabia. Considerable revisions of the Latin nomenclature have been made in recent years. Nevertheless, it is difficult to make clear statements in this respect. The most detailed explanation for the correct nomenclature existing at present was surely indicated by Schultze-Motel in 1986 [18]. The name of the genus of *Matricaria* L. dates back to the year 1753. Later *Chamomilla* S.F. Gray (1841), *Lepidotheca* Nutt. (1841), *Tripleurospermum* Schultz-Bip. (1844), *Sphaeroclinium* Schultz-Bip. (1844), *Gastrosulum* Schultz-Bip. (1844), *Gastrostylum* Schultz-Bip. (1844), *Rhytidospermum* Schultz-Bip. (1844), *Courrantia* Schultz-Bip. (1844), *Dibothrospermum* Knaf (1846), *Chamaemelum* Vis. (1847) non Adans. (1753), *Trallesia* Zumag. (1949), *Akylopsis* Lehm. (1850), *Otospermum* Willk. (1864), *Heteromera* Pomel (1874) were names used more recently. *Matricaria recutita* L. is to be considered as a type.

In the older statement by Bolkhovskikh et al. [3] the following species are united in the genus of *Matricaria* L.: *M. ambigua* (Lebed.) Kryl., *M. chamomilla* L., *M. discoidea* DC., *M. grandiflora*, *M. inodora* L., *M. maritima* L., *M. matricarioides* (Less.) Porter, *M. pubescens* (Desf.) Schultz-Bip., *M. suaveolens* Buch., *M. tchichatchewii* Boiss.

The genus of *Anthemis* L. includes the species of *A. alpina* L., *A. altissima* L., *A. arvensis* L., *A. austriaca* Jacq., *A. carpatica* Waldst. et Kit., *A. cotula* L., *A. cupaniana*, *A. iberica* Bieb., *A. jailensis* Zefir., *A. kelwayi*, *A. maritima* L., *A. monantha* Willd., *A. montana* Willd., *A. nobilis* (L.) J. Gray, *A. orientalis* (L.) Deg., *A. peregrina* L., *A. rigescens* Willd., *A. rudolphiana* Adams, *A. ruthenica* Bieb., *A. santi-johannis*, *A. sosnovskyana* Fed., *A. tigreensis* J. Gray ex A. Rich., *A. tinctoria* L., *A. trotzkiana* Claus ex Bunge, *A. woronowii* Sosn., *A. zyghia* Woronow.

The genus of *Chamomilla* is not indicated.

In the meantime nomenclatural knowledge could be actualized. Today the genus of *Anthemis* L. comprises about 210 species, thus representing one of the greatest genera of the chamomile relatives [3]. In Europe these include (among others):

- *Anthemis altissima* L. emend. Spreng. (syn. *Anthemis cota* L. emend. Vis.), being common in Europe, the Mediterranean area, Portugal, the Near East, and Central Asia
- *Anthemis arvensis* L., Field chamomile, spread over Europe, North Africa, Asia Minor, established in the eastern part of North America
- *Anthemis austriaca* Jacq., Austrian Field chamomile, being common in the eastern part of central Europe, southeast Europe, Anatolia, Armenia, the Caucasus Mountains, North Africa
- *Anthemis carpatica* Waldst. et Kit. ex Willd., spread over northern Spain, the Pyrenees, the Austrian Alps up to the eastern Carpathian Mountains and northern Greece
- *Anthemis cinerea* Panc., being common in the Balkans
- *Anthemis cotula* L., Stinking chamomile, being common in Europe, the Near East, North Africa, Canary Islands, established in North and South America, Australia
- *Anthemis cretica* L., spread over south Europe and the western part of the Czech Republic
- *Anthemis haussknechtii* Boiss. et Reut., being common in Syria, Iraq, Iran, Southern Anatolia
- *Anthemis marschalliana* Willd. (syn. *A. biebersteiniana* (Adams) K. Koch, *Chrysanthemum biebersteinianum* Adams), spread over the Caucasus Mountains and in the mountainous regions of Asia Minor
- *Anthemis plutonia* Meikle, being common in Cyprus as an endemic variety
- *Anthemis ruthenica* M. Bieb., Ruthenic Field chamomile, spread over Germany and Austria
- *Anthemis sancti-johannis* Turrill, being common in the southwestern part of Bulgaria
- *Anthemis tinctoria* L., Yellow chamomile, spread nearly all over Europe, western Asia, established in North America
- *Anthemis tricolor* Boiss., being common in Cyprus as an endemic variety
- *Anthemis triumfettii* (L.) DC., spread over Switzerland and southern Europe.

Zander [21] puts both True chamomile as *Chamomilla recutita* (L.) Rauschert (syn. *Matricaria recutita* L., *M. chamomilla* L. pro parte) and nonradial chamomile as *Chamomilla suaveolens* (Pursh) Rydb. (syn. *Matricaria matricarioides* (Less.) Porter pro parte, *M. discoidea* DC., *M. suaveolens* (Pursh) Buchenau non L., *Santolina suaveolens* Pursh) in a genus of its own: *Chamomilla* S.F. Gray. This classification was also used by Schubert and Vent [17]. But in Zander [22] this nomenclature is revised back to *Matricaria recutita* L.

According to Zander [21, 22] the genus of *Matricaria* L. (syn. *Tripleurospermum* Schultz-Bip.) comprises the following species:

- *M. capensis* hort. non L. → now *Tanacetum parthenium* (L.) Sch. Bip.
- *M. caucasica* (Willd.) Poir. → now *Tripleurospermum caucasicum* (Willd.) Hayek, spread over Bulgaria, Albania, the Caucasus Mountains, Asia Minor
- *M. inodorum* L. → now *Tripleurospermum perforatum* (Mérat) Laínz, spread over Asia Minor, Caucasus, Siberia, Amur, Sachalin, Kamchatka
- *M. maritima* L. → now *Tripleurospermum maritimum* (L.) W.D.J. Koch, spread over Iberic peninsula, France, central Europe, Poland
- *M. oreades* Boiss. (syn. *Chrysanthemum oreades* (Boiss.) Wehrh.) → now *Tripleurospermum oreades* (Boiss.) Rech., being common in Asia Minor and Syria
- *M. parthenioides* (Desf.) hort. → now *Tanacetum parthenium* (L.) Sch. Bip.
- *M. perforata* Mérat (syn. *M inodora* L., *M. maritima* L. ssp. *inodora* (K. Koch) Soo, *Chrysanthemum maritimum* (L.) Pers. ssp. *inodorum* (K. Koch) Vaar., *Chamaemelum inodorum* (L.) Vis.) → now *Tripleurospermum inodorum* (L.) Schultz-Bip., spread nearly all over Europe, the Caucasus Mountains, western Asia. With Schubert and Vent [17] this species is — among others — classified as Scentless chamomile, carrying the name of *M. maritima* L. (syn. *T. maritimum* (L.) Koch, *M. inodora* L.) and subdivided into the ssp. *maritima* and ssp. *inodora* (L.) Dostal. (see also Reference 2)
- *M. recutita* L. (see Section 3.2.1)
- *M. tchihatchewii* (Boiss.) Voss (syn. *Chamaemelum tchihatchewii* Boiss.) → now *Tripleurospermum tchihatchewii* (Boiss.) Bornm., being common along the coasts of Asia Minor

In the genus of *Chamaemelum* Mill. Zander [21] is only specifying the species of *Ch. nobile* (L.) All. (syn. *Anthemis nobilis* L. is an older name often still used in drug trade, *Ormenis nobilis* (L.) J. Gray ex Coss. et Germ.) spread over western Europe (up to Northern Ireland), established in south Europe and in the southern part of central Europe as well as in North Africa. A more detailed subdivision into the following species is made by Hänsel et al. [5]: *Chamaemelum eriolepis* (Cosson ex Maire) Benedi, *Ch. flahaultii* (Emberger) Benedi, *Ch. fuscatum* (Brot.) Vasc. (syn. *Anthemis fuscata* Brot., *A. praecox* Link in Schrader., *Chamomilla fuscata* (Brot.) Gren et Godr., *Maruta fuscata* Brot., *Ormensis fuscata* (Brot.) Schultz-Bip., *Ormensis praecox* (Link in Schrader) Briq. et Cavill., *Perideraea fuscata* (Brot.) Webb.), *Ch. mixtum* (L.) All. with var. *glabrescens* (Maire) Benedi, var. *aureum* Durieu in Bory et Durieu) Benedi, *Ch. nobile* (L.) All.

3.2.1 *MATRICARIA RECUTITA* L.

The best-known botanical name for True chamomile, also used in the pharmacopoeias, is *Matricaria recutita* (syn. *Matricaria chamomilla* L., *Chamomilla recutita* (L.) Rauschert). The Latin name of *recutitus* refers to the petals, meaning truncated, trimmed. The name of "chamomilla" may well originate from Dioscoride and Plinius the Elder who — due to the pomaceous odor — called the plant "chamaimelon." *Chamaimelon* means, more or less, "low growing apple tree" (Greek: chamai = low, melon = apple). Plinius the Elder wrote about "Chamaimelon quoniam odorem mali habet." Tabernaemontanus [19] also noted this applelike odor: "Die Chamillenblum/die bey uns in Teutschland vor sich selbst in Fruchtfeldern häuffig wächst/ist die rechte Chamillen oder Chamaemelum der Alten/die den Namen daher empfangen hat/dieweil sie reucht/wie ein lieblicher wolriechender Apffel." ("The chamomile flower frequently growing on fruit fields with us in Germany/is the true chamomile or chamaemelum of the ancients/which has got its name because it smells like a delicious apple.")

With Bolkhovskikh et al. [3] chamomile is classified as *Matricaria chamomilla* L. Zander [21], Schubert and Vent [17], and others call the True chamomile *Chamomilla recutita* (L.) Rauschert; Aichele and Schwegler [1] call it *Matricaria chamomilla* L.; with Hänsel et al. [5] the name of *Chamomilla recutita* (L.) Rauschert is preferred (syn. *Chamomilla meridionalis* C. Koch, *Ch. vulgaris* S.F. Gray, *Matricaria chamomilla* L. pro parte, *M. coronata* Gay ex Koch, *M. pusilla* Willd., *M. recutita* L., *M. suaveolens* L.).

In Schultze-Motel [18] a detailed compilation of the available nomenclatural literature is to be found. Here the name of *Matricaria recutita* L. is given primacy (1753). The following names may be regarded as derived: *Matricaria suaveolens* L. (1755), *M. chamomilla* L. (1755 and 1763), *M. chamaemilla* Hill (1761), *Leucanthemum chamaemelum* Lam. (1779), *Matricaria patens* Gilib. (1782), *Chamomilla patens* Gilib. (1782), *Matricaria tenuifolia* Salisb. (1796), *Chrysanthemum chamomilla* Bernh. (1800), *Ch. suaveolens* (L.) Cav. (1803), non (Pursh) Aschers. (1864), *Matricaria pusilla* Willd. (1807), *Chamomilla vulgaris* S.F. Gray (1821), *Matricaria courrantiana* DC. (1837), *M. pyrethroides* DC. (1837), *Chamomilla officinalis* C. Koch (1843), *Ch. meridionalis* C. Koch (1843), *Matricaria coronata* Gay ex Koch (1843), *M. kochianum* Schultz-Bip. (1844), *Courrantia chamomilloides* Schultz-Bip. (1844), *Chamomilla courrantiana* C. Koch (1851), *Matricaria bayeri* Kanitz (1862), *M. obliqua* Dulac (1867), *Chamomilla recutita* (L.) Rauschert (1974).

The habit of *Matricaria recutita* L. is relatively stable. In Hegi [7], four differing forms are distinguished:

1. f. *nana* Custer (= *monocephala* Junge = f. gracilis Chenévard): The plant is a dwarf plant and remains small (10 to 15 cm), not ramified, small leafed. Stem approx. 1 mm thick, capitulums small (8 to 15 mm in diameter).
2. f. *paleata* Abromeit: Capitulums with fine palets.
3. f. *eradiata* Rupr. (f. *subdiscoidea* A. Peter): The ligulate flowers are missing completely or are very short.
4. f. *coronata* (J. Gay) Coss. et Germ.: Widespread above all in southern Europe, with clearly developed pappus.

3.2.2 *Chamaemelum fuscatum* (Brot.) Vasc.

Syn. *Anthemis fuscata* Brot., *A. praecox* Link in Schrader., *Chamomilla fuscata* (Brot.) Gren et Godr., *Maruta fuscata* Brot., *Ormensis fuscata* (Brot.) Schultz-Bip., *Ormensis praecox* (Link in Schrader) Briq. et Cavill., *Perideraea fuscata* (Brot.) Webb.

3.2.3 *Chamaemelum nobile* (L.) All.

Syn. *Anthemis nobilis* L., *Ormenis nobilis* (L.) J. Gray ex Coss. et Germ., *Anthemis chamomilla-romana* Crantz, *Anthemis nobilis* L., *A. odorata* Lam., *Chamaemelum odoratum* Dod., *Chamomilla nobilis* God., *Leucanthemum odoratum* Eid Ap., *Lyonnetia abrotanifolia* Webb., *Matricaria nobilis* Baill., *Ormensis aurea* R. Loewe, *Ormensis nobilis* J. Gay.

3.2.4 *Anthemis arvensis* L.

Syn. *Anthemis agrestis* Wallr., *Chamaemelum arvensis* (L.) All.

3.2.5 ANTHEMIS COTULA L.

Syn. *Anthemis foetida* Lam., *A. psorosperma* Ten, *A. ramosa* Link, *A. heterophylla* Wallr., *A. abyssinica* J. Gay, *Chamaemelum cotula* (L.) All., *Ch. foetidum* Baumg., *Maruta cotula* (L.) DC., *M. foetida* (Lam.) S.F. Gray, *M. vulgaris* Bluff et Fingerh.

3.2.6 ANTHEMIS TINCTORIA L.

Syn. *Chamaemelum tinctorium* (L.) All., *Cota tinctoria* J. Gay ex Guss.

3.3 COMMON NAMES, SYNONYMS

3.3.1 MATRICARIA RECUTITA L.

German: Echte Kamille, Deutsche Kamille, Feldkamille, Gemeine Kamille, Hermel, Kamille, Kleine Kamille, Mägdeblume, Apfelkraut, Kummerblume, Mutterkraut, Kindbettblume, Hermännle, Helmchen, Lungenblume, Ramerian, Remey, Stomeienblume
English: True chamomile, common camomile, German chamomile, Hungarian chamomile, small camomile, horse gowan, wild camomile, camomile flowers, pin heads
French: Camomille, matricaire, camomille vraie, camomille allemande, camomilla commune, camomille ordinaire, camomille vulgaire, chamomille d'Allemagne, matricaire camomille
Dutch: Kamille
Italian: Camomilla, camomilla comune, capomilla
Portuguese: Camomila, camomilla des Alemães, camomilla vulgar, mançanilha, margaça das boticas
Spanish: Manzanilla, camomilla, manzanilla alemana, manzanilla común, manzanilla de Aragón, manzanilla dulce, manzanilla ordinaria, camamilda
Danish: Kamille
Swedish: Kamomill
Czech: Hermánek pravý, kamilka, rumancek kamilkový
Slovakian: Rumancek pravý
Hungarian: Orvosi székfü, kamilla, székfü
Polish: Rumaniek pospolity
Russian: Romaška aptecnaja, romaška obodrannaja
Turkish: Papatya
Indian: Babunphul
Persian: Babunah

3.3.2 CHAMAEMELUM NOBILE (L.) ALL.

German:Römische Kamille, Römische Hundskamille, Edel-Kamille, Gartenkamille, Doppelte Kamille, Große Kamille, Dickköpfe, Welsche Kamille, Dicke Gramille, Gaadekamille, Krampf-Kamille, Gartenchamille, Gemeine Chamille, Hemdknöpkes, Hemmerknebche, Kragenknebcher, Härmelchen, Johannisgnebel, Kathreinenbläume, Kragengebcher, Kühmelle, Mandl, Mönetli, Römischer Romey, Tüfelschrut, Wälschi Öpfelblümli, Zandelkraut
English: Roman chamomile, Noble chamomile, small camomile, ground apple, low camomile, whig plant, English camomile, Scotch camomile, May-then, sweet chamomile, double chamomile, white or low camomile
French: Camomille romaine

Dutch: Roomse kamille

Italian: Camomilla romana, camomilla odorosa, camomilla de Boemia, camomilla di Germania, camomilla inglese, camomilla nobile, camomilla ortense

Portuguese: Camomilla verdadeira, camomilla nobre, camomilla odorante, macela dourada, macela galega, camomila romana

Spanish: Manzanilla romana, manzanilla noble, manzanilla inglesa, manzanilla de Paris, manzanilla officinal

Catalonian: Camamilla romana

Basque: Bitxilora, lilibitxi, enbasabedarr, larranbillo

Danish: Romerske cameelblomster, Rómerske Kamille

Czech: Rmen sličný,   Heřmánek římský, paruman spanilý

Slovakian: Ruman rímsky (spanilý), paruman spanilý

Hungarian: Rómikamilla, római kamilla, nemes pipitér, római székfü

Polish: Koszyszek rumianku rzymskiego, Rumianek rzymski, Rumian szlachetny

Russian: Rimskaja romaška

Swedish: Ängsröllika

Turkish: Rumi papatya, Alman papatyasi

Arab: Babunj, Babunaj, Shajrat-ol-kafur

Indian: Babunike-phul, Shimedapu, Shimaichamantipu

Persian: Babunah

3.3.3 *ANTHEMIS ARVENSIS* L.

German: Acker-Hundskamille, Feld-Hundskamille, Ackerkamille, Afterkamille, Chrotechrut, Dreckkamille, Gensstöck, Hundsblume, Hundsdille, Korngöckala, Kretengosch, Krötendill, Pferdskamille, Rüenblaume, Schafkamille, Stinkpotsch, Taube Kamille, Wilde Kamille

English: Wild chamomile, corn chamomile, field camomile

French: Oeil-de-vache, camomille sauvage, fausse camomille

Dutch: Valse kamille

Italian: Camomilla senza odore, camomilla bastarda

Spanish: Manzanilla bastarda, manzanilla borde, manzanilla de los campos, manzanilla silvestre

Swedish: Åkerkulla

3.3.4 *ANTHEMIS COTULA* L.

German: Hundskamille, Stinkende Hundskamille, Kuhdille

English: Dog fennel, wild camomile, stinking chamomile, dog chamomile, stinking mayweed

French: Camomille des chiens

Italian: Camomilla mezzana

Portuguese: Macela fetida

Spanish: Magarza, manzanilla hedionda, manzanilla malagata

3.3.5 *ANTHEMIS TINCTORIA* L.

German: Färber-Hundskamille, Färberkamille, Rindsauge, Ochsenauge

English: Dyer's camomile, golden marguerite, ox-eye chamomile, yellow chamomile

French: Camomille de teintures

Russian: Pupavka krasilnaja

3.3.6 *Tripleurospermum inodorum* (L.) Schultz- Bip.

German: Bitterstock, Brautschleier, Falsche Kamille, Feinblättrichte Johannisblume, Feinblät-
   trige Wucherblume, Geruchlose Bertramwurz, Geruchlose Wucherblume, Hemdeknöppchen,
   Hummel, Hundskamille, Krötenstock, Pferdekamille, Wilde Kamille
English: Bachelor's button, barnyard daisy, corn feverfew, corn mayweed, false chamomile,
   false mayweed, scentless camomile, wild chamomile
French: Matricaire inodore
Danish: Falsk kamelblomst, hundeblomst
Swedish: Baldersbrå, knoppsörört
Hungarian: Pupa
Romanian: Musetel-prost, romanita-neadevarata

3.4 DESCRIPTION OF THE PLANT

3.4.1 Botany

True chamomile is an annual plant, grows 10 to 80 cm high, prefers sandy to loamy, mostly sour
fields and fresh ruderal places. Salty soils like in the Puszta or in Argentina are duly tolerated. The
plant has thin spindle-shaped roots that only penetrate flatly into the soil. The stem is in an upright
position, mostly heavily ramified, bare, round, and filled with marrow. The leaves are alternate,
double to triple pinnatipartite, with narrow-linear prickly pointed sections being hardly 0.5 mm
wide. The flower heads are placed separately, they have a diameter of 10 to 30 mm, and they are
pedunculate and heterogamous. The involucre is semispherical. The 26 to 48 involucral leaves are
arranged in one to three rows, they are ovoid to lanceolate upside down, green with a narrow
brownish membranous rim. The golden yellow tubular florets with five teeth are 1.5 to 2.5 mm
long, ending always in a glandulous tube. The plant flowers concentrically from the bottom to the
top. The 11 to 27 white ligulate flowers are 6 to 11 mm long and 3.5 mm wide, longer than the
involucre and bent back. The filamentous style shows two stigmata. The pollen has short prickles
and three hila. The receptacle is 6 to 8 mm wide, flat in the beginning and conical, cone-shaped
later, hollow (being a very important distinctive character) and without paleae. The fruit is a
yellowish brown to brown achene, 0.8 to 1.3 mm long, about 0.3 mm wide, slightly pressed together
and bent horn-like, attenuated at the base, truncated at an angle at the top, with four to five ribs
on the concave lower side. On the ribs small mucous glands are to be found, rounded on the back,
ribless, poorly punctured with glands on the outside, humid, mucous. Pappus is missing or can be
traced as a rim being hardly developed. The thousand kernel weight is 0.02 to 0.06 grams with
diploid forms and 0.04 to 0.12 grams with tetraploids (see Figure 3.1).

   True chamomile is very often confused with plants of the genera *Anthemis* and *Matricaria*.
Special attention has to be paid to confusions with *Anthemis cotula* L. For the differentiation from
A. cotula the following characters may particularly be taken into account: *A. cotula* has setiform
paleae, a receptacle with marrow, and a revolting smell. *A. arvensis* L. and *A. austriaca* Jacq. also
have prickly pointed paleae and a filled receptacle. Both species are, however, nearly odorless.
Matricaria maritima L. and *M. perforata* Mérat are odorless as well. These species have a filled
receptacle but no paleae. *Chamomilla suaveolens* (Pursh) Rydb. (syn. *Matricaria discoidea* DC,
M. matricarioides [Less.] Porter) has a typical chamomile scent, too, but has no ligulate flowers
and a compact growth can be noticed.

   After exclusion of the genus of *Chamaemelum* Mill. (syn. *Ormenis* Cass.) *Anthemis* L. is well
separable as a genus. Contrary to True chamomile the following generic characteristics may be
noted: the leaves are alternate, mono or double pinnatipartite with mostly linear-lanceolate to linear
sections. The laciniae are comb-shaped pinnatifid. The involucre is bowl-shaped to semispherical

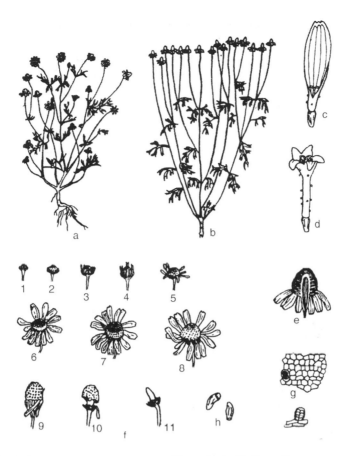

a. Scheme of a plant
b. Form of flowering horizon
c. Petal, ligulate floret
d. Tubular floret

e. Flower head with the cavity
f. 1–11: different stages of flower development and ripeness
g. Fragments of epidermis
h. Seeds

FIGURE 3.1 True chamomile (*Matricaria recutita* L.) in schemes [14].

or funnel-shaped. The involucral leaves are multiserial and blunt and have a dry membranous seam. The receptacle is flat or curved semispherically to conically, filled and mostly provided with leathery to membranous paleae. The white or yellow uniserial ligulate flowers are female or sterile, and sometimes they are missing. The numerous hermaphroditic yellow tubular florets often have a somewhat applanate and thickened tube with no pouchlike enlargement at the base. The achenes are oblong conical upside down to cylindrical and applanate from the front to the back, tetragonal with numerous (10 to 20) more or less distinct ribs. The pappus is a membranous collar or it is missing altogether.

Matricaria recutita L., *Chamaemelum nobile* (L.) All., and some *Cotula* species are morphologically closely related, so there is a danger of adulteration. The major palynological characteristics that proved useful to distinguish all three taxa were equatorial and polar diameters, colpi length, spine length, number of spine rows between colpi, and foraminate and non-foraminate sculpturing [20].

The independence of the genus of *Chamaemelum* Mill. — in comparison with the genus of *Anthemis* L. — is nowadays accepted on account of the following characteristics: *Chamaemelum* plants reach a height of 6 to 40 cm. The plants have many stems and are hairy. The long pedunculate flower heads have a diameter of 20 to 25 mm. The multiserial involucre is hairy with a dry membranous rim. The bottom of the inflorescence is conical. The paleae are broadly spatulate, curved, thin-skinned, the central nerve is often hairy. The marginal white ligulate flowers are female or sterile, the tongue has three laciniae. The yellow hermaphroditic tubular florets with five laciniae have a ring-shaped enlargement at the base. The brownish achenes are 1 to 1.6 mm long, roundish in the cross section, little applanate, smooth, without distinct ribs, but on the lower side they have three thin longitudinal stripes. The pappus is missing.

3.4.2   CHROMOSOME NUMBERS

Considering a basic chromosome number of n = 9 the diploid chromosome number 2n = 18 is predominant in nearly all species. An exception may well be *Anthemis tchichatchewii* with 2n = 4x = 36, and *A. carpatica* with both 2n = 4x = 36 and 2n = 6x = 54. Only a few species like *Anthemis maritima* and *Matricaria perforata* exhibit diploid and polyploid forms. Furthermore, B-chromosomes were found on the diploid stage [10] with *M. perforata*. In breeding forms of *Chamaemelum nobile* and *Matricaria recutita* polyploidizations were carried out so that now tetraploid breeding forms/varieties are available (see Chapter 5, Table 5.3.1).

3.5   DRUG NAMES

In numerous names of the medical literature the plural is used and the constituent parts of the plants are put in front (e.g., *Flos* or *Flores Chamomillae*). Linguistically this is not correct.

In the following the singular of the constituent part of the plant is therefore always used and the name of the constituent part of the plant's drug name is placed behind (*Matricariae flos*), as it is also done in the *European Pharmacopoeia* (1997), even if that was not always the case in the primary literature.

3.5.1   *MATRICARIA RECUTITA* L.

- *Chamomillae flos officinalis, Chamomillae flos* courant I or II variety, *Chamomillae vulgaris flos, Chamomillae vulgaris flos* fines, *Chamomillae vulgaris flos* pro balneo (bath chamomile), *Chamomilla, Matricariae flos, Chamomillae anthodium, Chamomillae cribratum, Chamomillae stramentum, Chamomillae herba, Chamomillae cum floribus herba*
- *Chamomillae aethereum oleum, Chamomillae aetheroleum, Chamomillae oleum, Matricariae aetheroleum*
- *Chamomillae tinctura*

3.5.2   *CHAMAEMELUM FUSCATUM* (BROT.) VASC.

- Chamaemelum fuscatum — flowers

The drug is not traded in Europe.

3.5.3   *CHAMAEMELUM NOBILE* (L.) ALL.

- Chamomillae romanae flos, Chamomillae hortensis flos, Chamomillae majoris flos, Chamomillae nobilis flos, Chamomillae odorati flos, Anthemidis flos, Leucanthemi (romani) flos, Chamomilla romana, Chamaemelum nobile hom., Anthemidis nobilis hom.
- Chamomillae romanae aethereum oleum, Chamomillae romanae aetheroleum, Chamomillae romanae oleum, Anthemidis aethereum oleum, Anthemidis oleum

3.5.4   *CHAMOMILLA SUAVEOLENS* (PURSH) RYDB.

- Matricariae discoideae herba, Matricariae suaveolens herba

TABLE 3.1
Chromosome numbers in the genera *Anthemis*, *Chamaemelum*,
***Matricaria* (according to [3], modified)**

Genus	Species	Chromosome number
Anthemis	*alpina* L.	18
	altissima L.	18
	arvensis L.	18
	austriaca Jacq.	18
	carpatica Waldst. et Kit.	36, 54
	cotula L.	18
	cupaniana	18
	iberica Bieb.	18
	jailensis Zefir.	18
	kelwayi	18
	maritima L.	18, 36, 54
	monantha Willd.	18
	montana Willd.	18
	orientalis (L.) Deg.	18
	peregrina L.	18
	rigescens Willd.	18
	rudolphiana Adams	16, 18
	ruthenica Bieb.	18
	santi-johannis Turill	18
	sosnovskyana Fed.	18
	tigreensis J. Gray ex A. Rich.	18
	tinctoria L.	18
	trotzkiana Claus ex Bunge	18
	woronowii Sosn.	18
	zyghia Woronow	18
Chamaemelum	*nobile* (L.) All.	18
Matricaria	*ambigua* (Lebed.) Kryl.	18
	recutita L.	18
	discoidea DC.	18
	grandiflora	18
	perforata L.	18, 36
	pubescens (Desf.) Schultz-Bip.	18
	suaveolens (Pursh) Rydb.	18
	tchichatchewii Boiss.	36

3.5.5 *Anthemis arvensis* L.

As a drug, the name Buphthalmi herba is often used. *Anthemis arvensis* has no medicinal value.

3.5.6 *Anthemis cotula* L.

Anthemis cotula is without any medicinal value. Sometimes the powder is used against insects. The drug is not traded in Europe. From time to time the flowers and the herb are traded in the Southwest United States as a substitute for chamomile, and in Russia they are used against worms.

3.5.7 ANTHEMIS TINCTORIA L.

Anthemidis tinctoriae flos is without any medicinal value. Due to the xanthophyl content, there has been an increased use as a natural yellow dye for textiles, particularly for wool [16].

3.6 PARTS USED

Chamomile is known and highly appreciated as a medicinal plant. However, chamomile is also of great importance as a foodstuff and as raw material for the cosmetic industry. Extracts, mono-drug, as well as tea mixtures are used both for pharmaceuticals and in the foodstuff sector. In the cosmetic sector extracts and essential oil are preferred. Depending on the use in the foodstuff, cosmetic, or pharmaceutical sector, different demands are made on the original drug, and according to the field of application very different trade qualities are offered worldwide.

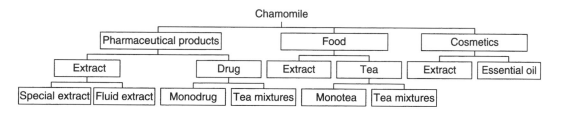

3.6.1 CHAMOMILE FLOWERS, MATRICARIAE FLOS

This drug is (mostly) used medicinally. The quality requirements are fixed by the pharmacopoeias of the different countries, and they can vary in wide ranges.

Almost complete flowers, flowers not fallen apart, or those fallen apart to a very small extent only with a small amount of fines are mostly called pharmaceutical chamomile. This quality is packed and delivered in cartons. Sometimes this drug is still harvested by hand. With mechanically harvested material a cost-effective sorting is necessary before or after drying. Mechanically harvested leaves and stems in addition to the flowers have to be sorted out. If the demands made on intact flower heads are not very high, they may be packed in pressed bales to save space.

3.6.2 CHAMOMILE FINES

Chamomile fines — in Hungary they are also handled under the name of Chamomillae cribratum — contain the tubular flowers of the chamomile flowers separated during the mechanical cleaning process, seeds and wing-petals. Depending on the content of essential oil, this material is either used for pharmaceutical purposes or as a foodstuff.

3.6.3 CHAMOMILE HERB WITH FLOWERS

This material comprises mechanically harvested chamomile flowers with a higher percentage of stems due to technological reasons. The quality requirements of the pharmacopoeias are usually not reached, but the material can easily be used as food teas.

3.6.4 CHAMOMILE HERB

Contrary to all other chamomile qualities this material is not produced by harvesting the flowers, but the flowering herb as such is cut and dried in one working operation and processed without any further cleaning or separation.

3.6.5 CHAMOMILE FOR EXTRACTION, INDUSTRIAL CHAMOMILE

This material comprises chamomile flowers, the evaluating properties of which are in accordance with the pharmacopoeia. Because of the mechanical harvest and its preparation, however, its external structure is in such a condition that this material is not suitable as pharmaceutical chamomile. From this quality medicinally used extracts are produced. Because of their high oil content flower heads fallen apart should not be regarded as worthless. If used in tea mixtures/blends a limitation of the fines is, however, justified due to the tendency of showing inhomogeneities.

3.6.6 CHAMOMILE ROOT

Roots of *Matricaria recutita* are used for pharmaceuticals of the anthroposophical therapy trend. Their active principles, especially the essential oil, differ considerably from the aerial parts [13].

3.6.7 CHAMOMILE OIL, MATRICARIA AETHEROLEUM

Chamomile oil is produced by distillation from fresh and dried flower heads and stems of the chamomile flowers. The essential oil of fresh flowers results in a higher yield, and the quality is said to be better.

3.6.8 CHAMOMILE FLUID EXTRACT

This product originated from flowers and has a content of ≥3% blue essential oil.

3.6.9 CHAMOMILE TINCTURE

Tinctures are mainly produced from fresh material for homoeopathic purposes.

REFERENCES

1. Aichele, D., Schwegler, H.-W. (1995) *Die Blütenpflanzen Mitteleuropas*, Vol. 4, Franck-Kosmos, Stuttgart, Germany, 528 pp.
2. Bär, B. (1995) *Analytik und Mikrobiologie des ätherischen Öls verschiedener Kamillenarten.* Dissertation, Univ. Berlin.
3. Bolkhovskikh, Z., Grif, V., Matvejeva T., Zakharyeva O. (1969) *Chromosome Numbers of Flowering Plants*, Nauka, Leningrad, Russia, 926 pp.
4. Carle, R., Fleischhauer, I., Fehr, D. (1987) *Dtsch. Apoth. Ztg.*, 127, 2, 451–457.
5. Hänsel, R., Keller, K., Rimpler, H., Schneider, G. (Eds.) *1992 Hagers Handbuch der pharmazeutischen Praxis*, 5th Ed., Vol. 4, Springer, Berlin, Heidelberg, New York, 1209 pp.
6. Heeger, E. F., Brückner, K. (1952) *Heil-und Gewürzpflanzen/Arten-und Sortenkunde.* 2nd ed., Berlin, 77–80.
7. Hegi, G. (1987) in Conert, H. J., Hamann, U., Schultze-Motel, W., Wagenitz, G. (eds.) *Illustrierte Flora von Mitteleuropa. Spermatophyta* IV, *Angiospermae Dicotyledones* 4, Part 4, 2nd ed., Parey Verlag, Berlin, Hamburg, p. 582.
8. Hegnauer, R. (1978) *Dragoco Report*, 24, 10, 203.
9. Hitchcock, A. S., Green, M. L. (1935) Species lectotypicae generum Linnaei. *International Rules of Botanical Nomenclature*, Jena, pp. 139–143.
10. Mulligan, G. A. (1959) Chromosome number of Canadian weeds II. *Canadian J. Bot.*, 37, 1, 81–92.
11. Oberprieler, C. (1997) Untersuchungen zur morphologischen und phytochemischen Variabilität zweier Hundskamillen-Arten (*Anthemis tricolor* Boiss. & *A. plutonia* Meikle *Compositae, Anthemideae*) aus Zypern. *Drogenreport*, 10, 17, 74.
12. Rauschert, S. (1974) *Folia Geobot. Phytotax.*, 9, 249–260.
13. Reichling, J., Bisson, W., Becker, H. (1984) *Planta Medica*, 4, 334.

14. Ruminska, A. (1983) *Rosliny lecznize. Podstawy biologii i agrotechniki (Heilpflanzen. Grundlagen der Biologie und Anbau).* 3rd Ed. Panstwaowe Wydw. Naukowe, Warszawa, p. 444.
15. Schilcher, H. (1987) *Die Kamille. Handbuch für Ärzte, Apotheker und andere Naturwissenschaftler,* Wiss. Verlagsgesell., Stuttgart, Germany, 152 pp.
16. Schweppe, H. (1992) *Handbuch der Naturfarbstoffe. Vorkommen, Verwendung, Nachweis,* ecomed, Landsberg/Lech, Germany, 800 pp.
17. Schubert, R., Vent, W. (Eds.) (1988) *Exkursionsflora, kritischer Band,* 7th ed., Volk und Wissen Verlag, Berlin, 811 pp.
18. Schultze-Motel, J. (Ed.) (1986) *Rudolf Mansfelds Verzeichnis landwirtschaftlicher und gärtnerischer Kulturpflanzen (ohne Zierpflanzen),* 2nd Ed., Akademie Verlag, Berlin, 1998 pp.
19. Tabernaemontanus, J. T. (1664) *New vollkommenlich Kräuter-Buch,* Jacob Werenfels, Basel, 1529 pp.
20. Uzma Nasreen, Khan, M. A. (1998) Palynological studies of *Matricaria chamomilla* L. (Babuna) and its related genera. *Hamdard Medicus,* 41, 4, 94–97.
21. Zander, R. (1994) *Handwörterbuch der Pflanzennamen,* 15th Ed., Ulmer, Stuttgart, Germany, 810 pp.
22. Zander, R. (2002) *Handwörterbuch der Pflanzennamen,* 17th Ed., Ulmer, Stuttgart, Germany, 990 pp.

4 Active Chemical Constituents of *Matricaria chamomilla* L. syn. *Chamomilla recutita* (L.) Rauschert

Heinz Schilcher, Peter Imming, and Susanne Goeters

CONTENTS

Chamomile contains a large number of therapeutically interesting and active compound classes. The most important ones are the components of the essential oil and the flavonoid fraction.

Apart from them, the following compound classes were detected and characterized: mucins, coumarins, phenol carboxylic acids (phenyl substituted carboxylic acids), amino acids, phytosterols, choline, and mineral substances.

The chamomile constituents are best categorized according to their lipophilicity [103]. The lipophilic fraction includes individual components of the essential oil, coumarins, methoxylated flavone aglyca, phytosterols, and "lipidic and waxy substances" [119]. The hydrophilic fraction consists of flavonoids, mucilage, phenyl carboxylic acids, amino acids, and choline.

4.1 MAIN ACTIVE CONSTITUENTS

4.1.1 THE ESSENTIAL OIL AND ITS CONSTITUENTS

The essential oil is present in all organs of the chamomile plant, with only the roots containing small quantities. The flowers and flower heads are the main organs of the production of essential oil. The composition of the essential oil in roots differs from that in flowers. The oil content changes during ontogenesis, reaching a maximum of 0.3–1.5% in flowers just before full bloom [103].

It is remarkable that chamomile flower oil mainly consists of sesquiterpene derivatives (75–90%) but only traces of monoterpenes [72]. The oil contains up to 20% of polyynes.

A recent study with two different chamomile cultivars reported the following rather typical values for capitula (flowers), shoot, and root as determined by gas chromatography (GC) and GC/mass spectrometry (MS): The major constituents of the oils of the capitula of the two cultivars were (E)-β-farnesene (4.9–8.1%), α-bisabolol oxide B (12.2–30.9%), α-bisabolol (4.8–11.3%), chamazulene (2.3–10.9%), α-bisabolol oxide A (25.5–28.7%), and cis-enyne dicycloether (4.8–7.6%). The shoot oil contained (Z)-3-hexenol (1.1–3.1%), (E)-β-farnesene (5.7–37.7%), germacrene D (0.5–14.6%), (E)-nerolidol (0.4–3.5%), spathulenol (0.5-3.5%), hexadec-11-yn-11, 13-diene (3.1–9.7%), and cis-en-yn-dicycloether (6.6–15.0%). The main constituents of the root oils were linalool (0.5–4.4%), nerol (3.5–16.6%), geraniol (1.2–9.0%), β-elemene (1.0–2.7%), (E)-β-farnesene (2.7–18.4%), spathulenol (11.9–9.4%), chamomillol (1.5–4.4%), τ-cadinol, τ-muurolol (0.5–6.4%), hexadec-11-yn-13,15-diene (0.3–6.2%), and cis-enyne dicycloether (4.7–5.3%). The shoot and root oils were found to be devoid of chamazulene. The presence of α-bisabolol and its oxides A and B as minor constituents in the shoot and root oils was established. α-Humulene, hexadec-11-yn-13,15-diene, phytol, isophytol, and methyl palmitate were detected for the first time in *Chamomilla recutita* [60].

The individual components are described in the following sections.

4.1.1.1 Matricin/Chamazulene

In 1863, the French chemist Piesse isolated a blue substance from the essential oil of chamomile. He characterized the compound as a hydrocarbon and called it azulene "à cause de sa couleur franchement bleue" [83].

The antiphlogistic effect of chamazulene had been known long before its constitution was found to be 1,4-dimethyl-7-ethyl-azulene in 1953 [68, 112]. It had first been assumed to be 1,4-dimethyl-7-isopropyl azulene [117]. The structure elucidation was actually done on chamazulene isolated from *Artemisia arborescens* L. [68]. The presence of chamazulene precursors had been reported for quite a long time until Cekan and co-workers finally isolated a substance. The structure of the compound that they called matricin was assigned as *(3S,3aR,4S,9R,9aS,9bS)*-4-acetyloxy-3a,4,5,9,9a,9b-hexahydro-9-hydroxy-3,6,9-trimethylazuleno[4,5-b]furan-2(3H)-one in 1956 [16, 17, 18]. Matricin is present in ligulate florets and tubular florets of chamomile only, but not in the bottom of the flower heads. In 1982, the constitution of matricin (see Figure 4.1) was confirmed by Flaskamp et al. using modern spectrometric methods [27].

The cycloheptene ring adopts a chair conformation, as was shown by ^{1}H nuclear magnetic resonance (NMR) data. The stereochemical assignment of matricin rests on these NMR data and a normal x-ray analysis of 4-epimatricin and of the adduct of 4-epimatricin with 3β-hydroxydihydrocostunolide, both isolated from *Artemisia arborescens* L. [1]. The absolute configuration was assigned on the basis of the assumption that 7-H always has α-orientation in guaianolides [2, 12, 20]. This assumption was shown to be true for matricin when Goeters et al. determined the absolute configuration of chamazulene carboxylic acid ([34, 47], *v.i.*). They also published a detailed analysis of the ^{1}H and ^{13}C NMR spectra of matricin and matricin epimers [34, 47].

FIGURE 4.1 Matricin [(−)-(3S*.3aR*.4S*.9R*.9aS*.9bS*–4-acetoxi-2.3.3a.4.5.9.9a.9b-octahydro-9-hydroxy-3-6-9-trimethylaculeno [4.5-b] furan-2-one]. (Formula and picture of 1.5 g pale yellow crystals isolated from chamomile by Mai Ramadan and Peter Imming, Marburg University, Germany [48].)

Matricin is very unstable and decomposes visibly by turning blue after a short time, particularly in aqueous solution [108, 109]. This color reaction also occurs during steam distillation of chamomile oil [103]. It is caused by decomposition to chamazulene. The transition of matricin to chamazulene can easily be demonstrated by treating a thin-layer chromatography (TLC) spot with steam [103]. The matricin content of chamomile varies considerably between cultivars. "Mabamille," a proazulene and (−)-α-bisabolol-rich cultivar of Martin Bauer GmbH and Robugen GmbH (both in Germany), contains up to 0.05% in dried flowers [39]. Of all chamomile preparations, the extract made with supercritical carbon dioxide has the highest matricin content (approx. 0.2%) [39]. Schmidt et al. published a procedure for the isolation of very pure matricin [105] and on the development of a matricin-rich chamomile preparation [80, 106, 107]. They also isolated and identified 8-desacetylmatricin, the product of the first degradation step of matricin [79].

The immediate precursor of chamazulene is chamazulene carboxylic acid (CCA). It is formed from matricin by the elimination of water and acetic acid and decarboxylates to chamazulene, probably through an electrocyclic reaction (see Figure 4.2).

Chamazulene carboxylic acid was first isolated by E. Stahl from chamomile and yarrow (*Achillea millefolium s.l.*) in 1954 [116]. Its constitution was confirmed by Cuong et al. using mass spectrometry and NMR. Apart from that, the compound was almost forgotten until in 2000, Imming recognized it to be a natural profen, constitutionally similar to synthetic antiphlogistic compounds

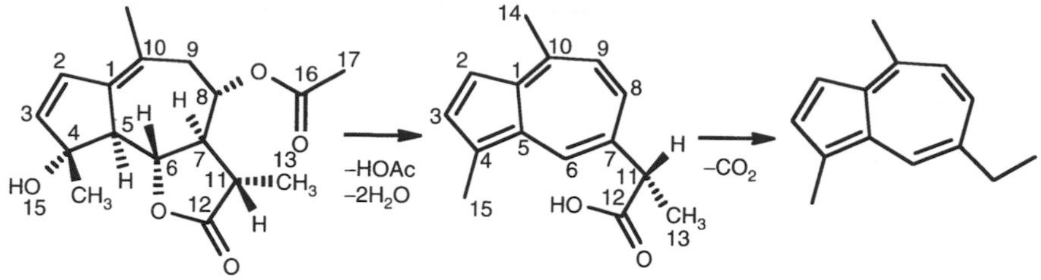

FIGURE 4.2 Hydrolytic degradation of matricin to chamazulene carboxylic acid and further decarboxylation to chamazulene.

like ibuprofen and naproxen. It was again isolated from a chamomile cultivar with a high proazulene content ("Mabamille"), extensively characterized (physico)chemically and shown to be more stable than originally reported, especially in neutral and weakly basic aqueous solutions [33]. In aqueous acid and in aprotic organic solvents, however, it rapidly loses carbon dioxide. Assuming first-order kinetics, the rate constant of decarboxylation k_1 was determined by ^{1}H NMR to be 1.70×10^{-3} in pure D_2O, 1.78×10^{-3} in pH 7.4 buffer, but 1.19×10^{-2} in pH 2.0 buffer. Its absolute configuration was found to be S by circular dichroism comparing with S- and R-naproxen [34, 47]. Its pK_a value (4.84) and log D (0.79 at pH 7.4) [33] are almost identical to ibuprofen, so CCA will be taken up equally well after oral application. The R-enantiomer was isolated from the Central American Stevia serrata [48].

An early paper claimed the presence of guaiazulene in chamomile [114], but this was never found again.

4.1.1.2 Bisabolols

In 1951, Šorm et al. isolated another essential constituent of chamomile oil, the monocyclic tertiary sesquiterpene alcohol (–)-α-bisabolol (INN: Levomenol) [114]. The constitution of bisabolol from chamomile was confirmed [40, 42] by comparing spectroscopic data with synthetic bisabolol already described by Ruzicka et al. [95, 96]. The isopropylidene structure of natural (–)-α-bisabolol was proved through infrared (IR) and nuclear magnetic resonance spectra [50].

This seems to be contradictory to Šorm et al., who determined bisabolol in chamomile oil to be a mixture of 85% and 15% of isopropylidene and isopropenyl isomers. As the two isomers were identified by ozonolysis, the isopropenyl isomer could have been formed during the ozonolysis. This is very likely in view of Naves' observation [77, 78] that a terminal isopropylidene group is found exclusively in aliphatic terpenes in plant metabolism, whereas the isopropenyl isomer is always an artifact.

As synthetic bisabolol has two intensive signals at 6.07 and 11.25 µm [42], the isopropenyl isomer is a by-product of the synthesis [50]. A 1:2 ratio of isopropylidene:isopropenyl determined for a commercially available synthetic mixture could not be confirmed in a later study [26].

Four optical isomers of α-bisabolol are possible. Three of them were isolated from different plants and distinguished because of their different optical rotation [49].

In 1977, Kergomard and Veschambre [57] determined the absolute configuration of (–)-α-bisabolol isolated from chamomile by stereoselective synthesis of the corresponding diastereoisomers and comparison of the NMR spectra. (–)-α-bisabolol has 5R,6S configuration (atom numbering as in β-bisabolol, isolated from cotton buds [69]). This result was confirmed by Knöll and Tamm [58].

The stereochemistry of the bisaboloids has been fully elucidated with the exception of bisabolol oxide C. In contrast to earlier assumptions, all steric centers of bisabolol oxides A and B, (–)-α-bisabolol and bisabolone oxide A have an S-configuration. The identical stereochemistry of all bisaboloids is also shown by the fact that some bisaboloids are interconvertible [26].

TABLE 4.1
Stereoisomers of α-bisabolol [15]

Isomer	Configuration	$(\alpha)_D^{20}$	Source
(+)-epi-α-bisabolol	1'R,2S	+ 67.4	Salvia stenophylla (purity 99.9%)
(–)-epi-α-bisabolol	1'S,2R	– 68.9	Myoporum crassifolium (purity 99.9%)
(+)-α-bisabolol	1'R,2R	+ 54.9	Populus balsamifera
		+ 52.6	(purity approx. 97.6%)
(–)-α-bisabolol	1'S,2S	– 57.7	Vanillosmopsis erythropappa (purity 99.9%)
		– 55.4	Matricaria chamomilla

(–)-α-Bisabolol from *Vanillosmopsis erythropappa* is identical with that from chamomile, showing the same stereochemistry (1'S,2S) [26]. The essential oil obtained from buds of the European *Populus balsamifera* [113] and the North American *Populus tacamahaca* [21] mainly consist of (+)-α-bisabolol. Both species are presumably identical. (+)-α-Bisabolol from *Populus balsamifera* is 1'R,2R configured and thus the optical antipode of bisabolol from chamomile.

Levogyrate α-bisabolol was also detected in the essential oil of *Myoporum crassifolium* [23, 81], making up more than 80% of the oil [23]. It had been found previously in small quantities in the essential oil of ylang-ylang (*Canaga odorata*) [58], neroli (*Citrus bigaradia*) [75], cabreuva (*Myrocarpus fastigiatus* and *M. frondosus* [76]), and lavender (*Lavandula spica*) [111] (quoted according to Reference 50).

Table 4.1 summarizes properties and origins of the four stereoisomers of α-bisabolol.

4.1.1.3 Bisaboloids

In 1951, (–)-α-bisabolol oxide A was isolated by Šorm et al. Many years later, Sampath et al. determined its structure [97] and isolated the isomeric bisabolol oxide B from chamomile [98]:

(–)-α-bisabolol oxide A ($C_{15}H_{26}O_2$): molecular mass 238, syrupy $[\alpha]_D = -42.2°$

(–)-α-bisabolol oxide B ($C_{15}H_{26}O_2$): molecular mass 238, $[\alpha]_D = -46.95°$

Two isomeric oxides of α-bisabolol, and consequently four cyclic structures, are possible (Figure 4.3): two containing a tertiary hydroxyl group and two with a secondary hydroxyl group.

They differ stereochemically at C-5 when numbered according to Reference 26 (previously C-2). The third chiral center in bisabolol oxide A and B has S-configuration [26].

During the isolation of the two liquid (–)-α-bisabolol oxides A and B, a small quantity of a crystalline substance was obtained and identified as (–)-α-bisabolol oxide C by Schilcher et al. (Figure 4.3) [104].

While testing chamomile material collected in Turkey, Hölzl and Demuth found a bisaboloid unknown so far [43]. Its constitution was determined by IR and NMR spectroscopy and through the products of oxidation and reduction found to be an α-bisabolone oxide (Figure 4.4) [44]. Bisabolol oxide A either can be transformed to it or can be obtained from it by reduction. The bisabolone oxide isolated from plant material displayed a different optical rotation compared to material prepared from (–)-α-bisabolol oxide A:

$$[\alpha]_D = -6,2° \text{ [97]}$$

$$[\alpha]_D = +3,5° \text{ [44]}$$

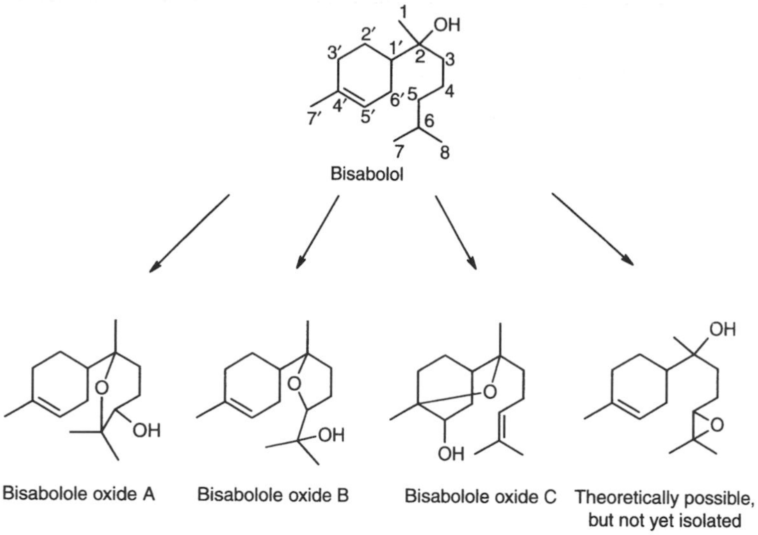

FIGURE 4.3 Products of the oxidation of α–bisabolol.

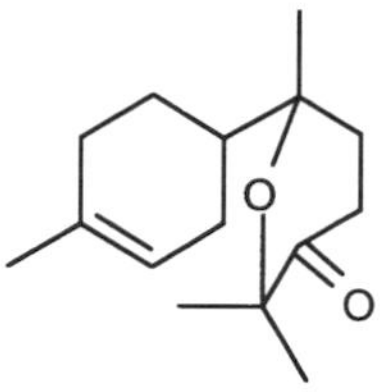

FIGURE 4.4 (–)-α-Bisabolone oxide A.

The nomenclature of the chamomile bisaboloids was reviewed according to the IUPAC guidelines and should be used as follows:

(–)-α-bisabolol:
(–)-(1'S, 2S)-6-methyl-2-(4-methyl-3-cyclohexene-1-yl)-hepten(e)-2-ol
(–)-bisabolol oxide A:
(–)-(1'S,3S,6S)-tetrahydro-2,2,6-trimethyl-6-(4-methyl-3-cyclohexen(e)-1-yl)-2H-pyran(e)-3-ol
(–)-bisabolol oxide B:
(–)-(1"S,2'S, 5'S)-1-methyl-I-[tetrahydro-5-methyl-5-(4-methyl-3-cyclohexen(e)-1-yl)-furan(e)-2-yl]-ethanol
(+)-bisabolone oxide A:
(+)-(1'S,6S)-tetrahydro-2,2,6-trimethyl-6-(4-methyl-3-cyclohexen(e)-1-yl)-2H-pyran(e)-3-on.

Bisabolol oxide A, B, and sometimes bisabolone oxide A are the main constituents of the essential oil. Schilcher [101] established a classification in chemotypes based on the composition of the essential oil (see Chapter 3, Plant Sources). In chamomile flowers, bisabolol oxide C is found in small quantities only or is chemically available from (–)-α-bisabolol by oxidation. Table 4.2 summarizes the results of different oxidation experiments, which led to varying quantities of bisabolol oxides A, B, and C [102, 104].

The photochemical tests were carried out in a quartz glass apparatus [102]. The rotation of the bisabolols was determined using long-wave UV light (mercury high-pressure lamp TQ 150 Hanau).

TABLE 4.2
Oxidation Stability of (-)-α-Bisabolol [65, 102–104]

Reaction Conditions	Main Components, Determined by GC
1. Bisabolol in hexane or heptane solution in brown glass containers with glass stopper	After 12 months: unchanged
2. (–)-α-bisabolol in ethanol 50% in white glass containers that were opened for 10 min once a week	After 12 months: ca. 70% (–)-α-bisabolol ca. 18% bisaboloids (mainly bisabolole oxide B) and unidentified other peaks (artifacts) After 18 months ca. 51% (–)-α-bisabolol ca. 25% bisabolole oxide B Remainder: not identified After 24 months: ca. 37% (–)-α-bisabolol ca. 31% bisabolole oxide B Remainder: not identified
3. (–)-α-bisabolol in Miglyol® 812 (Nobel Co.) — (see Schilcher, H. German Offenlegungsschrift Nr. 2331853, 22, 6, 1973) — in brown glass containers with glass stopper	After 24 months: ca. 98.5% (–)-α-bisabolol Remainder: not identified
4. Passage of purified air for 3 h without UV irradiation	ca. 85% (–)-α-bisabolol ca. 10% bisabolole oxide B Remainder: not identified
5. Passage of pure oxygen for 3 h without UV irradiation	ca. 68% (–)-α-bisabolol ca. 14% bisabolole oxide B Remainder: not identified
6. Irradiation of a bisabolol solution with visible light under exclusion of air for 6 h	Unchanged
7. Passage of pure oxygen for 5 h with UV irradiation	ca. 32% (–)-α-bisabolol ca. 42% bisabolole oxide B ca. 15% bisabolole oxide C, crystallized on wall of reaction chamber ca. 2% bisabolole oxide A Remainder: not identified
8. Passage of pure oxygen for 8 h with UV irradiation	Only traces of (–)-α-bisabolol and oxides B, C, and A left Reaction products had lower polarity and shorter GC retention times; mass spectra showed C_8 and C_7 fragments of mass 236

The increase of bisabolol oxide during the drying process was the object of further experiments. The plant material in this study consisted of flower heads of a chamomile variety being rich in bisabolol. The drying procedures were as follows [102, 103]:

1. Drying at room temperature in a dark and dry room (the relative atmospheric moisture was 35%)
2. Drying at 40°C
3. Drying at 50°C

TABLE 4.3

Relative Content of Constituents (GC) after Different Drying Procedure

	(–)-α-Bisabolol	Bisabolole oxide A	Bisabolole oxide B	Chamazulene
1	~34%	~ 4%	~ 20%	~ 12%
2	~34%	~ 4%	~ 20%	~ 12%
3	~24%	~ 5%	~ 28%	~ 10%
4	~17%	~ 8%	~ 33%	~ 7%
5	~7%	~ 26%	~ 19%	~ 1%

4. Drying at 60°C
5. Drying under conditions being similar to those of fermentation (e.g., layers of 10 cm height in a damp atmosphere)

Quantitative analysis of important constituents of the essential oil was done by gas chromatography. The results are summarized in Table 4.3.

Conditions similar to fermentation or high temperatures led to oxidation of bisabolol by atmospheric oxygen to the bisabolol oxides [102, 103]. The reactions during the drying process [102, 103] and the results of the oxidation experiments were evidence for the dependence of the content of bisaboloids on not only genetic or ecologic factors, but also on conditions of drying and storage of the plant material. Many analytical results found in the literature should be reinterpreted in the light of these findings [102, 103].

4.1.1.4 Other Terpenes

Motl et al. isolated the azulenogenic sesquiterpene alcohol, spathulenol [71]. Its constitution was determined by ^{1}H and ^{13}C NMR and by IR [55]. Its constitution was corroborated by regio- and stereoselective synthesis from (+)-aromadendrene [124]. In 1979, Lemberovics identified two farnesene isomers, viz., trans-β- and trans-α-farnesene [63]. According to her findings, β-farnesene appeared to be the main component, whereas α-farnesene was present in traces. Reichling et al. were not able to detect α-farnesene in the essential oil of chamomile flowers [89]. Further monoterpene hydrocarbons such as γ-terpinene, Δ^3-carene, and the sesquiterpene hydrocarbons α-cubebene, α-muurolene, and calamemene were detected by GC/MS [72].

Applying the same analytical procedure for a petroleum ether extract made from Czech chamomile flowers, Motl found a crimson fragrant azulene [72]. Its constitution was determined later [73] and found to be the aldehyde chamavioline (Figure 4.5). It had been synthesized previously as an intermediate of a chamazulene synthesis [74].

The constitution of matricarin was elucidated in 1978 [70] (Figure 4.5) and was corroborated by NMR data [67] and finally by a single-crystal x-ray analysis in 2002 [82]. The absolute configuration was not determined.

Motl et al. claimed [70] that not all active constituents of the petroleum ether extract have been identified yet. According to Stransky [118], the hydrocarbons found in chamomile flowers can be subdivided into n-alkanes, branched-chain alkanes, monoalkenes, terpenoid hydrocarbons, alkadienes, and aromatic hydrocarbons (see Figure 4.6).

In the essential oil of the root the sesquiterpene alcohol muurol-4-en-7-ol, also called chamomillol, was detected by Reichling et al. [90] (Figure 4.5). The authors also identified β-caryophyllene, cis-caryophyllene, and caryophyllene epoxide in this oil.

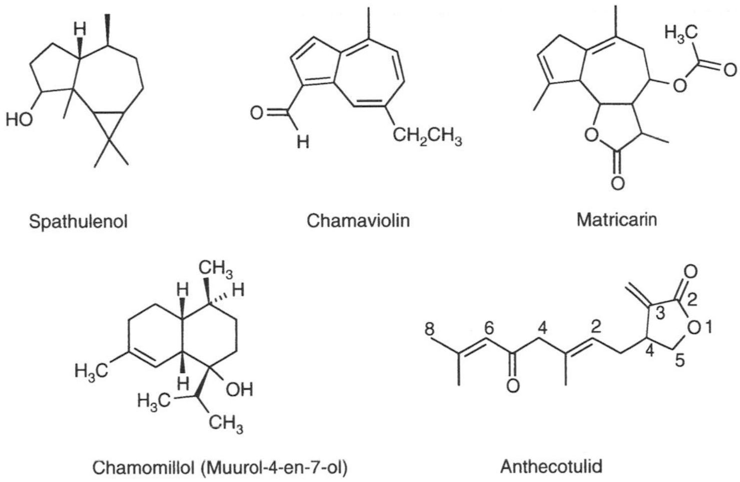

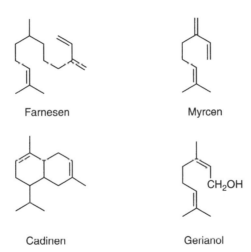

HO
H
Spathulenol
O
H
CH_2CH_3
Chamaviolin
H_3C
O
O
O
O
Matricarin
CH_3
H
H
H_3C
H
H_3C
OH
CH_3
Chamomillol (Muurol-4-en-7-ol)
8
6
4
2
3
2
O 1
4
5
O
O
Anthecotulid

FIGURE 4.5 Sesquiterpenes in chamomile flowers and roots. Anthecotulid is from *Anthemis cotula* (see text).

Farnesen
Myrcen
CH_2OH
Cadinen
Gerianol

FIGURE 4.6 Monoterpenes in chamomile flowers.

Yamazaki et al. reported the isolation of anthecotulide, a sesquiterpene lactone with an exocyclic methylene group (Figure 4.5), from chamomile collected in Argentina [130]. They claim to have isolated it in 7.3% (!) yield from aerial parts of *Matricaria chamomilla* L. (This number is sometimes wrongly quoted for the anthecotulide content of *Anthemis cotula*, but actually stems from the Yamazaki paper.) One of the authors of this chapter (P.I.) asked the curator of the Herbarium of the University of Texas, Austin, to redetermine the voucher specimen of Yamazaki et al. As suspected, it was actually *Anthemis cotula* L. [129]. Anthecotulide was originally isolated from *Anthemis cotula* L. (stinking mayweed, dog chamomile), where it is present in relatively high quantities [3, 9]. It is a very potent contact allergen [38]. Hausen et al. claim that it is present in variable but low levels in strains of the bisabolol oxide B chamomile type [38]. Bisabolol oxide B is the main constituent in the essential oil of this chemodem (chemical type), according to Schilcher [101, 103]. In all likelihood, the occasional presence of anthecotulide results from contamination of chamomile collections with *A. cotula* flowers and plants.

4.1.1.5 Spiroether (Syn. Enyne Dicycloether)

Two spirocyclic polyynes, the isomeric cis and trans enyne dicycloether, were found in the petroleum ether extract of chamomile flowers by Bohlmann et al. in 1961 [7] and reinvestigated in 1982, including a discussion of NMR data [8]. The cis spiroether (according to [114], *cis*-2-[hexadiyne]-(2,4)-ylidene]-1,6-dioxaspiro-[4,4]-nonene) is the major component in most plant specimens [102]. The trans spiroether was predominant only in certain commercially available chamomile flowers [102]. Both compounds were found in petroleum ether extracts and in freshly distilled essential oil [11].

The enyne dicycloethers were frequently accompanied by an aromatic compound with the molecular formula of $C_{10}H_{12}O_4$. Motl et al. [72] identified it as 2-hydroxy-4,6-dimethoxy-acetophenone by means of NMR spectroscopy. In the literature, this compound is also called xanthoxylin, brevifolin, or 6-methoxypaeonal. Motl et al. state that xanthoxylin is not separated from the spiroethers by thin-layer chromatography using usual solvent systems. Quantitative TLC (densitometry) may thus give too high quantities of spiroethers [72]. Unlike the enyne dicycloethers, xanthoxylin has no antiphlogistic effect.

Z-Enyne dicycloether
[(2Z)-2-(2,4-Hexadiynylidene)-
1,6-dioxaspiro[4.4]non-3-ene]

E-Enyne dicycloether
[(2E)-2-(2,4-Hexadiynylidene)-
1,6-dioxaspiro[4.4]non-3-ene]

4.1.1.6 Quantitative Composition of the Essential Oil

The approximate ratio of the main components in the essential oil is as follows [29, 102, 103]:

- Up to 15% chamazulene and its precursor matricin [116]
- Up to 33% bisabolols and bisabolol oxides [50, 102]
- Up to 45% trans-β-farnesene [14, 102]
- Up to 25% cis- and trans-isomers of the spiroethers [28, 102]

4.1.2 FLAVONOIDS

Flavonoid glycosides represent the major fraction of water-soluble components in chamomile. Apart from the glycosides, flavonoid aglyca were found in great variety among the lipophilic constituents. Chamomile flavonoids were recognized to be spasmolytic and antiphlogistic and are therefore of great interest.

Apigenin was the first flavone to be isolated from chamomile [85] in 1914. Its constitution, however, was elucidated as late as 1952 [115].

Lang and Schwand successfully isolated and identified apigenin-7-glucoside from ligulate florets and quercimeritin from tubular florets of chamomile [62].

Apiin (apigenin-7-[6"-*O*-apiosyl]-glucoside) had previously been found in white ligulate florets of some *Chrysanthemum* species and was again isolated from *Matricaria chamomilla*. Its constitution was elucidated by Wagner and Kirmayer in 1957 [125]. Six compounds separated by paper chromatography [123] were identified as apigenin glycosides through hydrolysis to apigenin.

Further constituents (e.g., luteolin-7-glucoside and patulitrin) were detected by Hörhammer et al., who also confirmed the presence of quercetin-7-glucoside and apigenin-7-glucoside [46]. The separation of 11 different flavonoid derivatives was successfully performed on polyamide plates. Less than six of them were apigenin glycosides [cf. 123].

Egyptian chamomile flowers were the subject of a qualitative analysis in 1963. They contained the flavonoids quercetin-3-rutinoside and quercetin-3-galactoside, but no luteolin-7-glucoside [22].

In 1965, the chromatographic separation of six to nine different apigenin glycosides was reported [99].

The first lipophilic chamomile flavone to be isolated and identified by Hänsel et al. in 1966 was chrysosplenitin (= 3,6,7,3'-tetramethyl-quercetagetin) [37].

Several papers were concerned with the analysis of the distribution of the flavonoids in different plant organs. Apart from chrysosplenitin and apigenin, three further apigenin glycosides were detected in ligulate florets, one of them apigenin-7-glucoside [128]. The tubular florets contained luteolin, luteolin-7-glucoside, and quercetin as well as numerous other flavonoids that were not identified yet. Surprisingly, apigenin and apigenin-7-glucoside were also observed in tubular florets, whereas luteolin glucoside could not be detected in ligulate florets (cf. [6]). Patulitrin, quercimeritrin, luteolin-7-glycoside, and apigenin-7-glucoside were detected in tubular florets of a new cultivar. In ligulate florets, only apigenin-7-glucoside was found [84].

In another report published in 1979, apigenin and apigenin-7-glucoside were again detected in ligulate florets. Two further glycosides were found and the presence of an isoflavone assumed.

According to a study published in 1979, the hydrolysis of flavonoids for the first time yielded the aglyca of isorhamnetin and chrysoeriol [88], apart from quercetin, patuletin, apigenin, and chrysosplenetin that had previously been found. In 1979 Kunde and Isaac, independently also Redaelli et al. [86], reported the isolation and identification of a new chamomile flavone, apigenin-7-(6''-*O*-acetyl-)-glucoside [61]. The aglyca apigenin, luteolin, patuletin, quercetin, and isorhamnetin were detected and the glycosides apigenin-7-glucoside, luteolin-7-glucoside, patulitrin, and quercimeritrin. They further reported the presence of flavonoid mono-glycosides and postulated a diacetylated apigenin-7-glucoside.

Kunde and Isaac classified chamomile flavonoids according to their polarity (Figure 4.7) [61]:

- "Lipophilic" flavone aglyca (e.g., methoxylated compounds)
- Hydroxylated flavone aglyca
- Acetylated flavone monoglycosides
- Flavone diglycosides

Systematic screening of a chloroform extract of Egyptian chamomile flowers [24] yielded further methylated flavones [37]. Apart from known compounds such as apigenin, patuletin, chrysoeriol, chrysosplenetin, and isorhamnetin, a number of thus far unknown methylated flavone aglyca such as jaceidin, chrysosplenol, eupatoletin, spinacetin, axillarin, eupaletin, and a 6-methoxy-kaempferol were discovered in chamomile. Especially noteworthy is the isolation of two methyl ethers of 6-hydroxy-kaempferol, eupaletin and 6-methoxy-kaempferol, since no kaempferol or kaempferol derivatives had been isolated from chamomile before [24] (Figure 4.8).

Again, the distribution of flavonoids in individual plant organs was studied and included organs studied earlier (ligulate florets and tubular florets, foliage, and roots) and additional organs such as involucral leaves, the receptacle, and the stems [6].

Comparing the glycoside pattern of tubular florets and ligulate florets as obtained from ethyl acetate extractions, the following conclusions can be drawn: on TLC plates, the ligulate florets showed uniformly dark violet spots that turned a greenish yellow color after spraying with natural product reagent. A total of five to seven substances were detectable. The spots obtained from the tubular florets showed yellow, green, and orange colors after spraying.

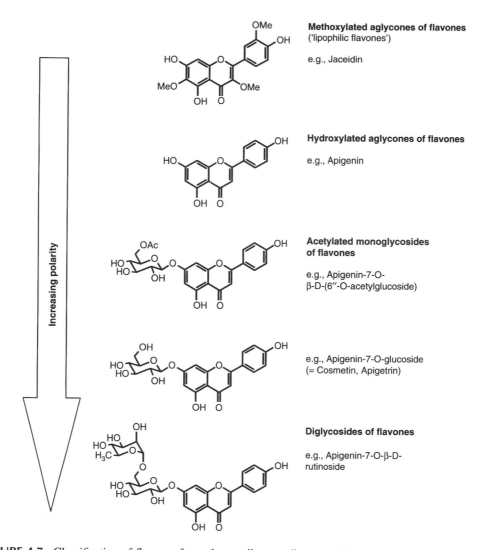

FIGURE 4.7 Classification of flavones from chamomile according to polarity [61].

In ligulate florets, the following flavonoids were found:

- Lipophilic chamomile flavones (for details see above, plus eupatoletin and spinacetin)
- Luteolin-7-glucoside
- Apigenin
- Apigenin-7-β-glucoside, apigenin-7-(6″-*O*-acetyl)-glucoside, apigenin-(6″-*O*-apiosyl-glucoside, apigenin-7-rutinoside)

After hydrolysis, the ligulate florets and tubular florets showed almost the same aglyca pattern except that patuletin and quercitin were missing in the ligulate florets, whereas apigenin was missing in the tubular florets.

According to the same report, receptacles, involucral leaves, foliage, and caulome contain mainly the same aglyca. Interestingly, oligomethylated aglyca are missing completely. The only aglyca to be found in all parts of the plant are isorhamnetin and luteolin with the exception of the root, which does not contain flavonoids at all [6, 128].

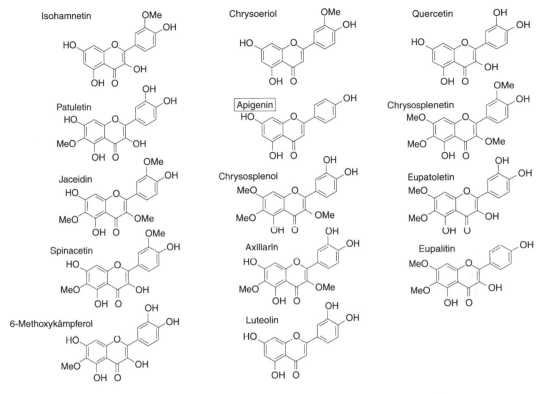

FIGURE 4.8 Flavone aglyca from chamomile flowers (according to Carle and Isaac [15]).

The following flavonoid glycosides were found in leaves by chromatographic methods:

Quercetin-7-glucoside, isorhamnetin-7-glucoside, luteolin-7-glucoside, chrysoeriol-7-glucoside, 6-hydroxyluteolin-7-glucoside, luteolin-7-rhamnoglucoside, and chrysoeriol-7-rhamnoglucoside [36]. No apigenin glycosides were observed; however, in another study [6] apigenin could be detected after hydrolysis.

Figure 4.8 shows a summary of flavonoid aglyca found in chamomile flowers. *Apigenin* is the most important therapeutically of them.

The presence of apigenin had not been proved unequivocally. Hölzl [41, 45, 122] radiolabeled apigenin and apigenin-7-glucoside through incorporation of $^{14}CO_2$ during biosynthesis in the ligulate florets. They measured the specific activity after selective ultrasound-assisted extractions with methanol and dichlormethane and proved that apigenin is in fact present in ligulate florets of chamomile. They isolated another acetylated glucoside, apigenin-7-β-D-(4″-O-acetyl) glucoside [122].

Carle et al. [13, 110] compared extracts from freshly collected wild plants with those from greenhouse plants. They assumed that apigenin is originally not present in fresh chamomile, but is the result of secondary enzymatic processes. This explains the great differences in the ratio of apigenin and apigenin-7-glycoside in the literature, which may also be due to different harvest conditions [110]. In the course of their work, they isolated flavone-glucoside-cleaving β-glucosidase from the protein fraction of flower heads, and in particular from freeze-dried ligulate florets, by means of fractionated ammonium sulfate precipitation and subsequent fast protein liquid chromatography (FPLC) on a Superose column. The enzyme has a relative molecular mass of 500,000 kD [64–66]. At its optimum temperature (37°C) and pH (5.6), this glycosidase showed a specificity for flavone-7-O-glucosides but did not hydrolyze α-glycosides or disaccharides. Agents carrying

TABLE 4.4

Flavonoid Glycosides and Aglyca in Chamomile Flowers

	R_1	R_2	R_3	R_4
Luteolin	H	H	OH	OH
Luteolin-7-glucoside	OGlu	H	OH	OH
Luteolin-4'-glucoside	H	H	OGlu	OH
Luteolin-7-rutinoside	OGlu-Rham	H	OH	OH
6-Hydroxy-luteolin-7-glucoside	OGlu	OH	OH	OH
Chrysoeriol	OH	H	OH	OCH_3
Chrysoeriol-7-glucoside	OGlu	H	OH	OCH_3
Apigenin	OH	H	OH	H
Apigenin-7-glucoside (Apigetrin)	OGlu	H	OH	H
Apigenin-7-(6"-*O*-acetyl)-glucoside	OGlu-ac	H	OH	H
Apigenin-7-(6"-*O*-apiosyl)-glucoside (Apiin)	OGlu-Apio	H	OH	H
Apigenin-7-rutinoside	OGlu-Rham	H	OH	H

sulfhydryl groups strongly inhibited the enzyme. They found that free apigenin is produced only after destruction of the cell compartments (e.g., after harvest), by enzymatic decomposition of glucosides [66]. During ontogenesis, an increase of the enzymatic activity and simultaneously the accumulation of flavonoids in flowers could be observed [66].

By HPLC with photodiode array detection and by thermospray liquid chromatography/mass spectrometry (TSP LC/MS), the isomeric 2"-, 3"-, and 4"-monoacetates and the 2"-, 3"-, and 3"-, 4"-diacetates of apigenin-7-glucoside could be separated and isolated from an *O*-acetyl-glycoside mixture [13].

Tables 4.4 and 4.5 summarize the flavonoid glycosides and corresponding aglyca identified so far.

4.1.2.1 Comparative Quantitative Tests of the Flavonoids

The quantities of (Z)- and (E)-2-β-D-glucopyranosyloxy-4-methoxycinnamic acids (GMCA) and apigenin glucosides of *Chamomilla recutita* were studied in progenies of selected tetraploid mother plants with significantly high and low contents of GMCA [94]. The relations between GMCA and apigenin aglycon content suggested that among the plants studied, groups of individuals with a high content of GMCA (6.4–9.2 mg/g dry weight) and high (4.1–5.1 mg/g dry weight) and medium (2.5–3.6 mg/g) contents of apigenin, as well as a group with a lower content of both compounds, could be distinguished.

Apigenin was also analyzed at two ploidy levels during a 3-year period. Higher percentages of apigenin were found in the ligulate florets of a diploid cultivar, in comparison with tetraploid plants. However, when the total apigenin in the anthodium was evaluated, tetraploid individuals accumulated significantly more flavonoid. Apigenin percentage in the ligulate florets was constant and not influenced by environmental conditions. Apigenin content was also found to change during inflorescence ontogeny. It represented the highest percentage of dry mass in young developing florets and anthodia of both cultivars. The total apigenin content of the anthodium, however,

TABLE 4.5
Flavonoid Glycosides and Aglyka, Especially Methoxylated Derivatives, in Chamomile Flowers

	R_1	R_2	R_3	R_4	R_5
Quercetin	OH	H	OH	OH	OH
Isorhamnetin	OH	H	OH	OH	OCH$_3$
Quercetin-7-glucoside (Quercimeritrin)	OGlu	H	OH	OH	OH
Quercetin-3-rutinoside (Rutin)	OH	H	OGlu-Rham	OH	OH
Quercetin-3-galaktoside (Hyperoside)	OH	H	OGal	OH	OH
Isorhamnetin-7-glucoside	OGlu	H	OH	OH	OCH$_3$
6-Methoxy-kaempferol	OH	OCH$_3$	OH	OH	H
Eupaletin	OCH$_3$	OCH$_3$	OH	OH	H
Patuletin	OH	OCH$_3$	OH	OH	OH
Patuletin-7-glucoside	OGlu	OCH$_3$	OH	OH	OH
Axillarin	OH	OCH$_3$	OCH$_3$	OH	OH
Spinacetin	OH	OCH$_3$	OH	OH	OCH$_3$
Eupatoletin	OCH$_3$	OCH$_3$	OH	OH	OH
Chrysoplenol	OCH$_3$	OCH$_3$	OCH$_3$	OH	OH
Chrysoplenetin	OCH$_3$	OCH$_3$	OCH$_3$	OH	OCH$_3$
Jaceidin	OH	OCH$_3$	OCH$_3$	OH	OCH$_3$

increased during flowering, although at later stages apigenin forms only a minor part of ligulate floret and anthodium dry mass [120].

While there are no qualitative differences between samples of individual "chamomile types" according to Reichling et al. [88], there are in fact considerable quantitative differences [102, 103]. They were established both in respect to the quantitative distribution of the aglyca and the total flavonoid content. Reichling et al. reported that a Bohemian chamomile variety, assigned as "K," and an Egyptian variety, assigned as "ART," contained much more quercetin than all other samples analyzed. Another Bohemian chamomile variety, assigned "C," had by far the highest apigenin content.

Repčak and Martonfi observed that the content of apigenin aglyca in ligulate florets and flower heads increases by polyploidization. In di- and tetraploid samples the amount of apigenin aglyca increased by about 15% [93].

Tests of about 100 samples performed by Schilcher [102] as well as the analysis of material obtained from 12 different origins, raised from seeds and cultivated under the same conditions, clearly showed that the total flavonoid content considerably differed, too. The flavonoid content of the samples tested ranged from 1.0–2.57%. Table 4.6 summarizes the results of these experiments. The total flavonoid content was determined photometrically [19]. This method is not suitable to obtain absolute values, but is sufficient for comparative analysis.

The results of Reichling and Schilcher demonstrate that it is important to determine the content of flavonoids in order to examine the quality of the plant material or preparation, especially with respect to the therapeutic significance of the flavonoids.

TABLE 4.6
Total Flavonoid Content of Chamomile from Different Locations [102, 103], Determined According to Reference 19

Origin of Cultivar	Location	Harvest Date	Total Flavonoids (%)
Wroclaw (Poland)	Kempten	1. harvest 22. 6. 1977	2.17
Sample No. 1	Kempten	2. harvest 30. 6. 1977	2.96
	Herrenberg	1. harvest 2. 8. 1978	2.33
	Herrenberg	2. harvest 14. 8. 1978	2.52
Niederrhein/Krefeld (Germany)	Kempten	1. harvest 23. 6. 1977	0.27
Sample No. 4	Herrenberg	1. harvest 24. 8. 1978	2.33
Bukarest (Romania)	Marburg	1. harvest 27. 6. 1977	2.13
Sample No. 2	Marburg	2. harvest 5. 7. 1977	2.71
	Marburg	3. harvest 13. 7. 1977	2.17
Aachen (Germany)	Marburg	1. harvest 22. 6. 1977	2.88
Sample No. 11	Marburg	2. harvest 27. 6. 1977	2.96
	Marburg	3. harvest 5. 7. 1977	2.80
	Kempten	1. harvest 23. 6. 1977	2.5
	Kempten	2. harvest 28. 6. 1977	1.42
	Kempten	3. harvest 12. 7. 1977	1.82
	Kempten	4. harvest 2. 9. 1977	0.17
Oldenburg (Germany)	Marburg	1. harvest 22. 6. 1977	2.21
Sample No. 12	Kempten	1. harvest 23. 6. 1977	1.98
Modena (Italy)	Marburg	1. harvest 22. 6. 1977	2.17
Sample No. 28	Marburg	2. harvest 27. 7. 1977	1.21
	Marburg	3. harvest 5. 8. 1977	1.30
	Kempten	1. harvest 23. 6. 1977	2.5
	Kempten	2. harvest 29. 7. 1977	1,0
Halle (Germany)	Marburg	1. harvest 22. 6. 1977	2.29
Sample No. 39	Marburg	2. harvest 27. 6. 1977	2.31
	Kempten	1. harvest 23. 6. 1977	2.17
	Kempten	2. harvest 29. 6. 1977	1.38
Bremen (Germany)	Marburg	1. harvest 22. 6. 1977	2.13
Sample No. 33	Herrenberg	3. harvest 24. 8. 1978	2.97
Oberstedten (Germany)	Herrenberg	3. harvest 24. 8. 1978	1.83
Argentina	Herrenberg	1. harvest 25. 6. 1980	2.8
	Herrenberg	3. harvest 28. 7. 1980	2,7
Spain	Herrenberg	1. harvest 22. 7. 1980	2.82
	Herrenberg	2. harvest 6. 8. 1980	2.75
	Herrenberg	3. harvest 12. 9. 1980	2.73
	Herrenberg	4. harvest 16. 9. 1980	2.29
Bratislava (Slovakia)	Herrenberg	1. harvest 1980	2.06
Tripleurospermum sp.	Herrenberg	2. harvest 1980	2.28
	Herrenberg	3. harvest 1980	2.5
Nijmegen (the Netherlands)	Herrenberg	1. harvest 1980	2.71
Tripleurospermum sp.	Herrenberg	2. harvest 1980	2.75

4.1.3 CHAMOMILE POLYSACCHARIDES, MUCILAGINOUS SUBSTANCES

In recent years, there has been an increased interest in chamomile polysaccharides, formerly called mucilaginous substances. In the literature, the content of this class of compounds varies considerably from 17% [56] to 10% [15] and even 3–5% [30, 31]. Polysaccharides are localized in the so-called "slime ribs" of chamomile flowers. The main chain of the polysaccharide consists of α-1->4 connected D-galacturone acid [15]. Other components of polysaccharides are as follows: xylose (about 21%), arabinose (about 10%), galactose (about 15%), glucose (about 7%), and rhamnnose (about 2%). In addition to these findings of Janecke and Weisser [53, 54], Schilcher identified fucose [102, 103].

Wagner et al. [126] characterized the crude mucilage of chamomile. They found it to be of high molecular mass (>500,000) and heavily branched with (1->4)-β-connected xylose moieties and a large amount of 4-O-methyl-glucuronic acid. In an *in vitro* test system, a significant stimulation of phagocytose was observed in the presence of chamomile polysaccharide [127]. The detection was done by means of phagocytose-chemoluminescence (CL). The test of the granulocytes was modified according to Brandt [10].

Recently Franz [31] could confirm the structure of the polysaccharide as a 4-O-methyl-glucurone oxylane, but not its activity toward phagocytose. He also identified a neutral fructane of medium molecular mass (3600) containing 74.3% fructose and 3.4% glucose (similar to inulin), and a strongly branched rhamnogalacturonane of medium molecular mass (93,000) consisting of 28% uronic acid, 3.2% protein (similar to pectin). They were found to have arabino-3,6-galactane glycoproteins as side chains. In an aqueous alcoholic chamomile extract preparation [31], only fructanes could be found. All three isolated chamomile polysaccharides showed remarkable antiphlogistic activity against mouse ear edema induced by crotone oil [29]. Further pharmacological tests are necessary. Nevertheless, a total extract is obviously superior to extracts containing the essential oil only because of the therapeutical relevance of the polysaccharides.

According to Janecke and Kehr [51, 52], the crude mucilage of chamomile has an exceptionally high mineral content. 100 g of dry mucilage extracted with hot-water extraction contained 18 to 29 g of ash. In measurements of the viscosity (see Chapter 10), the high mineral content has to be taken into consideration.

4.2 OTHER CONSTITUENTS

4.2.1 OTHER LIPOPHILIC CONSTITUENTS

4.2.1.1 Coumarins

Both the 7-methoxy-coumarin herniarin and the 7-hydroxy-coumarin umbelliferone (Figure 4.9) are of analytical and pharmacological interest. Herniarin is mainly present.

Analysis of material of different origins resulted in a range of 37.4–98.5 mg of herniarin [102] and 6–17.8 mg of umbelliferone both per 100 g of chamomile flowers. Both were detected in ligulate and tubular florets. The average content in ligulate florets was significantly higher [102, 103].

In herbal chamomile preparations sold in Italy, the preparations containing ligulate florets showed a higher content of coumarin when compared to other parts of the anthodia (flower heads). The ratio of umbelliferone:herniarin was <1:5 [121].

The content of herniarin decreased during anthodia ontogenesis in both diploid and tetraploid varieties, the tetraploid cultivar showing a higher concentration [91].

Umbelliferone was identified as a stress metabolite of the plant [92].

A Ukranian group claimed the isolation of another four coumarins: viz., coumarin itself, isoscopoletin, esculetin, and scopoletin from chamomile flowers [59].

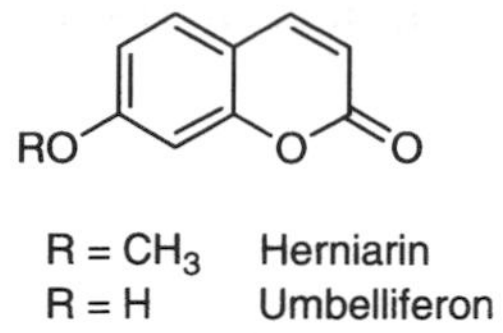

R = CH₃   Herniarin
R = H   Umbelliferon

FIGURE 4.9   Coumarins in chamomile flowers.

4.2.1.2   Ceraceous Substances

Streibel [118, 119] reported "lipidic and ceraceous substances" in a petrol ether extract of chamomile flowers. The composition was as follows:

Approximately 13.0% hydrocarbons
Approximately 16.0% simple aliphatic esters, sterol esters (phytosterols), and triterpenol
  esters
Approximately 3.0% triglycerides
Approximately 0.5% keto-esters
Approximately 6.0% esters containing acetylene

   Schilcher isolated a white ceraceous substance from steam distillate (using a glycerol bath for 6 h) by chromatography on silica gel [102, 103]. 1% of the compound was present in the dried drug. According to the ^{13}C spectrum, it was a straight-chain fatty acid with 30 carbon atoms.

4.2.2   OTHER HYDROPHILIC CONSTITUENTS

4.2.2.1   Phenyl Carboxylic Acids (Phenyl Substituted Carboxylic Acids)

Reichling et al. detected the following phenyl carboxylic acids chromatographically [88]: anisic acid, caffeic acid, syringic acid, and vanillic acid (Figure 4.10).

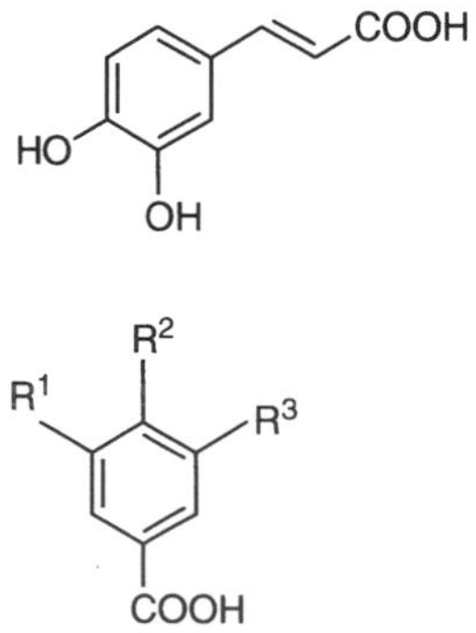

FIGURE 4.10   Arylcarboxylic acid derivatives.

Above: Caffeic acid
Below:

	R_1	R_2	R_3
Syringic acid	OCH₃	OH	OCH₃
Vanillic acid	H	OH	OCH₃
Anisic acid	H	OCH₃	H

Table 4.7 summarizes different contents of phenyl carboxylic acids observed [88]. In most of the tests anisic acid was used as reference.

TABLE 4.7
Phenylcarboxylic Acid Derivatives and Coumarins in Chamomile Flowers

	Coumarins		Phenylcarboxylic Acids			
Chamomile Sample	Herniarin	Umbelliferon	Anis Acid	Vanillic Acid	Syringic Acid	Caffeic Acid
Bulgarian chamomile	++	+	++	+	(+)	+
Bulgarian chamomile	+	+	++	+	(+)	+
Mexican chamomile	++	++	+	+	(+)	(+)
I, chamomile	+	+	++	++	+	+
II, chamomile	+	+	++	++	+	+
Bohemian chamomile, C	+	+	++	+	(+)	(+)
Bohemian chamomile, K	+	+	++	+	+	+
Egyptian chamomile, type Art	+	+	++	+	+	+
Egyptian chamomile, type HH	+	(+)	+	+	(+)	+
Egyptian chamomile, type M 77	++	+	++	+	(+)	(+)
Egyptian chamomile	+	+	+	+	(+)	+

Note: ++ = high concentration; + = low concentration; (+) = only trace detectable.

4.2.2.2 Other Constituents

Bayer et al. observed up to 0.3% choline [4, 5]. Choline is very likely to participate in the antiphlogistic activity of total extracts and aqueous preparations (infusions, etc.).

Graner found 6 amino acids in chamomile herb [35]. In 1970, Schilcher succeeded in detecting 13 different amino acids in fresh chamomile herb by paper chromatography. It is likely that L-leucine, DL-methionine, DL-α-alanine, glycine, L-histidine, and L-(+)-lysine are present, possibly also DL-threonine, DL-serine, and L-glutaminic acid [100]. Redaelli et al. found a "nitrogenous component" without reporting a further characterization.

REFERENCES

1. Appendino, G., Calleri, M., Chiari, G., Viterbo, D. (1985) *J. Chem. Soc. Perkin. Trans. II*, 203–207.
2. Barrero, A. F., Sánchez, J. F., Zafra, M. J., Barrón, A., San Feliciano, A. (1987) *Phytochemistry*, 26, 1531–1533.
3. Baruah, R. N., Bohlmann, F., King, R. M. (1985) *Planta Med.*, 51, 531–532.
4. Bayer, J., Katona, K., Tardos, L. (1958) *Acta Pharm. Hung.*, 28, 164.
5. Bayer, J., Katona, K., Tardos, L. (1958) *Naturwiss.*, 45, 629.
6. Becker, H. Reichling, J. (1981) *Dtsch. Apoth. Ztg.*, 121, 1285.
7. Bohlmann, F., Herbst, P., Arndt, Ch., Schönowski, U., Gleinig, H. (1961) *Chem. Ber.*, 94, 3193.
8. Bohlmann, F., Zdero, C. (1982) *Phytochemistry*, 21, 2543–9.
9. Bohlmann, F., Zdero, C., Grenz, M. (1969) *Tetrahedron Lett.*, 28, 2417–2418.
10. Brandt, L. (1967) *Scand. J. Haematol.*, Suppl. 2.
11. Breinlich, J. (1966) *Dtsch. Apoth. Ztg.*, 106, 698.
12. Calderón, J. S., Quijano, L., Gómez, F., Ríos T. (1989) *Phytochemistry*, 28, 3526–3527.
13. Carle, R., Dölle, B., Müller, W., and Baumeister, U. (1993) *Pharmazie*, 48, 304–306.
14. Carle, R., Fleischhauer, I., Fehr, D. (1987) *Dtsch. Apoth. Ztg.*, 127, 2451.
15. Carle, R. and Isaac, O. (1985) *Dtsch. Apoth. Ztg.*, 125 Nr. 43/Suppl. 1, 2–8.

16. Cekan, Z., Herout, V., Šorm, F. (1954) *Chem. Listy,* 48, 1071.
17. Cekan, Z., Herout, V., Šorm, F. (1954) *Collect Czechoslov. Chem. Commun.,* 19, 798.
18. Cekan, Z., Herout, V., Šorm, F. (1957) *Collect Czechoslov. Chem. Commun.,* 22, 1921.
19. Christ, B., Müller, K.H. (1960) *Arch. Pharm. Ber. Dtsch. Ges.,* 293, (65) 1033.
20. Cox, P. J., Sim, G. A., Herz, W. (1975) *J. Chem. Soc. Perkin. Trans. II,* 459–463.
21. Dull, G. G., Fairley, J. L., Gottshall, R. Y., Lucus, E. H. (1956–1957) *Antibiot. Ann.,* 682.
22. Elkiey, M. A., Darwish, M., Mustafa, M. A. (1963) *Fac. Pharm. Cairo Univ.,* 2, 107, ref. in Becker, H., Reichling, J. (1981) *Dtsch. Apoth. Ztg.,* 121, 1285.
23. Engelmann, E., Rauer, E. (1953) *Naturwiss.,* 40, 363.
24. Exner, J., Reichling, J., Cole, T. H., Becker, H. (1981) *Planta Med.,* 41, 198.
25. Ferri, S., Capresi P. (1979) *Atti. Soc. Toscana Sci. Nat. Pisa Mem.* Ser. B 86, 53, in Becker, H., Reichling, J. (1981) *Dtsch. Apoth. Ztg.,* 121, 1285.
26. Flaskamp, E., Nonnenmacher, G., Isaac, O. (1981) *Z. Naturforsch.* 36 b, 114, *Corrigendum* 36 b, 526.
27. Flaskamp, E., Zimmermann, G., Nonnenmacher, G., Isaac, O. (1982) *Z. Naturforsch.* 37 b, 508.
28. Franz, Ch., Hölzl, J., Vömel, A. (1973) *Acta Horticult.,* 73, 109–114.
29. Füller, E. (1992) Dissertation, University of Regensburg.
30. Füller, E., Blaschek, W., Franz, G. (1990) *Arch. Pharm.,* 323, 756.
31. Füller, E., Franz, G. (1993) *Dtsch. Apoth. Ztg.,* 133, 4224–4227.
32. Glichitch, L. S., Naves, Y. R. (1932) *Parfums de France* 10, 7 in O'Brian, K. G., Penfold, A. R., Werner, R. L. (1953) *Australian J. Chem.,* 6, 166.
33. Goeters, S. (2001) Dissertation, Marburg/Lahn, Germany.
34. Goeters, S., Imming, P., Pawlitzki, G., Hempel, B. (2001) *Planta Med.,* 67, 292–294.
35. Graner, G. (1965) *Präparative,* 1, 115.
36. Greger, H. (1975) *Plant. Syst. Evol.,* 124, 35.
37. Hänsel, R., Rimpler, H., Walther, K. (1966) *Naturwissenschaften,* 53, 19.
38. Hausen, B. M., Busker, E., Carle, R. (1984) 50, 229–234.
39. Hempel, B. (2001) Private communication (Robugen GmbH).
40. Herout, V., Zaoral, M., Šorm, F. (1953) *Collect Czechoslov. Chem. Commun.,* 18, 122.
41. Hess, S. (1989) Dissertation, Marburg, Germany.
42. Holub, M., Herout, V., Šorm, F. (1955) *Czechoslov. Farm.,* 3, 129.
43. Hölzl, J., Demuth, G. (1973) *Dtsch. Apoth. Ztg.,* 113, 671.
44. Hölzl, J., Demuth, G. (1975) *Planta Med.,* 27, 37.
45. Hölzl, J., Tschirsch, K. (1991) *Arch. Pharm.,* 324, 753.
46. Hörhammer, L., Wagner, H., Salfner, B. (1963) *Arzneim. Forsch.,* 13, 33.
47. Imming, P., Goeters, S., Pawlitzki, G., Hempel, B. (2001) *Chirality,* 13, 337–341.
48. Imming, P., Goeters, S., Ramadan, M. (2005) Unpublished results. See also Ramadan, M., Dissertation, Marburg/Lahn, Germany.
49. Isaac, O. (1979) *Planta Med.,* 35, 118–124.
50. Isaac, O., Schneider, H., Eggenschwiller, H. (1968) *Dtsch. Apoth. Ztg.,* 108, 293–298.
51. Janecke, H., Kehr, W. (1962) *Pharmaz., Ztg.,* 107, 432.
52. Janecke, H., Kehr, W. (1962) *Planta Med.,* 10, 60.
53. Janecke, H., Weiser, W. (1964) *Planta Med.,* 12, 528.
54. Janecke, H., Weiser, W. (1965) *Pharmazie,* 20, 580.
55. Juell, S. M. K., Hansen, R., Jork, H. (1976) *Arch. Pharm.,* 309, 458.
56. Kehr, W. (1960) Dissertation, Frankfurt/Main, Germany.
57. Kergomard, A., Verschambre, H. (1977) *Tetrahedron,* 33, 2215.
58. Knöll, W., Tamm, Ch. (1975) *Helv. Chim. Acta,* 58, 1162.
59. Kotov, A. G., Khvorost, P. P., Komissarenko, N. F. *Khimiya Prirodnykh Soedinenii* (1991), 853.
60. Kumar, S., Das, M., Singh, A., Ram, G., Mallavarapu, G. R., Ramesh, S. (2001) *J. Med. Arom. Plant Sciences,* 23, 617–623.
61. Kunde, R., Isaac, O. (1979) *Planta Med.,* 37, 124.
62. Lang, W., Schwandt, K. (1957) *Dtsch. Apoth. Ztg.,* 97, 149.
63. Lemberovics, E. (1979) *Sci. Pharm.,* 47, 330.
64. Maier, R., Kreis, W., Carle, R., Reinhard, E. (1991) *Planta Med.,* 57, Suppl. A 84.
65. Maier, R., Kreis, W., Carle, R., Reinhard, E. (1991) *Planta Med.,* 57, 297–298.

66. Maier, R., Kreis, W., Carle, R., Reinhard, E. (1993) *Planta Med.,* 59, 436–441.
67. Martinez, M. V., Munoz-Zamora, A., Joseph-Nathan, P. (1988) *J. Nat. Prod.,* 51, 221–228.
68. Meisels, A., Weizmann, A. (1953) *J. Am. Chem. Soc.,* 75, 3865–3866.
69. Minyard, J. P., Thomson, A. C., Hedin, P. A. (1968) *J. Org. Chem.,* 33, 909–911.
70. Motl, L., Repcak, M., Sedmera, P. (1978) *Arch. Pharm.,* 311, 75.
71. Motl, O., Felklova, M., Lukes, V., Jasikova, M. (1977) *Arch. Pharm.,* 310, 210.
72. Motl, O., Repcak, M. (1979) *Planta Med.,* 36, 272.
73. Motl, O., Repcak, M., Ubik, K. (1983) *Arch. Pharm.,* 316, 908.
74. Mukherjee, D., Dunn, L. C., Houk, K. N. (1979) *J. Am. Chem. Soc.,* 101, 251–252.
75. Naves, Y. R. (1934) *Parfums de France* 12, 61 in O'Brian, K. G., Penfold, A. R., Werner, R. L. (1953) *Australian J. Chem.,* 6, 166.
76. Naves, Y.R. (1948) *Helv. Chim. Acta,* 30, 278 in O'Brian, K. G., Penfold, A. R., Werner, R. L. (1953) *Australian J. Chem.,* 6, 166.
77. Naves, Y. R. (1946) *Perfume Record* 37, 120 Schilcher, H. (1973) *Planta Med.,* 23, 132.
78. Naves, Y. R. (1949) *Perfume Record* 40, 72 Schilcher, H. (1973) *Planta Med.,* 23, 132.
79. Ness, A., Metzger, J. W., Schmidt, P. C. (1996) *Pharm. Acta Helvet.,* 71, 265–271.
80. Ness, A., Schmidt, P.C. (1995) *Dtsch. Apoth. Ztg.,* 135, 3598–3610.
81. O'Brian, K. G., Penfold, A. R., Werner, R. L. (1953) *Australian J. Chem.,* 6, 166.
82. Parvez, M., Ahmad, V. U., Farooq, U., Jassbi, A. R., Raziullah, H. S. (2002) *Acta Cryst.,* E58, o324–o325.
83. Piesse, S. (1863) *Comptes Rend. hebdom. Séances Acad. Sciences,* 57, 1016.
84. Poethke, W., Bulin, P. (1969) *Pharm. Zentralh.,* 108, 733.
85. Power, F., Browning, H. Jr. (1914) *J. Chem. Soc.,* London, 105, 2280, in Becker, H., Reichling, J. (1981) *Dtsch. Apoth. Ztg.,* 121, 1285.
86. Redaelli, C., Formentini, L., Santaniello, E. (1979) *Herba Hung.,* 18, 323.
87. Redaelli, C., Formentini, L., Santaniello; E. (1979) *Int. Meeting on Medical Plant Reserarch, Budapest 1979,* Poster.
88. Reichling, J., Becker, H., Exner, J., Dräger, P. D. (1979) *Pharmaz. Ztg.,* 124, 1998.
89. Reichling, J., Beiderbeck, R., Becker, H. (1979) *Planta Med.,* 36, 322.
90. Reichling, J., Bisson, W., Becker, H., Schilling, G. (1983) *Z. Naturforsch.,* 38 c, 159.
91. Repcak, M., Eliasova, A., Ruscancinova, A. (1998) *Pharmazie,* 53, 278–279.
92. Repcak, M., Imrich, J., Franekova, M. (2001) *J. Plant Physiol.,* 158, 1085–1087.
93. Repcak, M., Martonfi, P. (1995) *Pharmazie,* 50, 696–699.
94. Repcak, M., Pastirova, A., Svehlikova, V., Martonfi, P. (2001) *Biologia (Bratislava, Slovakia),* 56, 455–457.
95. Ruzicka, L., Capato, E. (1925) *Helv. Chim. Acta,* 8, 259.
96. Ruzicka, L., Liguori, M. (1932) *Helv. Chim. Acta,* 15, 3.
97. Sampath, V., Trivedi, G. K., Paknikar, S. K., Bhattacharyya, S. C. (1969) *Indian J. Chem.,* 7, 100.
98. Sampath, V., Trivedi, G. K., Paknikar, S. K., Sabata, B. K., Bhattacharyya, S. C. (1969) *Indian J. Chem.,* 7, 1060.
99. Schäfer, J. (1965) *Wiss. Z. Karl Marx-Univ., Leipzig,* 14, 435.
100. Schilcher, H. (1970) *Planta Med.,* 18, 101–113.
101. Schilcher, H. (1973) *Planta Med.,* 23, 132.
102. Schilcher, H. (1985) *Zur Biologie von Matricaria chamomilla, syn. "Chamomilla recutita (L.) Rauschert,"* Research report 1968–1981, Inst. Pharmakognosie and Phytochemie of the FU, Berlin.
103. Schilcher, H. (1987) *Die Kamille — Handbuch für Arzte, Apotheker und andere Naturwissenschaftler.* Wissenschaftl Verlagsgesellschaft, Stuttgart, Germany.
104. Schilcher, H., Novotny, L., Ubik, K., Motl, O., Herout, V. (1976) *Arch. Pharm.,* 309, 189.
105. Schmidt, P. C., Ness, A. (1993) *Pharmazie,* 48, 146–147.
106. Schmidt, P. C., Soyke, B. (1992) *Pharmazie,* 47, 516–518.
107. Schmidt, P. C., Soyke, B. (1992) *Sci. Pharm.,* 60, 111–123.
108. Schmidt, P. C., Soyke, B. (1991) *Pharm. Unserer Zeit,* 20, 133–134.
109. Schmidt, P. C., Weibler, K., Soyke, B. (1991) *Dtsch. Apoth. Ztg.,* 131 (5) 175–181.
110. Schreiber, A., Carle, R., Reinhard, E. (1990) *Planta Med.,* 56, 179–181.
111. Seidel, C. F., Müller, P. H., Schinz, H. (1944) *Helv. Chim. Acta,* 27, 738.

112. Šorm, F., Nowak, J., Herout, V. (1953) *Chem. Listy,* 47, 1097.

113. Šorm, F., Vrany, M., Herout, V. (1952) *Chem. Listy,* 46, 364.

114. Šorm, F., Zaoral, M., Herout, V. (1951) *Collect Czechoslov. Chem. Commun.,* 16, 626–638.

115. Šorm, P., Zekan, Z., Herout, V., Raskova, H. (1952) *Chem. Listy,* 46, 308.

116. Stahl, E. (1954) *Chem. Ber.,* 87, 202, 205, 1626.

117. Stahl, E. (1954) *Naturwiss.,* 41, 257.

118. Stransky, K., Streibel, M., Ubik, K., Kohoutova, J., Novotny, L. (1981) *Fette, Seifen, Anstrichmittel,* 83, 347.

119. Streibel, M. (1980) Presentation, DFG Conference in Kiel, ref. in: *Seifen, Öle, Wachse,* 106, 503.

120. Svehlikova, V., Repcak, M. (2000) *Plant Biology* (Stuttgart, Germany), 2, 403–407.

121. Tosi, B., Romagnoli, C., Menziani-Andreoli, E., Bruni, A. *Int.* (1995) *J. Pharmacognosy,* 33, 144–147.

122. Tschirsch, K., Hölzl, J. (1992) *PZ-Wissenschaft,* 137, (5) 208–214.

123. Tyihak, E., Sarkany-Kiss, J., Vazár-Petri, G. (1962) *Pharmazie,* 17, 301.

124. Van Lier, F. P., Hesp, T. G. M., Van der Linde, L. M., Van der Weerdt, A. J. A. (1985) *Tetrahedron Lett.,* 26, 2109–2110.

125. Wagner, H., Kirmayer, W. (1957) *Naturwissenschaften,* 44, 307.

126. Wagner, H., Proksch, A., Riess-Maurer, I., Vollmar, A., Odenthal, S., Stuppner, H., Jurcic, K., Turdu, M., Yeur, Y. H. (1984) *Arzneim. Forsch./Drug Res.,* 34, 659.

127. Wagner, H., Proksch, A., Vollmar, A., Kreutzkamp, B., Bauer, R. (1985) *Planta Med.,* 50, 139.

128. Walther, K. (1968) Dissertation, Fak. Pharmazie FU Berlin.

129. Wendt, T. (2003) Private communication, Plant Resources Center, Herbarium, University of Texas, Austin.

130. Yamazaki, H., Miyakado, T., Mabry, T. J. (1982) *J. Nat. Prod.,* 45, 508.

5 Cultivation

Rolf Franke with cooperation of Jenö Bernáth, Tamer Fahmi, Norberto R. Fogola, Dušan Jedínak, Hans-Jürgen Hannig, Josef Holubář, Éva Németh, Viliam Oravec, Viliam Oravec, Jr., Miroslav Repčák, L'ubomir Šebo, Ivan Varga, and Eduardo Weldt S.

CONTENTS

5.1   ECOLOGICAL REQUIREMENTS

5.1.1   ORIGIN AND AREAS

The actual origin *of Matricaria recutita* L. is the Near East and south and east Europe. The species is to be found almost all over Europe, in western Siberia, Asia Minor, the Caucasus Mountains, Iran, Afghanistan, and India. After its introduction it also became common in North America, South America, New Zealand, and Australia [83]. It even appears in the coat of arms of Pehuajó (a provincial town in the Pampas about 500 km southwest of Buenos Aires), although chamomile is not found in a botanical statement containing the medicinal plants of Argentina found.

Good proof of the existence of chemodems has meanwhile been given. With regard to the sesquiterpene alcohols, original forms mostly show bisabololoxides. A form rich in $(-)\alpha$-bisabolol could be found in Spain. Nowadays the origin of chamomile flowers from wild collections can easily be determined by means of the chemical composition [26, 77, 80, Table 5.1.1].

Types rich in bisabolol are to be found endemically in Catalonia/Spain [14]:

- Bisabolol oxide A types originate from Egypt and central Europe (e.g., Hungary, Czech Republic, Slovakian Republic).
- Bisabolol oxide B types are from South American collections.
- Bisabolone oxide A types originate from southeast Europe and Turkey.
- Types poor in or free of matricine are to be found in Egypt, the Balkans (Romania and parts of Bulgaria), and Turkey; the types growing there are mostly those with yellowish-green oil [77].
- The composition of essential oil is obviously to a higher degree genetically determined than oil content. The oil content is more strongly influenced by environmental factors and shows considerable variation, even within a relatively small area (see Reference 89).

As a very high percentage of the material comes from cultivation, the required types are cultivated in various regions of the world, depending on what is needed.

5.1.2   SOIL

As far as the location is concerned, True chamomile is extremely tolerant and modest. It grows equally well in light and heavy soils of different scopes of reaction (from sour to neutral-alkaline). Wild locations are often sandy to loamy, mostly sour fields and fresh ruderal places. Salty soils

TABLE 5.1.1
Composition of Chamomile Oil (According to Frank, Data from Different References and Laboratories [23])

Identity number	Origin or type	Amount bisabolane	Bisabolol-oxide	Levomenol (Bisabolol) %	Bisa-bolol-oxide A %	Bisa-bolol-oxide B %	Bisa-bolon-oxide A %	Cham-azulene %	Cham-azulen% spectral-photom.	β-Farne-sene %	Spiro-ether %	Total amount essential oil%	Density	Optical rotation	Chemical type or comment	Ref.
1	Bulgaria	82.0	81.9	0.1	13.6	1.5	66.8	0.1		11.2	7.5	0.59				[70]
2	Bulgaria	77.6	77.5	0.1	13.1	1.3	63.1	0.1		12.2	12.0	0.60				[70]
3	Mexico	84.9	81.4	3.5	68.0	5.8	7.6	2.0		4.2	9.1	0.99				[70]
4	I-chamomile	86.5	84.4	2.1	73.4	6.0	5.0	1.0		3.4	9.1	0.91				[70]
5	II-chamomile	86.8	84.8	2.0	74.4	5.2	5.2	1.0		3.0	9.2	0.98				[70]
6	Bohemian chamomile c	76.0	73.5	2.5	60.4	5.8	7.3	9.7		5.1	9.5	0.76				[70]
7	Bohemian chamomile K	77.7	75.5	2.2	64.0	5.8	5.7	9.8		3.1	10.1	0.94				[70]
8	Egypt Type Art	81.9	79.6	2.3	65.5	8.6	5.5	3.5		3.4	11.5	0.90				[70]
9	Egypt Type HH	82.0	77.6	4.4	61.5	8.5	7.6	3.4		4.3	10.4	0.80				[70]
10	Egypt Type M77	81.6	78.8	2.8	64.6	7.4	6.8	2.1		5.5	10.2	0.88				[70]
11	Egypt	81.4	76.3	5.1	64.6	5.9	5.8	2.8		4.2	12.0	0.94				[70]
12	Ägypten	60.0	50.5	9.5	44.2	6.3		2.7	2.4		6.6	0.85			A	[80]
13	Ägypten	63.3	55.8	7.5	49.5	6.3		3.3	5.5		4.2	0.80			A	[80]
14	Egypt	63.3	56.0	7.3	50.2	5.8		6.3	5.0		10.3	0.80			A	[80]
15	Africa	48.9	41.2	7.6	36.5	4.8		2.4	2.8		8.0	0.65			A	[80]
16	Yemen	71.5	64.4	7.1	58.9	5.6		2.6	4.7		2.6	0.95			A	[80]
17	India	67.7	51.7	16.0	36.0	15.7		8.9	6.6		11.3	0.65			A	[80]
18	Japan	61.6	57.2	4.4	52.0	5.2		0.0	0.3		5.9	0.40			A	[80]
19	Frankonia	42.9	30.9	12.1	22.4	8.5		11.3	10.7		6.7	0.75			A	[80]
20	Self-cultivation	58.3	50.5	7.8	40.9	9.6		2.7	4.1		7.6	1.00			A	[80]
21	Czechoslovakia	48.1	40.6	7.5	35.9	4.7		11.1	7.5		8.8	0.80			A	[80]
22	Czechoslovakia	56.0	48.1	7.9	35.7	12.3		8.6	6.1		6.0	0.70			A	[80]
23	Bohemia	78.4	66.7	11.8	53.9	12.8		8.7	6.4		10.5	0.80			A	[80]
24	Hungary Variety I	58.9	50.6	8.3	40.7	9.9		15.2	14.8		7.5	1.00			A	[80]

No.														
25	Hungary Variety II	44.9	39.2	5.8	30.8	8.3		17.7	17.1	9.2	1.00	A	[80]
26	Poland	62.8	47.4	15.4	32.	15.0		5.4	5.9	9.3	0.42	A	[80]
27	Frankonia	52.3	39.2	13.1	25.2	14.1		9.5	10.3	4.8	0.36	A	[80]
28	Bohemia	56.5	48.4	8.1	39.1	9.3		6.7	7.5	5.7	0.52	A	[80]
29	Hungary Variety B1	46.9	39.7	7.2	31.7	8.0		13.4	13.5	5.2	0.78	A	[80]
30	Hungary Variety B2	41.2	34.8	6.4	26.2	8.6		10.7	10.4	4.4	0.66	A	[80]
31	Poland	39.8	32.1	7.7	23.8	8.3		8.6	0.0	5.0	0.40	A	[80]
32	Argentina	48.4	39.6	8.8	5.3	34.3		6.5	8.1	4.8	0.80	B	[80]
33	Argentina	74.0	61.0	12.9	8.8	52.3		5.4	7.7	4.1	0.70	B	[80]
34	Argentina	47.5	37.7	9.8	6.6	31.1		8.0	4.1	9.9	0.45	B	[80]
35	Buenos Aires	57.7	46.4	11.4	7.2	39.1		6.7	0.0	9.8	0.38	B	[80]
36	Yugoslavia A	56.2	42.3	14.0	24.2	18.1		3.7	2.1	10.7	0.35	D	[80]
37	Brazil	58.1	41.9	16.2	16.1	25.8		1.9	2.3	9.2	0.60	D	[80]
38	Yugoslavia B	28.5	20.1	8.5	10.4	9.6		3.0	3.2	10.3	0.32	D	[80]
39	Yugoslavia E	52.7	33.1	19.6	23.1	10.0		7.1	2.9	5.5	0.23	D	[80]
40	Poland	56.3	41.3	15.0	23.4	18.0		7.9	6.3	9.5	0.60	D	[80]
41	Egypt	85.5	81.5	4.0	67.5	7.5	6.5	5.0			0.85		[80]
42	Argentina	84.5	81.0	3.5	66.0	10.0	5.0	5.0			0.75		[80]
43	Bulgaria	85.1	85.0	0.1	13.5	1.5	70.0	0.1			0.60		[80]
44	Germany	81.5	71.5	10.0	54.0	15.0	2.5	16.5			0.80		[80]
45	Yugoslavia	76.8	64.8	12.0	25.8	31.5	7.5	19.0			0.90		[80]
46	Mexico	86.1	86.0	0.1	75.0	6.0	5.0	0.1			0.95		[80]
47	Czechoslovakia	82.5	79.5	3.0	65.0	7.0	7.5	12.5			0.80		[80]
48	Turkey	92.0	92.0	0.0	23.5	3.5	65.0	0.0			0.75		[80]
49	Hungary	84.5	77.0	7.5	63.5	8.0	5.5	10.0			0.85		[80]
50	Czechoslovakia	71.0	70.0	1.0	56.0	10.5	3.5	25.0			1.00		[80]
51	GDR	79.5	78.5	1.0	35.5	40.0	3.0	16.5			0.90		[80]
52	Poland	68.5	67.5	1.0	40.0	20.0	7.5	25.0			1.10		[80]
53	Hungary	67.5	65.0	2.5	52.0	8.0	5.0	27.5			1.00		[80]
54	Germany, Marburg 1974/1975	54.9	53.1	1.8	33.5	13.4	6.2	16.6	14.1	11.0	0.80		[80]

TABLE 5.1.1
Composition of Chamomile Oil (According to Frank, Data from Different References and Laboratories [23]) (continued)

Identity number	Origin or type	Amount bisabolane	Bisabolol-oxide	Levomenol (Bisabolol) %	Bisabolol-oxide A %	Bisabolol-oxide B %	Bisabolonoxide A %	Chamazulene %	Chamazulen% spectralphotom.	β-Farnesene %	Spiroether %	Total amount essential oil%	Density	Optical rotation	Chemical type or comment	Ref.
55	Germany, Rauischholzhausen 1975	70.6	69.8	0.8	59.8	3.1	6.9	8.1		11.5	8.1	0.80				[80]
56	Germany, Krefeld-Niederrhein 1975	74.1	65.9	8.2	54.3	4.6	7.0	4.3		6.5	14.0	0.90				[80]
57	Germany, Oldenburg 1975	62.6	60.1	2.5	49.2	9.3	1.6	2.9		12.2	20.7	0.75				[80]
58	Germany, Berlin 1974/1975	74.6	71.2	3.4	65.4	4.9	0.9	1.8		7.7	15.2	0.90				[80]
59	Italy, Modena 1974/1975	67.7	65.8	1.9	20.3	44.7	0.8	6.1		8.3	16.2	1.00				[80]
60	Argentina 1975/1976	66.6	56.9	9.7	3.3	53.2	0.4	9.4		14.1	7.2	0.90				[80]
61	Argentina 1980	58.8	24.4	34.4	3.4	20.6	0.4	12.5		7.1	17.0	0.90				[80]
62	Germany, Herrenberg 1979/1980	60.3	57.2	3.1	48.6	6.2	2.4	5.9		6.7	25.8	1.10				[80]
63	Chile 1974	25.0	22.6	2.4	5.3	12.4	4.9	7.3		2.1	45.1	0.60				[80]
64	Bohemia CSSR 1975	66.1	63.8	2.3	48.4	9.7	5.7	9.5		6.5	17.5	0.70				[80]
65	Turkey 1975	81.3	81.3	0.0	23.2	3.8	54.3	0.0		5.9	12.0	0.90				[80]
66	Hungary 1975	61.8	53.1	8.7	39.3	11.4	2.4	10.4		16.4	11.0	1.20				[80]
67	Bulgaria 1974	32.7	16.6	16.1	9.2	5.6	1.8	2.9		46.0	9.4	1.00				[80]
68	Argentina 1974	71.0	62.4	8.6	7.7	53.5	1.2	6.3		9.9	7.4	1.10				[80]
69	Argentina 1975	61.8	53.6	8.2	6.7	45.8	1.1	6.5		14.1	11.5	0.80				[80]
70	Argentina 1979	63.2	54.6	8.6	5.6	47.7	1.3	7.0		10.5	7.0	0.80				[80]
71	Bodegold 1975	72.9	71.6	1.3	32.2	34.6	4.8	11.3		4.2	11.3	1.00				[80]
72	Egypt	39.1	36.7	2.4	27.0	4.2	5.5	2.8			2.7				A	[12]
73	Egypt	43.3	40.8	2.5	29.0	5.8	6.0	1.5		20.0	6.0				A	[12]

No.	Sample										Density	n	Type	Ref.
74	Egypt	44.0	40.5	3.5	30.4	4.7	5.4	2.2	15.8	2.5			A	[12]
75	—	42.9	40.7	2.2	30.5	4.5	5.7	2.4	16.3	4.0			A	[12]
76	—	37.6	22.3	15.3	19.8	2.5	4.3	0.0	0.0				A/V	[12]
77	—	26.5	25.9	0.6	3.9	2.9	19.1	1.7	31.2	3.1			Bisabolonoxid-type	[12]
78	—	31.0	14.8	16.2	5.5	6.1	3.2	2.9	0.0	2.8			Bisabolol-type	[12]
79	Balkan	23.7	22.7	1.0	7.4	6.6	8.7	7.0	17.3	5.2			V	[12]
80	—	17.9	10.9	7.0	5.2	3.3	2.4	3.2	39.2	1.8			V	[12]
81	—	15.8	10.0	5.8	5.2	2.9	1.9	3.6	39.6	2.0			V	[12]
82	—	18.0	16.9	1.1	3.6	5.8	7.5	4.8	28.6	2.2			V	[12]
83	—	14.3	5.4	8.9	3.2	2.2		2.8	22.0	1.2			V	[12]
84	—	15.2	13.9	1.3	10.0	3.9		8.2	35.0	3.8			V	[12]
85	—	3.0	0.3	2.7	0.3	0.0		0.0					V	[12]
86	—	10.5	6.4	4.1	5.7	0.7		0.2						[12]
90	Degumille	20.5	2.1	18.4	0.6	1.3	0.2	3.6		0.7			Bisabolol-type	[12]
91	Degumille	43.2	10.7	32.5	4.1	5.4	1.2	14.2		2.8			Bisabolol-type	[12]
92	Degumille	25.1	5.2	19.9	1.7	2.8	0.7	5.9	36.6	1.6			Bisabolol-type	[12]
93	Degumille	32.8	6.8	26.0	3.0	3.5	0.3	11.3	16.7	8.5			Bisabolol-type	[12]
94	Fresh flowers	41.0	7.7	33.3	2.7	5.0		7.8	19.8					[13]
95	Drug	37.3	9.4	27.9	2.8	6.6		2.1	36.4					[13]
96	Chamomile Bleue Bulgarie (Adrian)	0.0	0.0								0.9096			[3]
97	Chamomile Bleue Bulgarie (Adrian)	0.0	0.0								0.9171	1.5031		[3]
98	Chamomile Bleue Bulgarie (Adrian)	0.0	0.0								0.9152	1.5032		[3]
99	Chamomile Bleue Bulgarie (Adrian)	10.4	2.8	7.6	2.8			4.7	45.4		0.898			[3]

TABLE 5.1.1

Composition of Chamomile Oil (According to Frank, Data from Different References and Laboratories [23]) (continued)

Identity number	Origin or type	Amount bisabolane	Bisabolol-oxide	Levomenol (Bisabolol) %	Bisabolol-oxide A %	Bisabolol-oxide B %	Bisabolon-oxide A %	Chamazulene %	Chamazulen% spectral-photom.	β-Farnesene %	Spiro-ether %	Total amount essential oil%	Density	Optical rotation	Chemical type or comment	Ref.
100	Chamomile Bleue Bulgarie (Adrian)	8.2	6.2	2.0	6.2			4.1		45.8			0.898			[3]
101	Chamomile Bleue Bulgarie (Adrian)	8.4	6.5	1.9	6.5			4.0		41.2			0.901			[3]
102	Chamomile Bleue Bulgarie (Adrian)	8.4	3.6	4.8	3.6			5.8		45.6						[3]
103	Chamomile Bleue Bulgarie (Adrian)	7.1	3.3	3.8	3.3			5.9		48.0						[3]
104	Chamomile Bleue Egypte	31.5	29.0	2.5	29.0			1.5		20.0						[3]
105	Chamomile Bleue Egypte	33.9	30.4	3.5	30.4			2.2		15.8						[3]
106	Chamomile Bleue Egypte	32.7	30.5	2.2	30.5			2.4		16.3						[3]
107	Chamomile Bleue Bulgarie	8.4	7.4	1.0	7.4			7.0		17.3						[3]
108	Chamomile Bleue Bulgarie	12.2	5.2	7.0	5.2			3.2		39.2						[3]
109	Chamomile Bleue Bulgarie	11.0	5.2	5.8	5.2			3.6		39.6						[3]
110	Chamomile Bleue Bulgarie	4.7	3.6	1.1	3.6			4.8		28.6						[3]
111	Chamomile Bleue Bulgarie	12.1	3.2	8.9	3.2			2.8		22.0						[3]
112	Chamomile Bleue Bulgarie	11.3	10.0	1.3	10.0			8.2		35.0						[3]
113	Chamomile Bleue Egypte (Adrian)	38.2	36.2	2.0	36.2			4.3		23.6			0.952			[3]
114	Chamomile Bleue Egypte (Adrian)	45.2	43.1	2.1	43.1			3.0		20.8			0.954			[3]

115	Chamomile Bleue Egypte (Adrian)	44.8	42.7	2.1	42.7	3.9	20.3	0.9631	1.5068	[3]
116	Chamomile Bleue Egypte (Adrian)	44.5	42.4	2.1	42.4	3.9	19.4	0.962	1.5093	[3]
117	Chamomile Bleue Egypte (Adrian)	43.9	41.9	2.0	41.9	2.8	23.3	0.958		[3]
118	Chamomile Bleue Egypte (Adrian)	43.2	41.6	1.6	41.6	2.7	25.4	0.954	1.5050	[3]
119	Chamomile Bleue Egypte (Adrian)	46.3	44.7	1.6	44.7	3.6	19.0	0.963		[3]
120	Chamomile Bleue Egypte (Adrian)	45.9	44.1	1.8	44.1	2.8	19.7	0.961		[3]
121	Chamomile Bleue Egypte (Adrian)	43.3	41.7	1.6	41.7	2.9	19.5	0.961		[3]
122	Chamomile Bleue Egypte (Adrian)	45.9	43.6	2.3	43.6	2.6	19.6	0.961		[3]
123	Chamomile Bleue Egypte (Adrian)	34.4	32.2	2.2	32.2	3.2	26.0	0.946	1.5063	[3]
124	Chamomile Bleue Egypte (Adrian)	41.2	39.2	2.0	39.2	3.1	24.2	0.950	1.5050	[3]
125	Chamomile Bleue Egypte (Adrian)	35.7	33.6	2.1	33.6	2.4	24.7	0.948	1.5030	[3]
126	Chamomile Bleue Egypte (Adrian)	40.2	38.1	2.1	38.1	3.0	20.8	0.957	1.5050	[3]
127	Chamomile Bleue Egypte (Adrian)	40.3	38.2	2.1	38.2	2.8	22.1	0.956	1.5050	[3]
128	Chamomile Bleue Egypte (Adrian)	35.5	33.4	2.1	33.4	2.8	26.8	0.946	1.5050	[3]
129	Chamomile Bleue Egypte (Adrian)	38.3	36.0	2.3	36.0	3.0	23.3	0.955	1.5064	[3]
130	Chamomile Bleue Egypte (Adrian)	46.2	44.0	2.2	44.0	4.1	19.2	0.963	1.5068	[3]
131	Chamomile Bleue Egypte (Adrian)	42.0	40.0	2.0	40.0	2.8	19.3	0.959	1.5062	[3]
132	Chamomile Bleue Egypte (Adrian)	47.1	45.0	2.1	45.0	4.3	20.3	0.958	1.5062	[3]
133	Chamomile Bleue Egypte (Adrian)	46.3	44.2	2.1	44.2	4.2	20.1	0.962	1.5084	[3]

TABLE 5.1.1
Composition of Chamomile Oil (According to Frank, Data from Different References and Laboratories [23]) (continued)

Identity number	Origin or type	Amount bisabolane	Bisabolol-oxide	Levomenol (Bisabolol) %	Bisabolol-oxide A %	Bisabolol-oxide B %	Bisabolon-oxide A %	Chamazulene %	Chamazulen% spectral-photom.	β-Farnesene %	Spiro-ether %	Total amount essential oil%	Density	Optical rotation	Chemical type or comment	Ref.
134	Chamomile oil, Germany 304093 March 1994	39.5	34.7	4.8	8.8	24.2	1.7	4.7		25.8	2.9					[36]
135	Chamomile oil, Germany 324509 July 1994	39.9	34.9	5.0	9.4	24.0	1.5	4.2		25.1	3.2					[36]
136	Dragoco, 1996, lab sample	66.6	66.1	0.5				23.4		2.9	0.0					[36]
137	Dragoco, 1984, prod. Batch	37.3	32.3	5.0				6.1		21.3	5.4					[36]
138	Dragoco, 1984, Argentina I	40.2	35.5	4.7				5.6		22.9	0.0					[36]
139	Dragoco, 1984, Argentina II	38.3	33.9	4.4				7.7		25.2	0.0					[36]
140	Dragoco, 1984, Egypt, Kato	53.4	51.3	2.1				3.1		22.7	0.0					[36]
141	Dragoco, 1984, prod. Batch	43.0	15.7	27.3				15.7		18.9	0.0					[36]
142	Dragoco, 1980, prod. Batch	46.3	40.0	6.3				6.8		22.6	0.0					[36]
143	Dragoco, 1978, prod. Batch I	49.0	42.5	6.5				11.5		7.5	0.0					[36]
144	Dragoco, 1978, prod. Batch II	41.5	35.0	6.5				10.0		8.0	0.0					[36]
145	Dragoco, 1975, prod. Batch	35.0	29.0	6.0				5.0		24.0	4.0					[36]
146	KMI-R99-1	62.0	61.6	0.3	52.4	4.4	4.8	13.0		2.3	19.7	0.66			Distilled from the drug	[11]

147	KMI-R99-2	50.5	1.5	49.0	0.3	1.1	0.0	14.4	3.2	26.0	0.55	Distilled from the drug	[11]
148	KMI-R99-3	55.2	54.7	0.5	17.4	25.0	12.3	12.7	1.7	22.6	0.43	Distilled from the drug	[11]
149	KMI-R99-4	49.6	5.6	44.0	0.6	5.0	0.0	19.6	2.2	22.6	0.25	Distilled from the drug	[11]
150	KMI-R99-5	59.0	0.9	58.1	0.9	0.0	0.0	17.3	3.8	14.4	0.38	Distilled from the drug	[11]
151	KMI-R99-6	59.8	59.8	0.0	42.3	12.9	4.6	10.6	2.9	16.3	0.53	Distilled from the drug	[11]
152	KMI-R99-7	52.5	13.9	38.6	5.7	7.8	0.5	19.3	2.9	20.0	0.40	Distilled from the drug	[11]
153	KMI-R99-8	53.6	53.3	0.4	29.8	11.9	11.7	13.2	2.1	24.3	0.51	Distilled from the drug	[11]
154	KMI-R99-9	49.6	8.1	41.5	3.8	4.3	0.0	14.0	3.6	27.8	0.62	Distilled from the drug	[11]
155	KMI-R99-10	49.9	9.3	40.5	1.8	7.5	0.0	18.1	4.7	23.3	0.49	Distilled from the drug	[11]
156	KMI-R99-11	48.5	11.6	36.9	7.8	3.5	0.4	17.6	4.4	23.2	0.54	Distilled from the drug	[11]
157	KMI-R99-12	60.4	56.6	3.8	46.3	6.8	3.5	11.2	3.2	21.2	0.27	Distilled from the drug	[11]
158	KMI-R99-13	56.9	3.8	53.1	2.1	1.7	0.0	10.4.	2.7	24.5	0.47	Distilled from the drug	[11]

TABLE 5.1.1
Composition of Chamomile Oil (According to Frank, Data from Different References and Laboratories [23]) (continued)

Identity number	Origin or type	Amount bisabolane	Bisabolol-oxide	Levomenol (Bisabolol) %	Bisabolol-oxide A %	Bisabolol-oxide B %	Bisabolol-oxide A %	Cham-azulene %	Cham-azulen% spectral-photom.	β-Farnesene %	Spiro-ether %	Total amount essential oil%	Density	Optical rotation	Chemical type or comment	Ref.
159	KMI-R99-14	56.0	55.4	0.5	33.0	11.9	10.6	15.2		2.6	21.4	0.53			Distilled from the drug	[11]
160	KMI-R99-15	47.6	3.9	43.7	0.3	3.6	0.0	14.7		4.9	27.7	0.46			Distilled from the drug	[11]
161	KMI-R99-16	51.9	3.4	48.5	0.0	3.4	0.0	15.8		3.1	24.9	0.40			Distilled from the drug	[11]
162	KMI-R99-17	48.2	0.0	48.2	0.0	0.0	0.0	20.9		2.7	20.4	0.30			Distilled from the drug	[11]
163	KMI-R99-18	54.6	54.2	0.4	25.7	18.6	9.9	12.0		1.9	24.1	0.60			Distilled from the drug	[11]
164	KMI-R99-19	50.6	48.4	2.1	16.2	20.9	11.4	13.9		1.8	27.0	0.50			Distilled from the drug	[11]
165	KMI-R99-20	64.1	63.0	1.2	48.8	9.9	4.3	10.2		2.7	19.6	0.49			Distilled from the drug	[11]
166	KMI-R99-21	55.2	45.7	9.6	25.5	18.1	2.1	8.2		5.4	25.6	0.43			Distilled from the drug	[11]
167	RP KMI 2000-1	65.8	63.0	2.7	42.7	15.7	4.6	11.3		2.3	13.5	0.36			Distilled from the drug	[11]

168	RP KMI 2000-2	46.6	1.4	45.2	0.0	1.4	0.0	21.3	2.2	25.0	0.27	Distilled from the drug	[11]
169	RP KMI 2000-3	52.8	12.3	40.5	0.0	12.3	0.0	15.5	2.9	24.1	0.47	Distilled from the drug	[11]
170	RP KMI 2000-4	66.0	57.6	8.4	41.9	12.1	3.6	10.7	3.1	17.9	0.50	Distilled from the drug	[11]
171	RP KMI 2000-5	60.0	59.4	0.5	47.9	3.4	8.2	12.8	2.4	20.7	0.44	Distilled from the drug	[11]
172	RP KMI 2000-6	55.2	1.4	53.7	0.0	1.4	0.0	18.4	2.4	19.4	0.27	Distilled from the drug	[11]
173	RP KMI 2000-7	61.1	1.5	59.6	0.0	1.5	0.0	19.1	2.4	14.0	0.30	Distilled from the drug	[11]
174	RP KMI 2000-8	59.1	45.0	14.1	23.2	19.9	2.0	10.1	6.3	21.5	0.25	Distilled from the drug	[11]
175	RP KMI 2000-9	48.3	13.9	34.4	8.9	4.4	0.6	17.3	4.6	25.0	0.33	Distilled from the drug	[11]
176	RP chamomile 2001-1	72.5	70.6	1.9	51.3	14.1	5.2	10.7	1.7	11.8	0.81	Distilled from the drug	[11]
177	RP chamomile 2001-2	51.3	1.1	50.2	0.0	1.1	0.0	21.6	3.9	20.7	0.52	Distilled from the drug	[11]
178	RP chamomile 2001-3	51.8	1.0	50.7	0.0	1.0	0.0	23.5	4.2	18.4	0.63	Distilled from the drug	[11]
179	RP chamomile 2001-4	66.7	63.0	3.7	51.4	5.6	6.0	12.8	3.3	15.0	0.55	Distilled from the drug	[11]

TABLE 5.1.1
Composition of Chamomile Oil (According to Frank, Data from Different References and Laboratories [23]) (continued)

Identity number	Origin or type	Amount bisabolane	Bisabolol -oxide	Levomenol (Bisabolol) %	Bisa-bolol-oxide A %	Bisa-bolol-oxide B %	Bisa-bolon-oxide A %	Cham-azulene %	Cham-azulen% spectral-photom.	β-Farnesene %	Spiro-ether %	Total amount essential oil%	Density	Optical rotation	Chemical type or comment	Ref.
180	RP chamomile 2001-5	62.1	0.6	61.5	0.0	0.6	0.0	21.2		3.0	12.0	0.43			Distilled from the drug	[11]
181	RP chamomile 2001-6	56.8	56.8	0.0	51.7	5.1	0.0	15.9		2.5	16.8	0.68			Distilled from the drug	[11]
182	RP chamomile 2001-7	69.9	63.7	6.2	38.1	22.4	3.2	9.6		4.9	14.8	0.58			Distilled from the drug	[11]
183	RP chamomile 2001-8	57.1	15.5	41.6	9.9	5.0	0.5	20.6		3.4	17.3	0.58			Distilled from the drug	[11]
184	RP chamomile 2001-9	56.1	11.1	45.0	4.7	6.4	0.0	18.0		3.3	20.4	0.63			Distilled from the drug	[11]

Traces are given with 0.1% to be able to grade them.

n.d. or not detectable are given with 0.0.

like in the Hungarian Puzsta or in Argentina are duly tolerated. As a matter of fact, however, chamomile has to be sown on a smooth and solidified field to make sure that the fine seeds (thousand kernel weight 0.02 to 0.06 g with diploid forms and 0.04 to 0.12 g with tetraploids) are not washed in the soils due to rainfall. The soil should therefore be prepared on a flat working level only; it should duly settle before sowing is done and it should be solidified by means of a roller. The seeds require a lot of humidity for germination and for a quick juvenile development.

5.1.3 FLOWERING

True chamomile flowers over a longer period and produces new flowers. For the cultivation this is taken into consideration by a multiple harvest of flowers. The flowering time can last about 50 to 65 days. It takes about 20 to 35 days before a flower is fully developed. The relatively short vegetation period of 150 to 180 days also allows a cultivation at higher altitudes up to 500 m.

5.2 METHODS OF CULTIVATION

Cultivation measures such as sowing time, fertilization, weed control, and harvest have to be arranged in such a way that the yield properties and those of quality fixed in the genotype ensure an optimum development.

5.2.1 CULTIVATION PROCEDURE

In 1956 Heeger still reported that with respect to cultivation chamomile had been worked on to a small extent only [40]. Over the past 40 years rapid development could be observed. Preference is given to chamomile being cultivated. In most countries extensive field cultivation with mechanized sowing, nursing, harvest, and processing has gained acceptance [13, 28, 38, 75, 84]. In nearly all companies specializing in cultivation, technological innovations have been tested and introduced repeatedly. Often technological solutions are quite specific to the companies and having been adapted to the locations and varieties were developed for the individual sections of production [62, 68]. Therefore, a compilation of a generally applicable sample technology does not seem to be recommended. Some experiences from various countries of cultivation are shown in the paragraphs to follow.

Presently three basic variants — being mostly complementary to each other — are applied: autumn sowing, spring sowing, and cultivation of several years' duration with self-sowing.

5.2.1.1 Autumn Sowing

In the Northern Hemisphere graduated autumn sowing is done at the end of the summer at an interval of about 8 to 14 days, from the beginning to the end of September. A precondition is, however, that the anticipated fruit is removed as quickly as possible. If sowing is done too early, this leads to premature formation of shoots [16]. Both plants developed too far as well as those that are too small die in winter. Chamomile should start the winter months at a stage of six to eight leaves. At this stage it is definitely cold resistant. Autumn sowing means that the best yields are achieved with the possibility of a long vegetative developmental and tillering phase within the physiological short day. Being largely independent from the sowing time, the flowering period starts with an achieved day length of about 17 hours (in central Europe at the end of May/beginning of June). Due to the fact that flowering generally starts at the same time, the possible cultivation area is strongly determined by the harvest and drying capacity. Even if chamomile is flowering over a longer period, producing new flowers again and again temporarily, the first harvest process has to take place according to exact timing. Thoroughly ripening flower heads favor plant aging. Overripe flowers strongly tend to fall apart during the drying and preparation process after the harvest [50]. Seeds dropping out mean a burden to the location in the years to follow, changing the spectrum of active principles [18, 24, 30, 81].

5.2.1.2 Spring Sowing

The advantage of spring sowing is the possibility of graduated steps of cultivation in order to be able to utilize the harvesting and drying technique to the maximum by temporally graduated harvest dates. In central Europe it is a usual practice to start sowing in March at an interval of about 14 days. It is, however, a fact that the yield and the homogeneity of the populations decrease the later sowing is done. The pressure of contamination by diseases and infestation by parasitic insects in the course of the summer months increases just as well [21].

5.2.1.3 Cultivation of Several Years' Duration by Self-Sowing

With extensive production, a cultivation of several years' duration is possible on the same field considering that the population is being formed by the chamomile seeds that dropped out. Additional seeds may also be sown. After the last economically justifiable harvesting passage, the population remains on the field for a few more days and the plants are mulched. The next step is a non-turning superficial soil preparation. From September onward the self-sown seeds germinate, which may be compared with a thick carpet. This procedure is similar to broadcast sowing without a drilling machine. The strong competitive capability of chamomile against weed, the possibility of utilizing a harrow and a chain harrow or curry-comb for weed control, and thinning out make the procedure a cheap alternative. Companies particularly in Germany, Argentina, Hungary, and in the Czech Republic have practiced this procedure on the same field for nearly ten years. In the total remaining area the weed is reduced by taking the autumn chamomile from the crop rotation. The yields to be expected and the other peculiarities of this procedure may be compared with those of the sowings done in autumn. This form of cultivation is not recommended for the flower production of breeding lines, especially with the cultivation of defined chemotypes, because of the uncontrolled procedure of the pollination as a germination of wild (i.e., nonbred) chamomile or a segregation with hybrid seeds may occur in the following year.

Especially with autumn sowing slightly humid locations are more favorable than regions with plenty of summer rains where chamomile tends to have a luxurious growth of herb and leaves; consequently it is less suitable for being harvested and shows a lower content of active principles. The development of chamazulene substantially depends on the duration of sunshine and the day temperatures from the time of the formation of flower buds.

It is reported repeatedly that chamomile is self-compatible [16, 19]. Experiences regarding cultivation of several years' duration and self-sowing point in the same direction. Precise indications concerning the value of different fruit rotations are missing but it is a fact that chamomile grows particularly well in fields free from weed; viz., after root crops such as potatoes and sugarbeet, maize, winter cereals, and leguminosae. Using fields for the production of chamomile for several years has also proved to be a success. Chamomile should generally be cultivated as final crop in the rotation without any N-fertilization. A high content of N_{min} in the soil leads to strong vegetative growth, makes the harvest more difficult, and can finally also lead to a lower content of active principles. Weeds cause problems for the cultures to follow. First of all a nonturning superficial soil preparation should in any case take place after the chamomile cultivation to stimulate the germination of chamomile seeds dropped out, to be followed by mechanical weed control in autumn. In course of the following year a culture is started after refertilization, being tolerant against herbicides combating chamomile.

Chamomile particularly absorbs the heavy metal cadmium from the soil [78], at least it concentrates due to the mobility in the flowers like with other flower drugs, e.g., St. John's wort

TABLE 5.2.1
Maximum Tolerable Content of Cadmium for Soils Planned for the Production of Chamomile in Central Europe (According to Plescher [66], Cd Maximum Value for *Matricariae flos* 0.2 mg/kg)

pH Value of the Soil	Max. Cd Content of the Soil in mg Cd/kg Soil, Dry Matter	
	Spring Sowing	Autumn Sowing
5.5	0.13	0.20
6.0	0.16	0.24
6.5	0.20	0.29
7.0	0.27	0.36
7.5	0.42	0.49

and flowering yarrow herb [67, 72]. As long as the maximum value of 0.2 mg Cd/kg of drug — a recommendation of the German Federal Ministry of Health since 1991 — is applicable, the maximum soil values indicated in Table 5.2.1 may be used for the choice of the location, depending on the individual pH value and on the method of cultivation [66]. If there are critical soil values a trial cultivation should always be used to determine whether the location is suitable.

5.2.2 NUTRIENT SUPPLY

Although chamomile is a nitrogen-loving plant species, it is, in view of the harvest mechanization, seen that when cultivating it the soil only contains small amounts of N_{min} and mostly it is done without any N-fertilization whatsoever [87]. Otherwise moist and leafy populations with a delayed flower maturity or "endless flowering" develop and with the application of harvest technique the species-typical appearance of foreign matter (percentage of leaves and stems) increases. Table 5.2.2 gives a general idea of recommendations for fertilization with respect to the cultivation of chamomile.

For the development of active principles, a good flowering, and sufficient firmness of stems, a good potassium supply is necessary [29]. On the other hand chamomile is delicate with regard to excessive quantities of P_2O_5. According to Franz and Kirsch [32] the nutrient ratio of nitrogen:potassium oxide should be 1:2. Fertilization by using potassium and if necessary also nitrogen takes place during the tillering phase, i.e., in October (at the Northern Hemisphere) or April (at the

TABLE 5.2.2
Recommendations for Fertilization Concerning the Cultivation of Chamomile in kg/ha

	N	P_2O_5	K_2O
Heeger [39] 1956	30–40	40–60	120–140
Schröder [82] 1965	40–80	36–45	80–120
Fink [22] 1978	60–80	20–30	70–100
Ebert [19] 1982	20–50	30–50	100–150
Ruminska [74] 1983	40–50	50–60	60–80
Traxl [88] 1986	30	50	80
Dachler, Pelzmann [16] 1989	40	50	100
Bomme, Nast[a] [8] 1998	17[a]	8[a]	22[a]
Bernath [7] 1993	40–60	40–60	50–70

[a] Nutrient uptake.

Southern Hemisphere) when sowing in autumn; if sowing is done in spring it takes place at a correspondingly later date. Organic fertilizers should by no means be used for chamomile, as the limits of microbial contamination could easily be exceeded. Organic fertilizers may only be used for the anticipated fruit.

With extensive cultivation the recommendations for fertilization based on N_{min} values ascertained are the basis for the determination of fertilizer requirements pertaining to the fields. The deprivation of nutrients has become the basis of fertilization to an increasing extent [65, 72]. (See Table 5.2.3.)

5.2.3 HARVEST

Even though chamomile is cultivated throughout the world, the collection from wild populations, particularly in central, eastern, and southern Europe, Greece, and Turkey has not lost its importance. There is a difference between "pure wild collection" as in Greece and Turkey and the organized harvest of wild populations, as is the case in Hungary for about 95% of the whole harvested quantity. From such wild collections or small cultivations and in gardens, chamomile is harvested at the period of full flowering, i.e., when most flowers are already open. Today, cultivation areas are mainly harvested automatically, although partly manual harvesting is still being practiced as well.

The harvest schedule plays an important role with regard to the quality of the product. Harvest of cultivated chamomile should be realized in the same way as for wild populations, i.e., when the major part of the flowering heads has already opened. The essential oil content of inflorescence

TABLE 5.2.3
Deprivation of Nutrients by Chamomile Flowers and Flowering Herb as well as Nutrient Content of the Herb Residues (According to References 48, 66, 72) (The deprivation of nutrients pertaining to the area is based on an average drug yield of 500 kg flower drug/ha.)

	Plant nutrients						
	N	P	P_2O_5	K	K_2O	Mg	MgO
Chamomile flowers, fresh							
Nutrient content kg/t fresh	4.2	0.9	2.1	4.5	5.4	0.4	0.6
kg nutrient in 3 t fresh material (kg nutrient/ha)	12.6	2.7	6.3	13.5	16.2	1.2	1.8
Chamomile flowers, dry							
Nutrient content kg/t drug	21.8–29.3	4.5–4.9	12.6	23.4–29.8	32.4	2.2–2.4	3.6
kg nutrient in 500 kg drug (kg nutrient/ha)	10.9–14.6	2.2–2.4	6.3	11.7–14.9	16.2	1.1–1.2	1.8
Flowering chamomile herb, fresh							
Nutrient content kg/t fresh	3.0	0.4	0.9	4.1	5.0	0,4	0.7
Flowering chamomile herb, dry							
Nutrient content kg/t drug	9.1–18.6	1.9–2.6	5.9	24.7–25.8	31.1	1.2–2.7	4.5
Herb without flowers, fresh							
Nutrient content in kg/t fresh	2.6	0.4	0.9	4.4	5.3	0.4	0.7
Herb without flowers, dry							
Nutrient content in kg/t drug	16.3	2.5	5.7	27.6	33.3	2.7	4.5

increases continuously starting from the creation of flowers and achieves its maximum when the ligulate flowers are in a horizontal position (see Section 5.5). As per Röhricht et al. [72] harvest is realized from the starting of the flowering period up to the period of full flowering, whereas two or three pickings are effected. As per Dachler and Pelzmann [17] the optimal time for harvest of the flowers is when a circle of tubular florets has already opened in the second third of the vaulted flowering receptacle.

In order to create an objective base for the determination of the optimal harvest time and not leaving this matter to the cultivators' intuition, the German Working Committee for Cultivation of Medicinal Plants proposed the following flowering index formula in the years 1971, 1972, 1973, 1977, and 1979. It is a compromise between an increasing yield of flowers, a decreasing content of essential oil, and changes in the composition of the essential oil [4, 30].

$$\text{flowering index} = \frac{V - Kn}{Kn + eB + V}$$

Kn = flower buds not yet flourished
eB = flowers ready to be harvested (flourished tubular flowers + ligulate flowers)
V = withering flowers

With this formula the point of time for harvest of the flowers is calculated from the relation of the number of withering flowers minus flower buds to the total amount of flowers (flower buds + flowers ready to be harvested + withering flowers). The first picking should take place when the flowering index reaches a value of –0.3 to –0.2. As the flowering period depends on the location, climatic conditions, and the variety of the flower and therefore ranges between several weeks up to 3 months (according to [19] eight to ten manual picking cycles are possible), this formula is a good way to determine the corresponding optimal time for harvest. In practical applications harvest takes place two or in rare cases three times.

In 1985 Franz [27] proposed the following flowering index formula:

$$I_K = \frac{IV - I}{I + II + III + IV} = -1 < I_K < +1$$

I_K = $\text{Index}_{chamomile}$
I   = flower buds
II = flowers ready to be harvested with a 50% open tubular flowers
III = flowers ready to be harvested with more than 50% open tubular flowers
IV = withered, decomposed flowering heads

The realization of the harvest of the flowering heads can take place in different ways, whereas the most labor-intensive method is manual picking. This manual picking is carried out in most countries only for small cultivation areas; in Egypt, however, it is still very popular and is applied almost exclusively (Figure 5.2.1).

Picking yield of freshly harvested short-stemmed chamomile flowers is about 3 to 5 kg/h. Slightly higher picking amounts can be obtained already with so-called chamomile picking combs. Today they are used (e.g., in Hungary) above all for organized harvest in wild populations, and have a capacity of about 50 to 150 kg fresh per day (see also Section 5.5).

Such a harvest with chamomile picking combs has already been carried out in Hungary well into the 1970s, as well for harvest on cultivated fields. Due to the higher plant and flower density, the picking yield was about 100 to 180 kg per day. Other procedures were the use of special forks

FIGURE 5.2.1 Handpicking in Egypt.

or comb shovels, applied in Hungary, similar to a scythe for harvesting the chamomile crop (Figures 5.2.2 and 5.2.3).

Finally, such manual methods are not practicable above all due to the labor time requirement of about 25 to 30 working days per ha for the production of larger quantities. This problem had

FIGURE 5.2.2 Comb shovel.

FIGURE 5.2.3 Harvest with comb shovels in Hungary.

FIGURE 5.2.4 Manually pushed picking cart.

to be solved through construction of mechanized harvesting machines. A first step in this direction was the use of manually pushed picking carts with two wheels (Figures 5.2.4 and 5.2.5).

Later on these picking carts were drawn by horses or tractors, which above all were applied in Argentina and again increased the yield per area.

Today the large-area industrial cultivation of chamomile worldwide uses automatic harvesting techniques with a picking yield of 200 to 300 kg/h and a capacity of about 3.5 ha per day, whereby (dependent on climate and crop-specific conditions) about 65 to 90% of all flowers are harvested. Since 1962 in Germany [19, 20] and since approximately 1975 as well abroad (e.g., in Argentina and Hungary), chamomile picking machines were used that pick the flowering heads on the field

FIGURE 5.2.5 Manually pushed picking carts in Argentina.

automatically and are self-propelling, as a cutter mounted in the front part of a tractor (see also Oravec et al., Chapter 5.6) or as combined harvesters (Figures 5.2.6 and 5.9.3).

For a competitive production of flower drugs a decrease of the high labor time requirement for the harvest is essential, whereby particular importance has to be attached to the outer quality of the harvested flowers. The demands on the harvest quality are high. The adherent stalk rests have to be as short as possible and the impurities of herb and other constituents the lowest possible. This can be achieved safely by manual picking. Such important countries of production as Italy [1, 5, 6], Argentina, Hungary [9], Russia [2], former Yugoslavia [9, 52], former Czechoslovakia [45], and former GDR [42, 71, 73] have developed and applied high-capacity harvesting machines with different picking systems.

FIGURE 5.2.6 Harvester.

Automatic harvest is to be carried out successfully in the case of a large-flowered chamomile variety, having their flowering heads almost at one level. The major part of the commercial drug originates from this automatic flower harvest, which entails *Pharmacopoeia's* monographs of chamomile flowers, describing chamomile to only "sometimes" have stem rests of 10 to 20 mm [79], are not applicable. If the harvested material is not processed after automatic harvesting, it is impossible to avoid stem rests with a length of 20 to 30 mm. With a corresponding postprocessing however, it is possible to produce chamomile of pharmaceutical quality. These not "sometimes" but almost always existing stem rests of 20 to 30 mm in length in no way disturb the medical benefit, insofar as the flowering head disposes of the required content of ingredients. Consequently this should appear in the *Pharmacopoeias* monographs as reality and therefore should be allowed [80].

Flourishing herb for production of *Chamomillae herba cum floribus*, that can be used for the production of tea in filter bags, mostly is harvested by the help of rotary mowers, forage harvesters, chaff cutters with blower, or combined harvesters [17, 19]. To obtain extraction quality for pharmaceutical use, the whole flowering heads are preferably harvested automatically. In this case, a posterior separation of stem rests is not necessary.

Another harvesting principle represent the machine developed in Argentina. Drums equipped with blades are used here. They run top down against a comb and shear the chamomile flowers. Here as well, industrial chamomile is preferably harvested due to the long flower stems. This principle of the rotating blades primarily has been developed for trailer machines that are drawn sideways behind the tractor. Later on these machines were developed further to self-propelling combined harvesting machines.

A third principle of mechanized harvesting is the picking procedure with picking combs. Here, machine widths of 2 to 6 meters of the width of the cutting unit are achieved. The picking drum is equipped with shifted comb strips, whereby long round tines in a wide distance or shorter sharp-edged tines with a lower interspace are used. The picking drum rotates against the driving direction, whereas the combs pass through the crop from the bottom up and separate the flowers from the plants. Adherent stalk rests are cut off with a cutting machine. For further cutting of the stalk parts the harvest is transported onto a swinging sieve with conveyor belts and then is collected in a silo. This is the basic principle of the machine of Ebert-Schubert [20], the Hungarian machines type "Szilasmenti" (see as well Section 5.5), the Slovakian machines type "VZR" (see as well Section 5.6), the Russian machines [2] and the machine "Linz III," developed in the years 1972 to 1978 in the GDR [71, 73] and built until 1988 [56]; the Linz proved itself in various countries (including Egypt), the quality and performance of which has not been achieved by other constructions so far. This is due to contra blades operated by cam discs behind the comb strips, that in contrast to other machines with picking combs lead mainly to an exact cutting of the flowering heads and clearly reduce the common tearing effect (Figure 5.2.7).

A fourth principle is the rotation of the picking drum in the driving direction, as realized with the Yugoslavian machine [9, 52]. The chamomile herb is grasped by the tines and then drawn into the machine. There, on the vaulted baffle plate a plant pad is formed that is repeatedly combed due to the high-revolution speed of the threshing drum. The combing effect results from a shifted arrangement of consecutive comb strips. The distance between the tines of the comb strips is wider than the diameter of the flowers and consequently inhibits a clogging of the picking drum.

The principle of immobile combs with appropriate separation of the flowers from the plant, e.g., by means of a modified scraper chain on top of the comb, similar to marigold harvesters [46] has been realized in the Italian *Chamaemelum* harvester [5, 6].

Summarizing, it can be noted the best principle is a picking drum with comb strips rotating against the driving direction. Unfortunately, at present there is no new, well-priced, and similarly effective machine on the market (see also [38, 57, 58, 59, 60]).

For after-harvest treatment there are two possibilities. In countries with high availability of manual work, e.g., Egypt, primarily drying is effected and afterward processing and sorting take place. In this way, drying often is realized in racks, e.g., from palm fronds (in Egypt) or on frames

FIGURE 5.2.7 Comb strips of the chamomile picking machine "Linz III."

covered with gauze (as was used formerly on a large scale in Argentina). During daytime these racks or frames are placed in the open air; at night they are piled (mostly) under the roof, in order not to take up humidity. In Argentina after 1970, in the mornings workers brought daily 35,000 racks outside for sun drying and back to the stockrooms in the evenings. Temporarily during season more than 5000 persons were occupied with harvesting and processing.

A further increase in production was only possible by use of modern picking machines. In 1974 the Argentine production amounted to 2000 tons. With such a high grade of mechanization of harvesting and processing as in Argentina and Germany, sorting and sieving is effected automatically before drying. If need be, after drying short stem rests are separated in a further step. That way, pure flower material is obtained. During season 60 to 80 tons of fresh plant material are processed per day, running continuously through the sorting machines to the belt dryers or discontinuous flatbed dryers and after drying are sorted by the help of transport conveyors and are freed from weeds.

In Poland, which has become an important country for the production of chamomile drug in recent years, such harvesting procedures are applied that by cutting the flower horizon after drying by threshing, and separation by sieving and sifting, lead to a production of chamomile pollen and industrial pollen.

5.2.4 Cultivation in Specific Countries

Cultivation areas in the different countries vary greatly. In France approximately 100 ha of Roman chamomile are cultivated, mostly used for perfumery. In Italy cultivation of Roman chamomile is about 25 ha, of which 7 ha is organic, and 170 ha *Matricaria recutita*, of which about 100 ha is organic. From Sweden cultivation of 10 ha, from Austria cultivation of 15 ha, from Bosnia-Herzegovina 40 ha [47], Bulgaria 100 ha [47] of True chamomile, whereas in the United Kingdom 80 ha Roman chamomile and 120 ha *Matricaria recutita* are cultivated. In Spain plants are collected rather than cultivated. This is the origin of the diploid bisabolol-rich form "Degumill." In Brazil the main production is concentrated in the southern region Mandirituba (Parana state).

Cultivation experiences in some counties are shown in the following sections:

Section 5.5: Production of chamomile (*Matricaria recutita* L.) in east and south European countries
Section 5.6: Cultivation experiences in Slovakia
Section 5.7: Growing varieties of chamomile in the Czech Republic
Section 5.8: Experiences with the cultivation of chamomile in Argentina
Section 5.9: Chamomile in Chile: Cultivation and industrialization
Section 5.10: Cultivation experiences in Egypt
Section 5.11: Cultivation in Germany

5.3 PLANT SELECTION AND BREEDING

5.3.1 BREEDING TARGETS AND TECHNIQUES

The composition of the essential oil with varying percentages of bisabolol, bisabololoxide, bisabolone, and matricine is fixed genetically [25]. Schick and Reimann-Philipp [76] remarked in 1957: "As far as breeding is concerned *Matricaria Chamomilla* has not yet been worked on." In 1950 only the two group varieties "Quedlinburger Großblütige Kamille" (Quedlinburg large-flowered chamomile) and "Erfurter Kleinblütige Kamille" (Erfurt small-flowered chamomile) were mentioned [41]. Often the stability, the resistance against diseases, the germinability, the flower yield, the fact that the individual flower heads are ripe at the same time, the homogeneous flowering horizon, the stability of the flower head, and consequently the suitability for a mechanical harvest are not sufficient [42, 85]. Since then a rapid development has been experienced. A number of chemotypes [80] with a varying content of matricine/chamazulene, (–)-α-bisabolol, spiroethers, and the bisabolol oxides A and B as well as bisabolone oxide A were selected. Varieties with very different breeding targets were bred. The content, especially the composition of the active principles, was worked on by precise selections [16, 32, 49, 80]. For nondestructive characterization of valuable substances in chamomile single plants, a near-infrared (NIR) spectroscopical rapid method is available [55].

Some of the present breeding targets are (sequence does not mean order of priority):

• High yield
• Large flowers
• Compact flower heads
• Stability of the flowers for a mechanical harvest with a low percentage of stems
• Regular growth with basilar ramification and many flowers (close proportion of herb:flowers)
• Narrow homogeneous flowering horizon
• Uniform flowering time, especially simultaneous ripening
• Good shooting capacity for a high yield with the harvest to follow
• Firm flower head with little inclination of disintegration and formation of fines
• High content of chamazulene (blue oil) >25% within the oil
• High content of (–)-α-bisabolol >30–50% within the oil
• Low or much bisabololoxide

- High content of total flavonoids >3%
- High content of essential oil (0.7–>1%)
- Desirable composition of the essential oil
- High germinability
- Resistance against powdery mildew and other diseases
- Stable stand

Particularly the working group led by Franz has made intense genetical investigations ([44, 53, 54] and others). By this research it was proved that medicinally relevant (–)-α-bisabolol is inherited recessively. The formation of the (–)-α-bisabolol oxides A and B is dominant over (–)-α-bisabolol, whereas (–)-α-bisabolone oxide A dominates (–)-α-bisabolol as well as the (–)-α-bisabolol oxide A and B [33, 34]. This is verified by using PCR-based marker techniques like Random Amplified Polymorphic DNAs (RAPDs) and Amplified Fragment Length Polymorphisms (AFLPs) as useful tools for the acceleration and improvement of the breeding process [88].

Nowadays quite different forms are available to cover various demands. It should be taken into consideration that not all results of these breeding works are admitted as varieties or not all of them can be purchased but often individual firms use them for their internal cultivation to produce special products. A short (incomplete) survey of released varieties and used breeding material is given in Table 5.3.1, in Table 5.3.2 some of this breeding material is classified according to its characteristics.

Most of these are realized in tetraploid varieties [10, 51]. When choosing the breeding place as well as the company for seed multiplication and maintenance breeding, this should be done with care. In case of a contamination of the soil by chamomile seeds (wild growing or cultivated populations some years ago) a crossing of tetraploid and diploid chamomile takes place. A clear separation by mere sifting of the (mostly) bigger tetraploid seeds is not possible. As a large amount of chamomile seeds is found in the soil of the cultivation areas and as there is no reproductive isolation in respect of wild chamomile, it is impossible to cover the next year's seed requirements from normal material of cultivation being mostly still the case with traded material from Egypt. The multiplication of seeds has to be regulated by taking suitable steps of organization and agrotechnical measures. For this purpose the basic material has to be produced from individual plants by partly taking *in vitro* multiplication steps. The seeds on hand are more or less treated as a hoe culture to make sure that no unwanted chamomile can contaminate the seeds. If the multiplication of seeds is done properly there is little danger of hybridization even with diploid chamomile, so that this supposed advantage of tetraploid chamomile does not seem to be relevant in the main cultivation areas.

With the breeding at a diploid stage of valence, in Germany for instance, the first step was a selection on an early flowering time on relatively shortened long-day terms. In the breeding pattern individual plant selection from more than 10,000 plants of an Argentinean population in connection with diallel crossings, backcrossings, and *in vitro* cloning was applied. By means of thin-layer chromatography a preselection of plants of suitable qualitative composition was made. After the *in vitro* phase the clones (adapted to substrate culture and recultivated) were planted on plots of comparison at two places in four repetitions, then they were tested on susceptibility to diseases and finally the oil content and the oil composition was determined. So the material could be restricted to a great extent.

After the calculation of suitability for combination according to method II, model II, on the basis of the parents and F_1- descendants kept in tissue culture further seven diallel crossings of the suitable clones were carried out, which were multiplied again by a tissue culture before. The crossing partners ceased flowering isolated from the other crossings. From each crossing about 15,000 plants could be cultivated, which were planted into fields of 500 m² each with a distance of normally several kilometers from each other. A hybridization of wild material was prevented by protective sowings such as by a broad belt of maize or cereals and by control of the surrounding

TABLE 5.3.1
Varieties and Breeding Material in Chamomile (*Matricaria recutita* L.)

Country	Variety/line
Austria	Manzana (4x)
Bulgaria	Lazur (4x) Bisabolol Sregez (17% chamazulene)
Brazil	Mandirituba
Chile	Manzanilla Primavera Puelche
Czech Republic	Bohemia (2x) Bona (2x) Goral (=Kosice II, 4x, high content oil, chamazulene, bisabolol) Lutea (4x) Novbona (2x) Pohorelicky Velkosvety
France	MA.VS.1 (ca. 1% oil, 25% chamazulene, 45% bisaboloxid A/B)
Germany	(Leipzig) Bodegold (4x) Camoflora (2x) Chamextrakt (2x) Degumill (2x) Euromille Mabamille (4x) Manzana (4x) Robumille (4x)
Hungary	Budakalszi 2 (=BK2,4x, high content oil, chamazulene) Soroksari 40 (2x)
Italy	Minardi (2x) Olanda (matricin)
Poland	Zloty Lan (4x, high content of oil, chamazulene, low bisabolol) Tonia Promyk (2x)
Rumania	Margaritar (4x) Flora (4x)
Slovak Republic	Bona (2x) Goral (=Kosice II, 4x, high content oil, chamazulene, bisabolol) Lutea (4x) Novbona (2x)
Slovenia	Tetra
Spain	Adzet (2x, bisabolol)

TABLE 5.3.2
Characteristics of Varieties and Breeding Material in Chamomile (*Matricaria recutita* L.)

Characteristics	Variety/line/origin
High matricin-/chamazulene	Olanda
	Sregez
	Turkey
Low matricin-/chamazulene	Egypt
	Turkey
High matricin-/chamazulene and bisabolol	Adzet
	Bona
	Camextrakt
	Degumill
	Goral
	Lutea
	Mabamille
	Manzana
	Novbona
	Robumille
High matricin-/chamazulene and bisabololoxid	Bodegold
	Bohemia
	Budakalaszi 2 [BK 2]
	Camoflora
	Flora
	MA.VS.1
	Pohorelicky Velkosvety
	Promyk
	Soroksari 40
	Tetra
	Tonia
	Zloty Lan
	Egypt
	Argentina (usually bisaboloxid B)
	Mexico
High matricin-/chamazulene and bisabolon	Lazur
	Turkey
	Bulgaria

fields on wild chamomile. Any unsuitable plants were removed by negative mass selection. With the crossing descendants from bisabolol-azulene parent plants, the bisabololoxide A-azulene parent plants and the bisabolol oxide B-azulene parent plants backcrossings were carried out with the maternal parent from the *in vitro* depot.

5.4 OTHER SPECIES

With regard to the propagation of *Chamaemelum nobile* (L.) All. and some species of *Anthemis* and *Matricaria* short notations are made in Chapter 3, Plant Sources. The main area of *Anthemis*

L. ranges from the Mediterranean area to central Asia. The genus is particularly rich in species and forms in the Balkans, Transcaucasia, Anatolia, and the Near East. The genus of *Chamaemelum* Mill. is mainly spread over the Mediterranean area, west Europe, the British Isles, France, the Iberian Peninsula, and North Africa, whereas the main distribution area of the genus of *Matricaria* L. is to be found nearly all over Europe (including Scandinavia) and northern Asia.

The chief countries of origin of traded material of *Chamaemelum nobile* (L.) All. were Germany, Belgium, England, France, Spain, and Italy, but also the United States and Argentina [43]. Nowadays the filled flowering form is almost exclusively cultivated for drug production. At present the main supplier countries are France, Belgium, England, Belorussia, Ukraina, Transcaucasia, Czech Republic, and Poland. More rarely the drugs originate from India, North America, Brazil, Argentina, Chile, Mexico, Germany, and Austria. Regarding *Chamaemelum nobile* (L.) All. the filled flowering form is the one almost exclusively cultivated in sunny altitudes for drug production. Sometimes the cultivars "weißköpfig" (white headed), "gelbköpfig" (yellow headed) [55], "doppia," "stradoppia" [69], and "flore pleno" [19, 40] are traded. With the pollination of the completely filled forms consisting of female ligulate flowers only, the descendants are segregating into all intermediate forms from filled to unfilled due to the high grade of heterozygosis. That is why in most cases the multiplication is done vegetatively by division of the root-stock [35, 40]. The culture can be utilized for about 3 to 4 years.

REFERENCES

1. Anonymous (1985) *Dossier mechanisation*. Institut Technique Interprofessionel des Plantes Medicinales, Aromatiques et Industrielles. Chemille, 83–105.
2. Anonymous (1987) (*Eine Maschine für das Sammeln von Apotheker-Kamillen*) (Russ.) Ministerium für landwirtschaftliche Maschinen der USSR. Taganrog, 1760.
3. Anonymous (1999) *PA/PH/Exp.13A/T* (99) 27 July 1999.
4. Arbeitsgemeinschaft für Arzneipflanzenanbau. *Jahresberichte 1971, 1972, 1973, 1977, 1979*. Lehrstuhl für Gemüsebau der Technischen Universität München-Freising.
5. Baraldi, G., Bentini, M., Guarnieri, A. (1985) Raccolta della *Anthemis nobilis*: primi resultati. *m & ma* **12**, 15–18.
6. Bentini, M., Guarnieri, A. (1988) Primi resultati di prove sulla raccolta meccania della camomilla romana (*Anthemis nobilis*). *Riv. di Ing. Agr.* No. 4, 193–197.
7. Bernáth, J. (1993) *Vadon termö és termeszetett*. Kioado, Budapest, 566 pp.
8. Bomme, U., Nast, D. (1998) Nährstoffentzug und ordnungsgemäße Düngung im Feldanbau von Heil- und Gewürzpflanzen. *Z. Arznei. Gewürzpflanzen.*, **3**, 82–90.
9. Brkic, D., Lukac, P., Babic, T., Sumanovac, L. (1989) *Ispitivanje adaptiranog zitnog kombajna u herbi kamilice*. Trogir, Aktualni zadaci mehanizacije poljoprovrede, 213–221.
10. Bundessortenamt (Ed.) (1996) *Beschreibende Sortenliste 1996-Heil-und Gewürzpflanzen*, Landbuch-Verlag, Hannover, Germany, 137 pp.
11. Bundessortenamt (1999–2001), different origins for registration.
12. Carle, R., Fleischhauer, I., Fehr, D. (1987) Qualitätsbeurteilung von Kamillenölen. *Dtsch. Apoth. Ztg.*, **127**, 47, 2451–2457.
13. Carle, R., Gomaa, K. (1992) Technologische Einflüsse auf die Qualität von Kamillenblüten und Kamillenöl, *Pharm. Zt. Wiss.* **137**, 2, 71–77.
14. Carle, R., Isaac, O. (1985) *Dtsch. Apoth. Ztg.*, **125** (Suppl. I), 2–8.
15. Correa, C. Jr. (1995) Mandirituba: new Brazilian chamomile cultivar. *Horticultura Brasileira*, 13 (1), 61.
16. Dachler, M., Pelzmann, H. (1989) *Heil-und Gewürzpflanzen. Anbau-Ernte-Aufbereitung*. Österreich. Agrarverlag, Vienna, 244 pp.
17. Dachler, M.; Pelzmann, H. (1999) *Arznei-und Gewürzpflanzen. Anbau, Ernte und Aufbereitung*. 2nd Ed. Österreichischer Agrarverlag, Klosterneuburg, 351 pp.

18. Drangend, S., Paulsen, B. S., Wold, J. K. et al. (1994) Experiments on chamomile in South East Norway in 1993, in *NJF Utredning/Rapport*, Mikkeli, Finland, **91**, 36–39.

19. Ebert, K. (1982) *Arznei-und Gewürzpflanzen. Ein Leitfaden für Anbau und Sammlung.* 2nd Ed., Wiss. Verlagsgesell., Stuttgart, Germany, 221 pp.

20. Ebert, K., Schubert, H. (1962) Kamillenernte voll mechanisiert. *Dtsch. Apoth. Ztg.*, **102**, 6, 167–168.

21. Erfurth, P., Plescher, A. (1983) Zum Auftreten bakterieller, pilzlicher und tierischer Schaderreger an Heil-und Gewürzpflanzen. *Nachrichtenblatt für den Pflanzenschutz in der DDR*, **1**, 19–22.

22. Fink, A. (1978) *Dünger und Düngung.* Chemie, Weinheim.

23. Frank, B. (2004) Personal communication.

24. Franz, C. (1979) Content and composition of essential oil in flower heads of *Matricaria chamomilla* L. during its ontogenetical development. *Herba Hungarica*, **18**, 317–321.

25. Franz, C. (1982a) Genetische, ontogenetische und umweltbedingte Variabilität der Bestandteile des ätherischen Öls von Kamille, in Kubeczka, K. H. (Ed.) *Ätherische Öle. Analytik, Physiologie, Zusammensetzung.* Georg Thieme, Stuttgart, New York, pp. 214–224.

26. Franz, C. (1982b) *Dtsch. Apoth. Ztg.*, **122**, 1413.

27. Franz, C. (1985) in *Chamomile Symposium in Industrial and Pharmaceutical Use*, June 27–29, 1985, Trieste, Italy.

28. Franz, C. (1986) Züchtung und Anbau — Chancen für die Qualität pflanzlicher Arzneimittel. *Pharmazeut. Zeitung*, **11**, 611–616.

29. Franz, C., Hölzl, J., Kirsch, C. (1983) Einfluβ der Stickstoff-, Phosphor- und Kalidüngung auf Kamille (*Chamomilla recutita* (L.) Rauschert, syn. *Matricaria chamomilla* L.) II. Beeinflussung des ätherischen Öls. *Gartenbauwissenschaft*, **1**, 17–22.

30. Franz,C., Hölzl, J., Vömel, A. (1978) Variation in the essential oil of *Matricaria chamomilla* L. appending on the plant age and stage of development. *Acta Horticulturae*, **73**, 229–238.

31. Franz, C., Isaac, O., Kirsch, C. (1985) Neuere Ergebnisse der Kamillenzüchtung. *Dtsch. Apoth. Ztg.*, **125**, 20–23.

32. Franz, C., Kirsch, C. (1974) *Gartenbauwiss.*, **39**, 9.

33. Franz, C., Wickel, I. (1979) *Planta Medica*, **36**, 281.

34. Franz, C., Wickel, I. (1980) *Planta Medica*, **39**, 287.

35. Freudenberg, G., Caesar, R. (1954) *Arzneipflanzen. Anbau und Verwertung.* Parey, Berlin, Hamburg, 204 pp.

36. Hammerschmidt, P. (2002) Personal communication, February 13, 2002.

37. Hannig, H.-J. (1991) Qualitätsanforderungen der EG und der Arzneimittelprüfrichtlinie — Züchtung von Arzneipflanzen, in *Mitt. APV Pflanzliche Arzneimittel — Aktuelles zu Qualität, Wirksamkeit, Unbedenklichkeit*, Königswinter, October 28–30, 1991.

38. Hecht, H., Mohr, T., Lembrecht, S. (1992) Mechanisierung der Blütendrogenernte. *Landtechnik* **47**, 276–281.

39. Heeger, E. F. (1946) Die Kamille. *Pharmazie*, **1**, 211.

40. Heeger, E. F. (1956) *Handbuch des Arznei- und Gewürzpflanzenanbaus. Drogengewinnung.* Deutscher Bauernverlag, Berlin, 775 pp.

41. Heeger, E. F., Brückner, K. (1950) *Arten-und Sortenkunde der deutschen Heil-und Gewürzpflanzen.* Vol. **1**, Berlin.

42. Herold, M., Pank, F., Menzel, E., Kaltofen, H., Loogk, E., Rust, H. (1989) Verfahrenstechnische Entwicklungen zum Anbau von *Chamomille recutita* (L.) Rauschert und *Calendula officinalis* L. für die Gewinnung von Blütendrogen. *Drogenreport*, **2**, 2, 43–62.

43. Hoppe, H. A. (1958) *Drogenkunde.* **7**. Ed. Cram, de Gruyter & Co., Hamburg, 1229 pp.

44. Horn, W., Franz, C., Wickel, I. (1988) Zur Genetik der Bisaboloide bei der Kamille. *Plant Breed.*, **101**, 307–312.

45. Isaac, O. (1992) *Die Ringelblume. Handbuch für Ärzte, Apotheker und andere Naturwissenschaftler*, Wiss. Verlagsgesell., Stuttgart, Germany.

46. Kaltofen, H. (2002) Personal communication.

47. Kathe, W., Honnef, S., Heym, A. (2003) *Medicinal and Aromatic Plants in Albania, Bosnia-Herzegovina, Bulgaria, Croatia and Romania.* Bundesamt für Naturschutz/Federal Agency for Nature Conservation, Bonn, 200 pp.

48. Kerschberger, M. et al. (1997) in *Anleitung und Richtwerte für Nährstoffvergleiche nach Düngeverordnung*, Jena.

49. Kirsch, C., Franke, R. (1993) Neue Ergebnisse der Kamillenzüchtung. *Herba Germanica*, **3**, 97.

50. Letschamo, W. (1991) Vergleichende Untersuchungen über die nacherntetechnisch bedingten Einflüsse auf die Wirkstoffgehalte in der Droge bei Kamille-Genotypen. *Drogenreport*, Sonderausgabe, 128–134.

51. Letschamo, W. (1992) *Ökologische, genetische und ontogenetische Einflüsse auf Wachstum, Ertrag und Wirkstoffgehalt von diploiden und tetraploiden Kamillen*, Chamomilla recutita *(L.) Rauschert*. Dissertation, University of Giessen, Germany, 171 pp.

52. Martinov, M., Tesic, M., Müller, J. (1992) Erntemaschine für Kamille. *Landtechnik*, **47**, 10, 505–507.

53. Massoud, H., Franz, C. (1990a) Quantitative genetical aspects of *Chamomilla recutita* (L.) Rauschert. *J. Ess. Oil Res.*, **2**, 15–20.

54. Massoud, H., Franz, C. (1990b) Quantitative genetical aspects of *Chamomilla recutita* (L.) Rauschert. II. Genotype: environment interactions and proposed breeding methods. *J. Ess. Oil Res.*, **2**, 299–305.

55. Melegari, M., Albasini, A., Pecorari, P., Vampa, G., Rinaldi, M., Rossi, T., Bianchi, A. (1988) *Fitoterapia*, **59**, 449–455.

56. Menzel, W. (2002) Personal communication.

57. Mohr, T. (2002) Untersuchung und Weiterentwicklung einer Erntemaschine für Heilpflanzenblüten. *Z. Arznei. Gewürzpflanzen*, **7**, Sonderausgabe, 196–202.

58. Mohr, T., Hecht, H. (1996) Entwicklung einer Pflückmaschine. Ernte von Echter Kamille, (*Chamomilla recutita* (L.) Rauschert), Ringelblume (*Calendula officinalis* L.) und Johanniskraut (*Hypericum perforatum* L.). *Z. Arznei. Gewürzpflanzen, (Sonderheft)* 68–77.

59. Mohr, T., Hecht, H., Eichhorn, H. (1996) Vergleichende Untersuchungen einer verbesserten Pflückmaschine zur Gewinnung von Blütendrogen der Echten Kamille (*Chamomilla recutita* (L.) Rauschert), Ringelblume (*Calendula officinalis* L.) und Johanniskraut (*Hypericum perforatum* L.). *Drogenreport*, **9**, 14, 15–23.

60. Mohr, T., Hecht, H., Eichhorn, H. (1996) Vergleichende Untersuchungen einer verbesserten Pflückmaschine zur Gewinnung von Blütendrogen (Teil 2). *Drogenreport*, **9**, 15, 5–9.

61. Müller, J. (2003) Stand und Forschungsbedarf bei der Erntetechnik von Arznei-und Gewürzpflanzen. *Z. Arznei. Gewürzpflanzen*, **8**, 2, 56–60.

62. Müller, J., Köll-Weber, M., Kraus, W., Mühlbauer, W. (1996) Trocknungsverhalten von Kamille (*Chamomilla recutita (L.)* RAUSCHERT), *Z. Arznei. Gewürzpflanzen*, **1**, 104–110.

63. Pank, F., Wettrich, K., Rust, H. (1997) Rationalisierung von Produktionsverfahren der Arznei- und Gewürzpflanzen. Ökonomische Effekte am Beispiel von Fenchel (*Foeniculum vulgare* MILL.) und Kamille (*Chamomilla recutita* (L.) Rauschert). *Z. Arznei. Gewürzpflanzen (Sonderheft)*, 25–30.

64. Pfeffer, S., Krüger, H., Schütze, W., Schulz, H. (2002) Schnelle Erfassung von Qualitätsparametern in Kamillenblüten mit Hilfe der Nah-Infrarotspektroskopie. *Drogenreport*, **15**, 28, 29–32.

65. Plescher, A. (1997) Akkumulation von Cadmium in Echter Kamille (*Chamomilla recutita* (L.) Rauschert) und Johanniskraut (*Hypericum perforatum* L.). *Drogenreport*, **10**, 17.

66. Plescher, A. (1997) Verfahrenstechnische Entwicklungen zum Anbau von Kamille (*Chamomilla recutita* Rauschert). *Z. Arznei. Gewürzpflanzen*, **4**, 193–201.

67. Plescher, A., Pohl, H., Vetter, A., Förtsch, U. (1995) Übergang von Schwermetallen aus dem Boden in Arznei-und Gewürzpflanzen. *Herba Germanica*, **3**, 116–125.

68. Plescher, A., Stodollik, A. (1995) Tendenzen im Pflanzenschutz bei nachwachsenden pharmazeutisch genutzten Rohstoffen. *Mitt. Biol. Bundesanstalt Landw.*, **310**, 109–118.

69. Pomini, L. (1972) *Riv. Ital. EPPOS*, **54**, 627–630.

70. Reichling, J. et al. (1979) Vergleichende Untersuchung verschiedener Handelsmuster von Matricariae flos. *Pharm. Ztg.*, **124**, 41.

71. Rimpler, R. (1972) Gutachten Nr. 121: Kamillenblütenerntemaschine. Potsdam-Bornim, Zentrale Prüfstelle für Landtechnik Potsdam-Bornim des Staatlichen Komitees für Landtechnik und MTV.

72. Röhricht, C., Mänicke, S., Grunert, M. (1997) Der Anbau von Kamille (*Chamomilla recutita* [L.] Rauschert) in Sachsen. *Z. Arznei. Gewürzpflanzen*, **2**, 135–146.

73. Rühlicke, A. (1978) Gutachten Nr. 353: Kamillenpflückmaschine LINZ III. Potsdam-Bornim, Zentrale Prüfstelle für Landtechnik Potsdam-Bornim des Ministeriums für Land-, Forst-und Nahrungsgüterwirtschaft.

74. Ruminska, A. (1983) *Rosliny lecznize. Podstawy biologii i agrotechniki (Heilpflanzen. Grundlagen der Biologie und Anbau)*. 3rd ed. Panstwaowe Wydw. Naukowe, Warsaw, 550 pp.

75. Salomon, I. (1992) Chamomille: a medicinal plant. *Herb, Spice and Med. Plant Digest*, **10**, 1–4.

76. Schick, E. R. and Reimann-Philipp, R. (1957) Die Züchtung von Heilpflanzen. *Züchter*, **27**, 7, 337.

77. Schilcher, H. (1973) *Planta Medica,* **23**, 132.
78. Schilcher, H. (1978) Influence of herbicides and some heavy metals on growth of *Matricaria chamomilla* L. and the biosynthesis of essential oil. *Acta Horticulturae — Spices and Medicinal Plants,* **73**, 339.
79. Schilcher, H. (1981) *Pharm. Ztg.,* **126**, 2119.
80. Schilcher, H. (1987) *Die Kamille. Handbuch für Ärzte, Apotheker und andere Naturwissenschaftler,* Wiss. Verlagsgesell., Stuttgart, Germany, 152 pp.
81. Schreiber, A., Carle, R., Reinhard, E. (1990) On the Accumulation of Apigenin in Chamomile Flowers. *Planta Medica,* **56**, 179–181.
82. Schröder, H. (1965) Die Düngung von Sonderkulturen, in Linser, H. (Ed.) *Pflanzenernährung und Düngung,* **III/2** Springer, Vienna.
83. Schultze-Motel, J. (Ed.) (1986) *Rudolf Mansfelds Verzeichnis landwirtschaftlicher und gärtnerischer Kulturpflanzen (ohne Zierpflanzen),* **2**. Ed. Akademie Verlag, Berlin, 1998 pp.
84. Seitz, P. (1987) Arznei-und Gewürzpflanzen in der DDR. *Deutsch. Gartenbau,* **51**, 3040–3046.
85. Svab, J. (1969) Untersuchungen über den Einfluß der ökologischen Faktoren auf den Öl-und Chamazulengehalt von Kamilledrogen. *Herba Hungarica,* **8**, 91–96.
86. Traxl, V. (1986) *Pestovanie liecivných rastlín (Züchtung von Arzneipflanzen).* Vydavatelstvo OBZOR, Bratislava, 142 pp.
87. Vrany, J. (1968) (Can you cultivate Chamomile?) (Slovak.) *Naše liecive rastliny,* **5**, 170.
88. Wagner, C., Marquard, R., Friedt, W., Ordon, F. (2001) Untersuchungen zur genetischen Diversität der Echten Kamille (*Chamomilla recutita* (L.) Rausch.) mittels PCR-basierter Markertechniken. *Z. Arznei. Gewürzpflanzen,* **6**, 216–221.
89. Wogiatzki, E., Marquard, R. (2002) Gehalte und Zusammensetzung des ätherischen Öls von Wildkamillen von 25 Standorten in Griechenland. *Drogenreport,* **15**, 27, 49–53.

5.5 PRODUCTION OF CHAMOMILE (*MATRICARIA RECUTITA* L.) IN EAST AND SOUTH EUROPEAN COUNTRIES

JENÖ BERNÁTH AND ÉVA NÉMETH

5.5.1 INTRODUCTION

Medicinal and aromatic plants, especially for self consumption, have been produced in the territory of Hungary and other east and south European countries for many centuries. However, until the end of the 19th century the cultivation of medicinal and aromatic plants was carried out on the "garden" scale in a limited area [1]. The intensification of the production only started in the early years of the 20th century, when the processing of the raw plant including industrial oil distillation was started as well.

The shortage of medicines, teas, and spices at the time of World War I and afterward drew much more attention to the production and utilization of medicinal and aromatic plants. Both the collection of the wild populations and the cultivation of some selected species started at that special period, due to the enlarging local and export demand. This specialization took place spontaneously, effected by different biological, economical, and social factors.

Natural occurrence of *Matricaria recutita* in Hungary is an example of how the regional specialization could have happened for the utilization of actual species of indigenous flora [2]. As a result of the increasing west European demand (especially that of Germany) the Great Plain of Hungary became an important indigenous region of chamomile production. The *Chamomillae flos* became a well-known Hungarian product sold on the world market with high success. From a socio-economical point of view the formation of this special region was promoted by the abundance of labor. According to the data of trade companies, during the harvest of chamomile flowers as much as 15,000 to 20,000 people are involved in the collection today [2].

As a result of the modern scientific and industrial achievements the demand for raw chamomile changed from both a qualitative and a quantitative point of view [3, 18, 29]. However, the traditional wild collection of plant material is practiced even today; many efforts were made in the past four decades to establish economical cultivation [5, 12, 14, 27, 32]. At present, collection and large-scale cultivation methods are practiced in east and south European countries in parallel, having biological and economical advantages or disadvantages. This section discusses the utilization of indigenous flora and the effective methods of cultivation.

5.5.2 PRODUCTION OF CHAMOMILE FROM INDIGENOUS POPULATIONS

5.5.2.1 Distribution of Chamomile

5.5.2.1.1 *Distribution in East and South European Countries*

Chamomile is a species of east Mediterranean origin, occurring naturally in all the above-mentioned regions of Europe [21, 23, 30]. However, it is proved by botanical investigations that its distribution may vary from region to region depending on the ecological conditions of the actual habitats. Based on the Hungarian analysis different types of growing areas have to be distinguished. In the western part of Hungary (Transdanubia) as well as in the Hungarian Central Chain located in the north, the chamomile grows in sporadic populations being a constituent of the ruderal plant communities. The individuals show robust habit, forming a large number of branches, many relatively large flowers. In contrast, the dwarf type, with a limited number of flowers, is very frequent in the great Plain of Hungary where the plants grow on the ruderal places as well, occupying sodic fields, lay lands, and sowings in abundance. These morphological features are reflected in the differentiation of the chemical characteristics of the populations, which will be discussed later.

5.5.2.1.2 *Soil Conditions*

Analyzing the natural occurrence of the populations of chamomile, it is obvious that it is present in the poor sodic fields, lay lands, and sowings in great abundance. In spite of physiological optimum measured at pH 7 the plant grows well up to pH 9. However, the statement that chamomile should prefer the poor sodic field must be a false one, which was refuted even by the first cultivation experiences. Based on the results of Kerekes [13] the high tolerance of certain species can be explained by the outstanding salt accumulation ability of the root. The plant can accumulate sodium salts in the root cells in amounts up to 10 mg/g. The high salt concentration helps the plant in water uptake when its level in the soil is under the limit, which is not available for other members of the plant community [11]. The interspecific competition ability of the plants under extreme condition is supported by this physiological feature.

5.5.2.1.3 *Coenological Aspects*

It is obvious from Section 5.5.2.1.2 that chamomile shows good interspecific competition under extreme conditions. However, this advantage due to the salt accumulation is lost in other plant communities. To clear up the competition regularities of chamomile, more detailed coenological investigations are required. It would have both practical and theoretical importance. From a practical point of view, the presence of related species (*Matricaria matricarioides, M. inodora, Anthemis austriaca, A. cotula*, etc.) spoils the quality of the collected drug, as was proved by Romanian experts.

The investigation carried out by Mathé [21] can be considered as an outstanding one in this respect. Analyzing the coenological aspects of chamomile at ten different regions of Hungary, the most frequent associated plant species were pointed out. The result is summarized in Table 5.5.1. It is obvious that the related species *Matricaria matricarioides* and *M. inodora*, which may decrease the drug quality, can be present in the majority of habitats used for wild collection.

5.5.2.2 Chemical Diversity of Wild Populations

The chemical diversity of wild-growing chamomile populations had been studied by numerous scientific groups worldwide. In Hungary the first investigations were carried out by Rom [26] at the beginning of the large-scale commercial utilization of Hungarian chamomile populations. The high

TABLE 5.5.1
The Most Frequent Plant Species Associated with Indigenous Chamomile Plant Communities at Ten Different Districts of Hungary [20]

Accompanying Plant Species	Frequency of the Distribution in Different Communities (%)									
	1	2	3	4	5	6	7	8	9	10
Polygonum aviculare	■	■	■			■			■	■
Capsella bursa-pastoris	■	■	■		■				■	■
Malva neglecta	■	■				■			■	■
Lepidium ruderale	■		■		■					
Matricaria matricarioides	■								■	
Lepidium draba	■		■	■		■				■
Taraxacum officinale			■	■		■			■	
Matricaria inodora							■	■		
Chenopodium album	■									
Festuca pseudovina		■								
Plantago media		■								
Lolium perenne		■						■		
Sisymbrium sophia						■				
Ranunculus arvensis			■		■					
Achillea millefolium			■	■						
Bromus mollis					■	■				
Poa angustifolia										■
Plantago major						■				■
Plantago lanceolata						■			■	
Erigeron canadensis										■
Arctium lappa								■		
Potentilla anserina	■									
Agropyron repens		■								
Ballota nigra		■								
Vicia sepium			■							
Consolida regalis			■							
Artemisia monogyna				■						
Poa bulbosa				■						
Adonis aestivalis				■						
Hordeum hystrix					■					
Cerastium anomalum					■					
Puccinellia distans					■					
Rorippa kerneri					■					
Lepidium perfoliatum					■					
Camphorosma annua					■					
Apera spica-venti						■				
Poa annua						■				
Xanthium stumarium						■				
Rorippa silvestris								■		
Rumex crispus									■	
Urtica urens									■	
Centaurea cyanus										■

20–40% ☐   40–60% ▨   60–80% ▦   80–100% ■   20–40% ☐   40–60% ▨

1. Plain in Northwestern Hungary, 2. Danubian basin, 3. Territory east of the River Tisza, 4. District of Rivers Krös-Tisza, 5. District of Rivers Krös-Maros, 6. Nyirség, 7. Bodrog River district, 8. Plain of Bereg, 9. Hungarian Central Chain, 10. Transdanubia

chemical diversity of the species was proved even by the first data. To fulfill the requirement for collecting equalized and standardized drug, more detailed analysis were needed on the chemical diversity of indigenous populations. In the period 1959–1961 500 wild chamomile plants were sampled by Mathé [21]. However, the accumulation of essential oil and its chamazulene content were affected by conditions of the year, and also regional differences were established. The frequency of populations of low pro-chamazulene content is very high in the regions, located to the east from the River Tisza (region 3 in Table 5.5.1). In this region about one third of populations accumulate chamazulene in a range as low as 0–29 mg%, and the presence of the pro-chamazulene-free individuals is very common, too. In contrast, in the region of the Bodrog River (northeast Hungary) and Transdanubia, 25–27% of populations accumulate 100 mg% or higher amount of pro-chamazulene.

According to recent investigations [36] at 12 collection areas of Hungary, populations accumulating typically chamazulene (10–20%), α-bisabolol (30–50%) or bisabolol-oxid (30–50%) could be completely distinguished (Table 5.5.2). Samples collected in Danube-Tisza Mid Region could be characterized by highest accumulation of chamazulene (up to 20% of oil). According to Mathé [21] the compositional characteristics of the populations may be affected by the climatic conditions and the soil properties, too.

In the same experiment, the flavonoid concentrations had also been checked. The quantity of apigenin-7-glucoside proved to be outstanding in the populations originating from the Great Hungarian Plain (1.8–2.8 mg/g), while samples collected in Transdanubia could be characterized by lower levels (2 mg/g).

The chemical diversity of the chamomile populations was justified in the Bulgarian flora as well. It was proved by the investigations of Peneva et al. [23], analyzing populations for presence of essential oil, chamazulene content, and flavonoids during the period 1982–1983. Some of their data are presented in Table 5.5.3.

In the experimental field, chamazulene was found even in the populations that had been characterized as chamazulene-free local races. In fact, the flavonoid content did not change in the course of the introduction. However, by the opinion of the authors the local populations of chamomile show low productivity of essential oil, which makes them unsuitable as a starting material for enlarging production. By the more detailed investigations of Stanev et al. [30], four different chemotypes of chamomile were identified in Bulgaria. Analyzing the samples collected from 29 regions of Bulgaria, two main groupswere separated: populations accumulating chama-

TABLE 5.5.2

Main Components of the Wild Growing Chamomile Populations in Different Regions in Hungary [36]

Number of Habitats	1	2	3	4	5	6	7	8	9	10	11	12
Essential oil content	0.7	0.8	0.7	0.6	0.6	0.5	1.0	1.3	0.6	0,5	0,9	1,3
Farnesene-a	3.4	4.5	2.4	3.2	3.1	2.5	1.8	3.0	2.8	2,6	3,6	2,6
Farnesene-b	0.0	0.0	0.0	3.2	0.7	2.7	0.2	0.0	0.0	1,8	0,0	0,0
Bisabolol-oxid-II	10.7	11.9	8.2	13.8	11.5	18.4	15.4	7.6	9.7	10,1	12,2	13,0
α-Bisabolol	13.0	16.9	13.3	46.3	44.5	30.9	29.2	19.3	10.5	17,5	22,3	9,3
Chamazulene	8.7	5.1	7.0	6.8	9.7	11.0	16.3	20.1	11.0	13,0	10,0	12,5
Bisabolol-oxid-I	41.3	34.8	50.1	3.5	5.9	7.3	24.0	35.2	44.9	38,9	22,8	41,4
En-in-dicycloether	17.6	21.6	14.8	15.8	16.1	17.7	10.1	13.8	18.0	13,3	25,3	18,2
Others	5.2	5.2	4.3	7.4	8.3	9.5	3.0	1.0	3.1	2,8	3,7	3,0

Habitats: 1. Jászberény, 2. Csincse, 3. Poroszlo, 4. Hortobágy, 5. Püspökladany, 6. Bakonyszeg, 7. Felsökerek, 8. Akaszto, 9. Szabadszallas, 10. Apajpuszta, 11. Nikla, 12. Somogytarnoca

TABLE 5.5.3
Essential Oil and Flavonoid Content of Bulgarian Chamomile Populations of Different Origin under Natural and Field Conditions [23]

Origin of Population	Essential Oil (%)		Flavonoids (%)	1	2	3	4	5	6	7
	Original	Introduced								
Razgrad	0.533	0.320	0,318	+	+		+		+	+
Mezdra	0.850	0.360	0.195	+	+	+	+	tr	+	
Roman	0.675	0.240	0.239	+	+	+	+	+	+	+
Karlovo	0.566	0.320	0.250	+	+		+		+	+
Stara Zagora	0.680	0560	0.298	+	+	+	+	+	+	+
Dbenye	0.800	0.720	0.345	+	+	+	+			+
Kazanlik	0.400	0.271	0.309	+	+	+	+		tr	+

Flavonoids: 1. quercetagenin, 2. ferulic acid, 3. apigenin-7-O-glycoside, 4. luteoline-7-O-glycoside, 5. hyperosid, 6. rutin, 7. apigenin

zulene and populations characterized by the presence of bisabolol. In the chamazulene type the chamazulene content ranges between 4.0 and 10.3% of the oil. Populations with cultivar "Lazur," especially in southeastern Bulgaria, accumulated threefold higher amounts of α-bisabolol (73.7%). At the same time a great variation in the bisabololoxide A and bisabololoxide B content was found in the natural populations. It is confirmed by the data that Bulgarian chamomile populations are rich in bisabololoxide A, and its amount varies slightly (the variation coefficient calculated to the presence of bisabololoxide A is 21%). From the practical point of view the southeastern Bulgarian population could have a special importance in selecting plant material producing high amounts of α-bisabolol.

5.5.2.3 Harvest of Wild Populations

The cultivation of chamomile is practiced all over the world; however, the collection of the wild population did not lose its importance, especially in the east and south European countries. In Hungary about 95 percent of 500 to 700 t of *Chamomillae flos* produced yearly comes from the wild population. For collection the regions are chosen where the chamomile plant forms a compact stand on relatively large fields. These territories in Hungary are as follows: territories east to the River Tisza, district of Rivers Körös-Tisza, and districts of Rivers Körös-Maros, Nyirség, and the Bodrog River. In Figure 5.5.1 the distribution of the main collection regions of chamomile are shown from the data of a Hungarian merchant company.

The collection area is much smaller compared to natural distribution of the plant, which can be explained by practical consideration to make economical production.

In Hungary the blossom of wild chamomile starts in the second half of April and continues till the beginning of June. However, large yearly variability can be found in the time of flowering, which is affected by the weather and ecological conditions. The earliest flowering usually occurs in poor sodic fields, which is followed by the populations distributed in showings. The flowering of plants shows 5 to 8 days delay in heavy clay soils comparing to the sandy ones.

The wild populations of chamomile should be harvested when most of the flowers are already opened (at the time of the full blossom). If collection is done before the optimal time the yield will be small and large amount of unripe flower heads will depreciate the quality. As a result of the late harvest, the flower heads fall into pieces in the course of postharvest processing, which depreciates the quality, too. For choosing the proper time of harvest the essential oil content of the drug has to be taken into consideration. It is supported by both practical and scientific observations; the

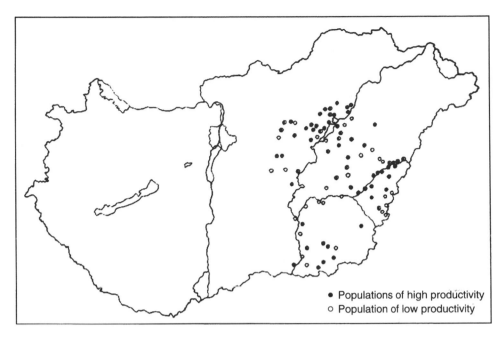

FIGURE 5.5.1 Distribution of the main collection regions of chamomile in Hungary from the point of view of economic production [15].

essential oil content of the inflorescence increases continuously from the time of budding and reaches its maximum when the ligulate flowers stand in a horizontal position. In the course of the ripening afterward, the essential oil content and the amount of chamazulene decreases again. However, the essential oil content may depend on the weather conditions. For collection warm and sunny days are recommended by the agencies responsible for purchasing.

In the wild populations the inflorescences are collected exclusively with a chamomile comb (Figure 5.5.2).

FIGURE 5.5.2 Harvest of inflorescences of indigenous chamomile populations in Hungary with the aid of a special chamomile comb.

An experienced worker can gather 50–150 kg fresh flower in a ten-hour shift. At the time of combing, the stems and large leaf segments should be removed, which will avoid quality problems. According to the data of purchasing firms 20,000 to 25,000 people can participate in the collection of Hungarian chamomile during the harvest period.

After the collection, the chamomile flower has to be dried immediately. The traditional methods, drying on balks and sheds etc., are only used for making drug for home consumption. The collected raw material is transported to the processing centers of the purchasing firms, which are usually equipped with up-to-date drying machines. The drying procedure is the same as for the cultivated crop and is discussed later.

5.5.3 CULTIVATION OF CHAMOMILE

5.5.3.1 Selection of Land

Chamomile is a characteristic plant of poor sodic fields, ruderal places, lay lands, and sowings. However, the statement that chamomile should prefer poor soil conditions is a false one. The selection of the land depends on many factors and changes country by country. In Hungary, because of economical reasons, the poor sandy soils are recommended, which are unsuitable for cultivation of other agricultural crops [10]. Based on the Rumanian experiences, the plant can be grown on the light neutral sandy soils, sodic fields, and other soils of low fertility [25]. Yugoslavia's poor or fertile soil conditions are utilized [14]. In Slovenia the production method of chamomile has been developed for the farms, which are specialized for hop cultivation [37]. Hop grows there in the form of monoculture on large and small fields. The monoculture of hop has an adverse effect on the soil fertility. Between the two plantations of the hop, cultivating other plant species including chamomile is recommended. In accordance with this situation the production of chamomile has developed on heavy clay and also on light sandy soils, where the hop gardens are spread. However, as an effect of heavy clay soil the plant stand may become an inhomogenous one, showing vegetative habit.

There have been some efforts to cultivate chamomile under extreme conditions. It was reported by Yugoslavian scientists [31] that chamomile grows well on waste lands, especially on recultivated fields around the Pljevlja coal mine.

5.5.3.2 Cultivation Methods

5.5.3.2.1 Soil Preparation

It is agreed by different authors that chamomile can be cultivated in monoculture. This type of culture suits the biological characteristics of the species, being a self-sowing plant under natural conditions in sodic and ruderal places. According to recent experiences in Hungary [33], it can be cultivated for 4 to 5 years, or even longer, with proper agronomic techniques (not only with self-sowing), in the same field. The field should be changed when the presence of the resistant plant species in the weed flora becomes dominant. The advantage of not changing chamomile fields frequently is to avoid the appearance of chamomile plants in the following culture for years.

One of the preconditions of the effectual cultivation of chamomile is the proper preparation of the seedbed. As a first step of the cultivation, the stable of the predecessor plant should be turned over immediately, with disc. Ploughs are not recommended, except when the stable cannot be turned over by disc because of the weather conditions. The field should be worked over with disc tillers and rushing rollers until the soil is smooth, free of lumps, well compacted, and hard. If the plant is cultivated in monoculture, after the harvest of chamomile the stem residues should be removed by mobile chaffing machines. This action is followed by stubble stripping and rolling.

5.5.3.2.2 Sowing

The optimum time for sowing chamomile in this region is the end of August or the first half of September. This statement has been supported by Slovenian [37], Hungarian [33], and Yugoslavian

TABLE 5.5.4
The Effect of Sowing Time on the Yield of Dry Inflorescences of Chamomile [12]

Time of Sowing	Yield of Inflorescences (kg/ha)		
	1958	**1959**	**1960**
Autumn sowing			
13–16 July	284	400	532
3–6 August	366	462	378
24–27 August	362	576	364
14–17 September	322	593	390
5–8 October	214	646	198
	110	580	342
Average of autumn sowing	274.6	542.8	367.3
Spring sowing			
29 February–15 March	256	188	346
14 March–1 April	181	68	308
28 March–15 April	13	91	196
11 April–29 April	40	107	202
25 April–12 May	60	103	79
9 May–27 May	11	93	0
Average of spring sowing	93.5	108.3	205.2

[14] experiences. The advantage of the autumn sowing was first proved by Kerekes [12]. The results of his investigations, which were carried out for three years, are summarized in Table 5.5.4.

It is supported by the data that the optimum time for sowing is the first half of September. The spring sowing could be suggested only exceptionally, and the success of spring sowing depends on weather conditions and the type and state of soil. In the case of spring sowing the earliest time possible should be proposed (second half of February or first half of March); however, the yield is less compared to any of the autumn variation. There is only a single contradictory result, achieved by Gasic et al. [8], in the evaluation of time of sowing. Investigating 17 chamomile cultivars in Yugoslavia the essential oil content was significantly higher if the plants were grown by spring sowing. It is in contrast to the majority of the experiences as well as to the result of Franz [7], who found that plants sown in the autumn reached a higher oil content due to slower development at the time of blossom. However, in spite of the observation of Gasic et al. [8], in this region the optimum time for sowing is autumn. Autumn sowing avoids the adverse effects of the dry springs, which are very common in that region.

For sowing chamomile seed pulvis is used in the majority of cases. The pulvis consists of 20–30% of seeds, and about 70–80% of parts of the inflorescence, which help to make a homogenous sowing. About 50 kg/ha of pulvis is used. In the cases of monoculture good results were attained with ploughing of chamomile fields after harvest and sowing only 25 kg/ha pulvis in Reference 37. According to Hungarian experiences in the case of monoculture no sowing is necessary at all. The seeds germinate well after suitable turning and rolling. Because the seed of chamomile prefers light for germination, the pulvis should be spread on the soil surface in 12–15-cm rows. Depending on the local technology and ecological conditions other row distances are recommended; for instance, 12–36 cm in Yugoslavia [14], or up to 40 cm in Romania if the field is weedy and a path is required for mechanized weed killing [25]. In order to facilitate uniform sowing on the surface, the coulter of the sowing machine should be adjusted according to the usual row distance of cereals, and should be locked in the raised position. After sowing, a Cambridge roller should be used; it helps to prevent the blowing away of small seeds and presses them to the soil surface, helping the water uptake. A seed-covering harrow must not be used after sowing chamomile.

5.5.3.2.3 Nutrient Supply

Chamomile grows in the sodic soils or in poor ruderal habitats under natural conditions. According to the former studies [11] chamomile takes 53 kg of N, 85 kg K_2O, and 21 kg P_2O_5 from the soil for production of 1.0 t of inflorescence. According to Hungarian cultivation experiences, the chamomile gives good yield on the medium quality soils without any fertilization, because of the limited nutrient requirement of species. In Slovenia chamomile production is located to the hop production areas, where the plant is alternated with maize and wheat, which are fertilized with 50 kg nitrogen, 100 kg phosphorous, and 150 kg potassium per ha. For this reason only 13 kg/ha nitrogen is proposed to spread over at the time of vegetation [37]. There are experiences with the adverse effect of high amounts of fertilizers, too, causing intensive vegetative growth of plants, which results in decreasing yield as an effect of the plant leaning to the ground.

However, in the loose sandy soils or on eroded sands fertilization is required for a good yield. In exceptionally poor soils, application of 20–30 kg/ha N and 20–30 kg/ha P_2O_5 is recommended. On eroded sands, with neutral pH, the dosage of P_2O_5 can be raised up to the 40- to 60-kg level and nitrogen top-dressing in a dosage of 30–40 kg may be effectual.

The cultivation of chamomile in monoculture cannot be economical without regular application of fertilizers. In monoculture the nutrient reserve of soil decreases to the critical level by the third year; in consequence, chamomile starts to disappear from the field. To manage successful monoculture both basic fertilization (in autumn) and top-dressing in spring is necessary. From the second year 10–12 kg/ha N, 60–70 kg/ha P_2O_5, and 50–70 kg/ha K_2O must be applied before wintering and 40–60 kg/ha N in top-dressing form.

There is a special case when top-dressing with N fertilizers should be suggested. In east and south European countries chamomile may be damaged by late frosts. As a result of low temperature the development of plants stops, and the leaves become yellow. In this situation the nitrogen top-dressing in a dosage of 30–40 kg/ha will help the plants through the critical period.

5.5.3.2.4 Care of Plants

In the cultivation of chamomile chemical weed control is solved [32]. The former manual mechanical methods have no importance and are practiced on small parcels only. Maloran (chlor-bromuron), the most effective herbicide, should be applied in April with 3–4 kg/ha dosage. The optimum time for application is when the plants are in the two- or three-leaf stage and the weeds are just starting to sprout. Chamomile tolerates Afalon (linuron) in 3–4 kg/ha dosage. This herbicide is suggested for Yugoslavian [14] and Slovenian [37] farmers, too, but in a much lower (1–2 kg/ha) dosage. The effectiveness of such a low amount could be uncertain; however, it may contribute to avoid damaging flowers, which could happen under disadvantageous weather conditions. In monoculture the amount of tolerant weeds increases continuously. In that case special herbicide treatments are required. For instance in Hungary the most dangerous monocotyledonous weed is *Bromus tectorum*. It can be controlled by spraying Kerb (propizamid) in a dosage of 2 kg/ha in the late autumn. Against the tolerant dicotyledonous weeds, the application of Sys 67 Prop (2,4-D) has to be considered.

5.5.3.2.5 Harvest

Chamomile in the indigenous populations should be harvested when the majority of the flowers have already opened. The essential oil content of the inflorescence increases continuously from budding, and reaches its maximum when the ligulate flowers take the horizontal position.

The harvesting process of chamomile depends on the purpose: whether plain flowers or a mixed product (flowers with 10–20% leaf and stem material) are to be gathered. The latter serves chiefly as raw material for oil distillation.

The flowers should be gathered when they are in full blossom. Up to the 1970s gathering was done mainly by hand, with the help of a long-handled "flower comb." By this method a very good quality drug can be produced, both from the wild-growing plants, and also from the cultivated ones. From 100 to 150 kg of flowers were collected by experienced workers from the cultivated stand in 10-hour shifts. In the case of the cultivated plants, which can be characterized by more robust habit, the stem parts should be removed more frequently from the flower comb, as the amount of green parts torn by the tool is much more than in the case of plants growing under sodic conditions. However, turning to large-scale cultivation, manual harvesting (because of manpower input of 25–30 workdays per ha) became simply unfeasible. This problem had to be solved by the construction of mechanized harvesters.

One of the possibilities was the application of the well-known Ebert-Schubert chamomile harvester. This type of machine produced excellent quality drug, but its output was low. In the 1980s, the majority of countries belonging to this region of Europe started to develop their own type of harvester [14, 19, 32, 37]. For instance, in Hungary a new and effective combine-harvester system was invented by the team of Cooperative "Szilasmenti." It was a complementary, high-capacity system, eliminating manual labor. The equipment consisted of a flower picker and a sifter for gathering 8–10 t of raw material per day. By this harvester raw material both for *flos* production, consisting of 50–60% of plain flower, and for essential oil distillation, coming from the sifter-top, were made simultaneously. The expected yield in Hungary is 0.5–2.0 t/ha of fresh flower, from which 0.1–0.5 t/ha dry drug can be produced. Much higher dry yield is reported from Romania, Slovenia, and Yugoslavia, which are 0.6–1.0 t, 0.5 t, and 0.4–0.5 t on average, respectively.

A different type of harvesting method was developed for collection of chamomile biomass used for isolating active agent complexes either by distillation or chemical extraction. The silo harvester (E 280) proved to be the most suitable for this purpose. The cutting level of the machines should be adjusted below the average height of the flowers, avoiding high amount of stem in the harvested material. The green stem parts impose an extra load on the essential oil distillation. The mixture of flowers and stems, especially that was gathered with mobile ensilage machines, should be immediately taken to the distillation vessel. By this mechanized harvesting method 4.0–8.0 t/ha of flower-stem mixture can be harvested. The yield varies widely because a second harvest of the crop becomes possible under suitable weather and edaphic conditions.

Further, remarkable progress was achieved recently by the elaboration of distilling containers, considerably reducing manpower input and operating at a higher output level.

5.5.3.2.6 Drying

The drying is one of the most important postharvest processes of chamomile. Its importance had been recognized even at the beginning of the large-scale cultivation of the plant [35]. There is no doubt that the flower — collected either by flower comb or mechanized harvester — should be transported to the drying location at once. The continuous transportation of the collected chamomile is a basic organizational task. If the delivered flowers cannot be put into the drier immediately, they should be spread in the shade for temporary storage for several hours.

For reducing the cost of drying, the collected flower should be screened before drying. A motorized chamomile screen of 7–12 mm mesh size can be used for this procedure.

For drying chamomile, either natural or artificial methods can be applied. In the case of large-scale cultivation, artificial drying is the only reliable method; it results in more aesthetic and high-quality drug, which is not easily disintegrated to dust. Different types of driers can be used for dehydration of chamomile. The most frequent type is counterflow driers (belt conveyor and tunnel types) managed at 40–70°C. However, different kind of driers can be used successfully as well, as it is proved by the Slovenian example [37]. There the harvested chamomile flower is transported to hop drying-house and put on drying meshes. The thickness of stratum is 8 cm, and the drying lasts from 8–10 hours at 40°C.

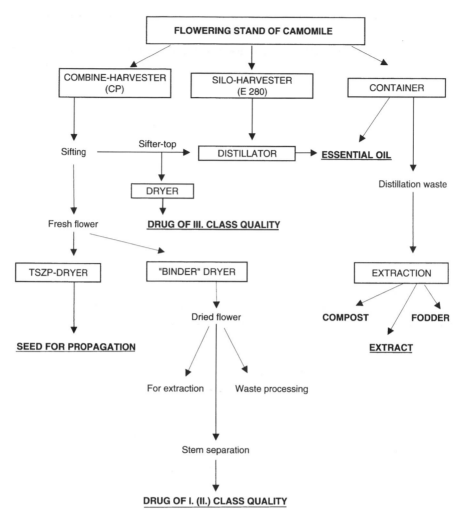

FIGURE 5.5.3 Complex processing system employed for the large-scale production of chamomile in Hungary [34].

5.5.3.3 Postharvest Processing

To make an economic large-scale production of chamomile, a complex processing system has to be employed. Such a system was worked out by Sváb [34] for the Hungarian conditions. It is demonstrated by the scheme in Figure 5.5.3.

5.5.3.4 Cultivars

The importance of growing high-quality chamomile cultivars has been emphasized by many authors. However, the productivity of the chosen material and its chemical composition are affected by the local ecological conditions and the cultivation method, as well. That is the reason that the majority of the countries in the above-mentioned regions have their own cultivars.

The selection of chamomile was started in the early 1960s in Hungary [28]. As a result of this breeding work two cultivars were registered and used for large-scale cultivation. The main characteristics of these cultivars are as follows:

"Budakalászi 2" (BK 2): It is a well-known Hungarian cultivar, used as a control population in different parts of the world, even proposed for commercial production [4, 6, 16, 17]. The cultivar is a tetraploid form and has large flowers [28]. It has a relatively long vegetation period, producing

TABLE 5.5.5
Changes of the Chemical Characters of Selected Cultivars during Three Vegetation Cycles (1980–1982) in the Experiment Carried out in Backi Petrovac, Yugoslavia [9]

Cultivar or Population	Essential Oil Content (%)		Ratio of Main Components in %				
	1st harvest	2nd harvest	(–)-α-bisabolol	Chamazulene	Bisabolol-oxide A	Bisabolol-oxide B	Farnesene
ZA-2							
min./max.	0.71–0.74	0.51–0.59	5–10	10–15	40–50	10–15	5–10
Ljubljana							
min./max.	0.65–0.89	0.48–0.66	5–10	10–20	40–50	10–15	5–10
Banat							
min./max.	0.36–0.48	0.34–0.46	20–30	10–20	10–20	35–40	5–15

a high yield of first-quality flower. Its essential oil content is 0.7–0.9% on average, and it contains 12–20% of chamazulene and about 10% of (–)-α-bisabolol.

"Soroksári 40": The cultivar "Soroksári 40" is a diploid form selected from local populations [13]. Its vegetation period is considered to be moderately long, producing high flower yield. Its essential oil content is about 0.8–1.1%. The ratio of chamazulene is 16–19% and that of the (–)-α-bisabolol 6–10%.

The Romanian cultivars were selected from Polish polyploid variety "Zloty Lan" [22].

"Margaritar": This is a polyploid cultivar [24] selected from Polish variety "Zloty Lan." The essential oil content is about 0.7% with 14% of azulene.

"Flora": The cultivar was created by repeated individual selection combined with self-pollination of the valuable elite of the Polish polyploid variety "Zloty Lan." It has higher flower production capacity compared to "Margaritar" and the essential oil content is higher, too, up to 0.8%. The ratio of azulene is about 15%. Another advantage of this cultivar is its higher tolerance against the infection of *Peronospora leptosperma* and *Erysiphe cichoracearum*.

The chamomile production in Bulgaria is based on a single cultivar, "Lazur." It was selected from the local wild growing chamomile populations [38].

"Lazur": "Lazur" is a tetraploid form, selected from local Bulgarian flora. Its essential oil content is about 0.7%. From the components of the essential oil the presence of chamazulene (24.37%), bisabolol oxide (19.42%), and bisabolol (12.38%) has to be mentioned as a characteristic feature of the cultivar.

In Yugoslavia different types of chamomile cultivars and populations were tested for getting starting material of high productivity for the large-scale cultivation [9]. Some of their results are shown in Table 5.5.5.

It is obvious from the data that the essential oil content of cultivars depends on both the time and the year of the harvest. It can also be concluded that external factors could induce quantitative modifications in the composition of essential oil; however, the basic chemical characteristics of the populations did not change.

REFERENCES

1. Bernáth, J. (1994) Dagli orti all industria storia di una tradizione. *Erboristeria Domani* (October 1994) **10** (175) 29–30.
2. Bernáth, J. (1996) Situation report on the Hungarian medicinal and aromatic plant section. Traffic report. 97 pp.
3. Carle, R. and Isaac, O. (1987) Die Kamille — Wirkung und Wirksamkeit. *Z. Phytotherapie*, **8**, 67–77.

4. Carle, R., Seidel, F., and Franz, C. (1991) Investigation into seed germination of *Chamomilla recutita* (L.) Rauschert. *Angewandte Botanik.*, **65** (1–2), 1–8.

5. Correa, C. Jr. (1995) Mandirituba: new Brazilian chamomile cultivar. *Horticultura Brasileira*, **13** (1), 61.

6. Falistocco, E., Menghini, A., and Veronesi,F. (1996) Osservazioni cariologiche in *Chamomilla recutita* (L.) Rauschert. *Atti del convegno internazionale: Coltivazione e miglioramento di piante officinali*, Trento, Italy (2–3 June), *Proceedings*, 459–463.

7. Franz, Ch. (1980) Content and composition of the essential oil in flower heads *of Matricaria chamomilla* L. during its ontogenetical development. *Acta Horticulturae*, **96**, 317–321.

8. Gasic, O., Lukic, V., and Adamovic, D. (1991) The influence of sowing and harvesting time on the essential oils of *Chamomilla recutita* (L.) Rausch. *J. Ess. Oil Res.*, **3**, 295–302.

9. Gasic, O., Lukic, V., Adamovic, R., and Durkovic, R. (1989) Variability of content and composition of essential oil in various camomile cultivars (*Matricaria chamomila* L.). *Herba Hung.*, **28** (1–2), 21–28.

10. Hornok, L. (1978) *Gyógynövények termesztése és feldolgozása (Cultivation and Processing of Medicinal Plants)*. Mezõgazdasági Kiadó, Budapest, 356 pp.

11. Hornok, L. (1992) *Cultivation and Processing of Medicinal Plants*. Akadémiai Kiadó, Budapest, 338 pp.

12. Kerekes, J. (1966) Kamillatermesztési kísérletek (Experiments for the production of chamomile). *Herba Hung.*, **5** (2–3), 141–147.

13. Kerekes, J. (1969) *Gyógynövénytermesztés (Production of Medicinal Plants)*. Mezõgazdasági Kiadó, Budapest.

14. Kisgeci, J. and Adamovic, D. (1994) *Gajenje lekovitog bilja*. Nolit, Beograd. 185 pp.

15. Lenchés, O. (1993) *Matricaria recutita* L., in Bernáth, J. (Ed.) *Wild Rowing and Cultivated Medicinal Plants*. Mezögazda Kiadó, Budapest, 566 pp.

16. Letchamo, W. (1996) Developmental and seasonal variations in flavonoids of diploid and tetraploid camomile ligulate florets. *J. Plant Physiol.*, **148**, 6, 645–651.

17. Letchamo, W. and Gosselin, A. (1996) High quality camomile for North American commercial processing. *Acta Horticulturae*, **426**, 593–600.

18. Mann, C. and Staba, J. (1986) The chemistry, pharmacology and commercial formulations of chamomile, in Craker, L. and Simon, J. (Eds.) *Herbs, Species and Medicinal Plants*. Oryx Press, Phoenix, AZ, pp. 239.

19. Martinov, M., Tesic, M., and Müller, J. (1992) Erntemaschine für Kamille. *Landtechnik*, **47** (10), 505–507.

20. Mathé, I. (1963) A kamilla (*Matricaria chamomilla* L.) magyarországi termõhelyei és hatóanyag-vizsgálata (Hungarian growing areas of chamomile and examination on active agents). *Kísérletügyi Közlemények*, Kertészet, 56/C, 11–26.

21. Mathé, I. (1979) *A kamilla (Matricaria chamomilla L.)* (The chamomile). *Magyarország kulturflórája*. 6. 79 pp.

22. Paun, E., Verzea, M., Dumitrescu, A, Barbu, C., and Ungureanu, N. (1996) Breeding research on medicinal and aromatic plants — A survey of the Romanian experience. *Int. Symposium, Breeding Reserach on Medicinal and Aromatic Plants* (June 30–July 4, 1996), Quedlinburg, *Proceedings*, 136–140.

23. Peneva, P.T., Ivancheva, S.I., and Terzieva, L. (1989) Essential oil and flavonoids in the racemes of the wild camomile (*Matricaria recutita*). *Rastheviedni Nauki*, **26** (6), 25–33.

24. Plugaru, V. (1996) Cercetari privind metodologia obtinerii de seminte la specia musetel (*Matricaria chamomilla* L.). *Herba Rom.*, **13**, 25–33.

25. Rácz, G., Rácz-Kotilla, E., and Laza, A. (1984) *Gyógynövényismeret (Knowledge on Medicinal Plants)*. Ceres Könyvkiadó, Bucharest.

26. Rom, P. (1930) Adatok a chamomilla összehasonlíto vizsgálatához *(Data on comparison experiments of chamomile). M.Gyógyszertud. Társ. Ért.*, **4**, 4–5.

27. Salamon, I. (1996) Large scale cultivation of chamomile in Slovakia and its perspectives. *Coltivazione e miglioramento di piante officinali, Trento* Italy, *Proceedings*, 413–416.

28. Sárkány, S. (1965) A kamilla (*Matricaria chamomilla* L.) nemesítése (Breeding of chamomile). *Herba Hung.*, **4**, 125–168.

29. Schilcher, H. (1987) *Die Kamille*. Wissenschaftliche Verlagsgesell., Stuttgart, Germany, pp. 99.

30. Stanev, S., Zheljazkov, V., and Janculoff, Y. (1996) Variation of chemical compounds in the essential oil from some native forms of chamomile (*Chamomilla recutita* L.). *Int. Symposium, Breeding Research on Medicinal and Aromatic Plants* (June 30–July 4, 1996), Quedlinburg, *Proceedings*, 214–217.

31. Stepanovic, B., Jovanovic, M., and Knezevic, D. (1989) Ispitivanje mogucnosti gajemja lekovitog I aromaticnog bilja na jalovistu rudnika uglja "Pljevlja." *Zemljiste Biljka*, **38** (1), 57–62.

32. Sváb, J. (1983) Results of chamomile cultivation in large-scale production. *Acta Horticulturae*, **132**, 43–47.

33. Sváb, J. (1992) German camomile, in Hornok, L. (Ed.) *Cultivation and Processing of Medicinal Plants*. Akadémiai Kiadó, Budapest, 246–254.

34. Sváb, J. (1997) Personal communication.

35. Sváb, J., Tyihák, E., and Rápoti, J. (1966) A magyar kereskedelmi kamillával végzett szárítási kiérlet (Drying experiment with Hungarian chamomile). *Herba Hung.*, 5 (1), 31–35.

36. Sztefanov A., Szabó K., and Bernáth J. (2003). Comparative analysis of Hungarian *Matricaria recutita* (L.) Rausch. populations. *J. Horticult. Sci.*, in press.

37. Wagner, T. (1993) Camomile production in Slovenia. *Acta Horticulturae*, 244, 476–478.

38. Zheljazkov, V., Yankuloff, Y., Raev, R.Tc., Stanev, S., Margina, A., and Kovatcheva, N. (1996) Achievements in breeding on medicinal and aromatic plants in Bulgaria. Int. Symposium, Breeding Research on Medicinal and Aromatic Plants (June 30–July 4, 1996), Quedlinburg, *Proceedings*, 142–146.

5.6 CULTIVATION EXPERIENCES IN SLOVAKIA

VILIAM ORAVEC, VILIAM ORAVEC, JR., MIROSLAV REPČÁK, L'UBOMÍR ŠEBO, DUŠAN JEDINAK, AND IVAN VARGA

5.6.1 INTRODUCTION

Chamomile has long been one of the most important medicinal plants cultivated in Slovakia. Its cultivation started in the beginning of the 1950s in the former Czechoslovak Republic. Diploid variety Bohémia with a high content of chamazulene and α-bisaboloxide A and B was sown. In 1957 the tetraploid variety Pohoelický Velkokvetý with similar characteristics, as far as efficacious compounds are concerned, was bred, but this variety was restricted because of the high degree of disintegration.

Chamomile is a plant with a wide growing range and can be grown in the Slovak Republic almost everywhere. It grows best in warmer areas protected from wind with plentiful sunshine and mean yearly precipitation ranging from 550 to 800 mm. Soils rich in nutrients and humus, heavy to mild, mold to luvisol character are the most suitable. After almost 40 years of experience, the crops reached the required level of market production; the cultivation of chamomile in beet and potato regions was proved to be the most suitable.

With regard to the initial slow growth of chamomile, it is necessary to choose the foregoing agricultural plant that leaves the land weed-free and in a good state. From this point of view root crops, peas, beans, and mustard are the most suitable; corn is also good. Clover, clover and grass mixtures, dill, coriander, and caraway are not suitable at all. It is also possible to grow chamomile in monoculture until the weed flora resistant to the herbicides permitted is formed.

The second precondition of successful chamomile cultivation is a high-quality soil preparation. In one-year chamomile cultivation, the stubble is ploughed over by medium-deep ploughing (180–220 mm) immediately after the foregoing agricultural plant harvest. Soil surface is adjusted by cold-crusher and harrow so that it was smooth enough, clod-free, and hardened. In monocultural growing, after the harvest of chamomile, it is necessary to remove the rest of the herb by a postharvest cutting machine and then a break-up and rolling follows [7].

Simultaneously with the medium-deep ploughing, fertilizers are applied. To produce 1 ton of drug from 1 ha the following average doses of pure nutrients are recommended in the ratio 1N:0.14P:2.05K for 1 ha: 60 kg N, 10 kg P, 142 kg K. Only one third of nitrogenous fertilizers is applied before sowing, the remaining two thirds are applied in two doses after overwintering and the second weeding [4].

Chamomile can be sown at any time of year on the soil surface, because the germination requires the presence of light. However, in order to reach a high crop and to be able to dry the harvested biomass, the sowing is limited to two periods.

Autumn sowing from August 15th to September 15th is recommended for the regions with regular autumn rainfall and frosts coming after October 20th. Chamomile sown in this period germinates quickly, takes roots well, and grows in the period of humid and warm-enough weather.

Spring sowing is suitable for all regions with the exception of dry and warm ones, where total rainfall for April and May does not reach 50 mm and day temperatures exceed 15°C. Spring sowing is applied to approximately 30% of production area in order to ensure continuous harvest up to August 15th. The sowing takes place from March to the end of April, and five to six harvests are reached.

The quantity of the seeds sown depends on the seed quality and varies from 1.5 to 2.5 kg/ha. In monoculture a self-seeding occurs, thus the quantity of seeds for sowing applied is lower. Chamomile is sown, dependent on the subsequent mechanization, to rows 30–45 mm apart.

From the point of view of mechanized harvest and drug quality, overriding attention should be paid to the purity and the state of health of the stand. Herbicides are effective in the struggle against weeds; however, they must be applied correctly. The most important damage to chamomile is caused by sucking insects, above all by *Erophyes convolvens* and representatives of *Thysanoptera, Heteroptera,* and *Homoptera* [3].

Cultivation areas of chamomile in Slovakia have varied from 150 to 400 ha in the last ten years, according to market demands. These demands are much influenced by the amount of chamomile collected from natural resources, which reached 30 to 50%. Up to 1990 only one firm was a monopoly bulk buyer and processor of chamomile. Export is recorded directly by producers for west European countries (Table 5.6.1).

The present need of chamomile drug is insufficient for the inland market in Slovakia. Chamomile flower is imported by processing pharmaceutical firms. On the other hand, producers export their product directly, or by means of commercial firms to west European markets under relatively advantageous conditions. In order to intensify producers' activities in the field of medicinal plants,

TABLE 5.6.1
Purchase of Chamomile by Liečivé Rastliny (Medicinal Plants) Malacky Division, Export in Tons and Prices

Year	Wild	Cultivated	Export	Price, 1st class (€)
1986	46	5	—	2.13
1987	46	15	—	2.13
1988	28	15	—	2.25
1989	62	14	—	2.25
1990	34	15	—	2.50
1991	20	7	14	2.75
1992	26	—	34	3.00
1993	34	—	39	3.00
1994	18	5	45	3.25
1995	19	16	27	3.50

TABLE 5.6.2
Economy of Model Firm of Agricultural Company ROZKVET Nová Ľubovňa at the Complex Exploitation of Chamomile

Year	Overall Takings	Overall Expenses	Area [ha]	Euro (€) Profit	Takings per 1 ha	Expenses per 1 ha	Profit per 1 ha
1983	4,150	2,312	5	1,538	830	522	308
1984	13,250	8,950	13	4,300	1,019	688	331
1985	11,478	8,626	15	2,852	765	575	190
1986	32,673	23,323	22	9,350	1,485	1,060	425
1987	33,425	25,600	30	7,826	1,114	853	261
1988	36,250	24,350	31	11,900	1,169	785	384
1989	55,200	18,550	43	36,650	1,284	431	852
1990	51,625	25,750	60	25,875	860	429	431
1991	42,500	34,750	82	7,750	518	424	95
1992	59,325	47,200	94	12,125	631	502	129

the interest group Rumanček (Chamomile) was found in 1996. It is a nonprofit association of both trade and juristic persons that represents the interests of improvers, cultivators, and processors of medicinal, aromatic, and tonic plants. The objective of the association is the enlargement of area and the rise of efficiency of breeding, cultivation, and processing of medicinal, aromatic, and tonic plants both for inland market and export, which should reach 20% of domestic production. The next goal of the association is to ensure the coordination in the field of breeding, cultivation, research, and production of final products from medicinal, aromatic, and tonic plants and to reach 2500 ha of cultivation area in the Slovak Republic, including the area for chamomile up to 500 ha.

The overall picture of chamomile cultivation gives us its economic evaluation. It was applied to a model firm in the Agricultural Farm Rozkvet in Nová Ľubovňa village in the years 1983–1992. This agricultural firm completely exploited the chamomile biomass for drug, seeds, essential oils, and extracts. Adequate prices and state appropriation for the years 1991–1992, when there was a disproportion caused by the growth of outlays and fall of prices and a substantial reduction of appropriations, had a favorable influence on economic results. The negative influence of prices and appropriation inputs changed the orientation of the market with respect to the export of chamomile drug above. The conditions for prospective quick growth of chamomile cultivation also within the criteria of alternative cultivation are given by long-term absence of pesticides in soil, fertilizers being applied only in minimum doses, and mechanical cultivation (Table 5.6.2).

5.6.2 Research, Breeding, Seed Growing, and Varieties

5.6.2.1 Research

State research of medicinal plants in the former Czechoslovakia was a dominant task mostly of the institutions situated in Slovakia. From the beginning of complex research, Slovakofarma Hlohovec was a co-ordinator. Ing. Ivan Varga, present director for science and research of Slovakofarma, was the responsible coordinator of these activities in the years 1980–1997.

Chamomile research started in 1976 at the Department of Experimental Botany and Genetics of the Faculty of Science of P. J Šafárik University in Košice (FS PJSU) by a partial task force on the "research of chamomile cultivation in soils with high salt content." The goal of the research was to find the most suitable complex chamomile farming technology aimed at the obtaining the maximum crop of high-quality drug [5].

The research continued from 1980 to 1990 and was expanded by another research workplace: Agricultural Farm in Nová Ľubovňa. It was aimed at the solution of problems of large-scale cultivation of this medicinal plant, harvest, the development of a wide-space chamomile harvester, postharvest arrangement by presorting of plant biomass, and also the sorting and stalk removing of the dry drug. It also built up a shop for scientific and technological development and piece production of machines for harvest, postharvest arrangement, and technological processing of chamomile.

5.6.2.2 Breeding and Seed Growing

Simultaneously with cultivation research, the research collectives dealt with breeding and preparing the material for registration for a variety tests.

In the course of 20 years the collective of breeders of the Department of Experimental Botany and Genetics of FS PJSU in Košice, Agricultural Farm Rozkvet, and the firm Vilora bred up to four varieties of chamomile. Breeding work consisted principally of breeding of indigenous Spanish chemotypes with local varieties and was aimed at maintenance of chamazulene content on the original level, but reaching high content of α-bisabolol to the detriment of α-bisaboloxide A and B. This was reached in the period evaluated. Prospective work in this field is aimed at coumarine and flavonoid compounds [11].

In order to maintain or improve the qualitative characters of chamomile essential oil, maintenance breeding and seed growing on the principles of production process are carried out [12].

5.6.2.2.1 Production Process Scheme

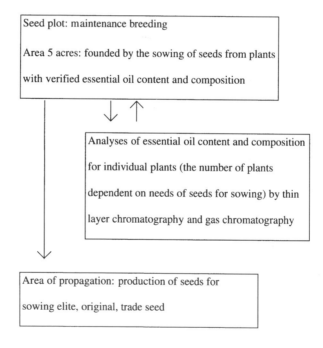

5.6.2.2.2 Methods

Chamomile is grown on seed plots with a 5-acre area. Seed plots are founded by sowing the seeds of plants in which essential oil contents and composition fulfills the standards of the drug quality. With this objective individual plants are analyzed by thin-layer chromatography, essential oil composition is studied, and plants with detectable amounts of α-bisaboloxides A and B are eliminated. The propagation areas are founded from the seeds produced on seed plots.

FIGURE 5.6.1 Crop of chamomile.

Chamomile anthodia disintegrate during the manipulation after harvest and drying. The mass is sieved; receptacles, ligulate florets and disc florets, dust, and flower and seed fragments are removed. Sorted seeds are dispatched packaged.

5.6.2.3 Varieties

At the present four varieties of chamomile are permitted in Slovakia.

Bona (National Variety Book ČSFR 1495/1984)

Reproduction concerned the rise of content in (−)-α-bisabolol essential oil at the expense of bisaboloxides, keeping a high level of chamazulene. Reproduction started in 1975 through the cross-breeding of wild-growing material from Spain with the variety Bohemia and following selections on the screening basis for individuals with high content of (−)-α-bisabolol in dichlormethan extracts by reduction with thin-layer chromatography and gas chromatography. The variety testing was executed from 1980 up to 1983.

The early diploid variety was of a smaller size with middle-green leaves and middle-sized reductions. The content of volatile oil equals 0.9%, chamazulene content in essential oil 16.1% and (−)-α-bisabolol 35%. During maintaining breeding (−)-α-bisabolol increased in essential oil by 42.9%.

Novbona (National Variety Book ČR reg. No 3052 and SR ev. No 3332)

The variety Bona served as the basic material. The breeding took place in Nová L'ubovňa from 1983 to 1990. By the selection through evaluation of the chemotype, inside the variety population, the part of plants with a high content of (−)-α-bisabolol and chamazulene increased.

The early diploid variety consisted of up to small plants, bright green fine leaves, small up to middle size of flower level, middle average of reduction, without tongue-shaped flowers, *and* including tongue-shaped flowers. In the variety population the (−)-α-bisabolol chemotype is represented by 94%. The content of essential oil in the drug equals 0.9%. The essential oil contains 18.0% chamazulene and 46.1% bisabolol.

Goral (National Variety Book ČSFR 1888/1990)

As basic material for the polyploid induction through colchicining in 1978, this was used in newbreeding and later registered as Bona. In the following years the population was selected on the basis of analysis for chromosome numbers and through chemotype screening individuals with appropriate characteristics. On the basis of the variety examined from 1986 to 1988, the variety was agreed on in 1989.

In comparison with diploid varieties, the variety distinguishes itself by increased breakdown. The mesh oversize up to 2 mm is 26% in comparison with 10% in the case of the variety Bona. The variety population represents itself as a mixture of chemotypes (35% (–)-α-bisabolol and 65% bisabololoxide). The content of essential oil is 1.1% and chamazulene in essential oil is 24.5% but the content of (–)-α-bisabolol 24% does not exceed the sum of other materials of the bisabolol type. Goral is favored by the farmers because of good harvest characteristics.

Lutea (National Variety Book ČR reg. No 3051 and SR reg. No 3333)

The basic material for breeding the variety of the bisabolol type is the variety Goral. In 1987 and 1988 under laboratory conditions and in strong isolation, selected individuals were cultivated on the basis of chemotype screening. In 1989 and 1990 the breeding was implemented under field conditions in Nová L'ubovňa.

Middle early tetraploid variety consisted of middle high plants, middle green central leaves, middle height of flower level, high detracted average without tongue-shaped leafs and with tongue-shaped leaves. The variety has a stable chemotype composition of the population (over 92% of (–)-α-bisabolol). The content of essential oil in the drug equals 1.2%, chamazulene in essential oil is represented by 21.2%, and (–)-α-bisabolol 43.3%. The breakdown is on the level of diploid varieties [9].

Values of secondary metabolites in the period 1991–1995 can be studied in Figure 5.6.2 (the varieties Bona and Novbona) and Figure 5.6.3 (varieties Goral and Lutea) [9].

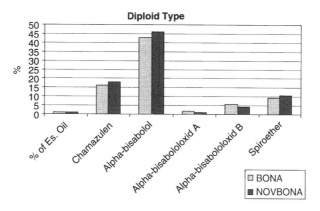

FIGURE 5.6.2 Values of secondary metabolites in diploid varieties (1991–1995).

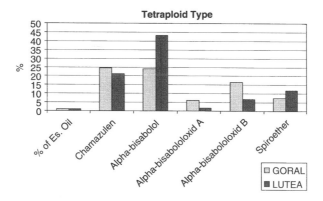

FIGURE 5.6.3 Values of secondary metabolites in tetraploid varieties (1991–1995).

5.6.3 PRINCIPLES OF QUALITY: DRUG DESCRIPTION

The criteria for quality classes of chamomile flowers are governed by the Branch Standard 86 62 11.

Criteria	Quality, max. %			
	1	2	3	4
Overripe, crushed flower heads passing through the sieve III (2 mm)	8	16	30	40
Flower heads with stem longer than 20 mm and with leaves	7	10	15	20
Flower heads with stems longer than 40 mm	1	2	3	4
Bare receptacles and undeveloped flower heads	Unique	8	12	16
Inorganic impurities	0.5	1	1.5	2
Loss by drying	14	14	14	14
Ash	13	14	15	16
Ash insoluble in HCl solution	4	5	6	7
Content of oil min.%	0.4	0.4	0.3	0.3

In special cases of the use of chamomile flowers, *Flos chamomillae*, for pharmaceutical needs, the evaluation was performed according to the *Czechoslovak Pharmacopoeia* IV 1984 and 1987 [2, 3].

Synonym: *Flos chamomillae vulgaris* is the dried flower head of *Matricaria recutita* L. species chamomile. It must contain min. 0.4% oil and min. 0.035% chamazulene calculated to guayzulene (1,4-dimethyl-7-isopropylezulene-C15H18-Mr198.31).

Identity Tests

Microscopy: methodical procedure as with DAB10 standard.
CHAMOMILE FLOWER (*Matricariae Flos*)
The chamomile flower consists of flower sets *Matricaria recutita* with the min. content of 0.4% (V/m) blue volatile oil.

Properties

The drug has an aromatic agreeable aroma. The opened flower heads consist of a single covering, which has the shape of one to three rows of leaves, conical lengthwise, eventually half-round bed (young flower heads), 12 to 20 edge-placed tongue-shaped flowers with white tongues, as well as several dozen central yellow tubular flowers.

Identity Test

A. The covering leaves are egg-formed or spear-formed having bright brown-gray edges. The flower bed is mostly conical and hollow without chaffs. The crown of tongue-shaped flowers consists of a single basal bright yellow or bright brownish and yellow tubular part coming over to white wide egg-shaped tongue. The crown of the tubular flowers is yellow, increasing in height, where it becomes five-pointed having on the base bright brownish to brown colors.

B. The flower heads will be separated into proper parts. Follows test with microscope using chlorhydrate R solution. The outer epidermis of bed flowers has a skin edge composed of one layer of radial prolonged cells and a central zone of chlorophyll-containing structure. Over the structure can be found the epidermis with lengthwise-shaped cells with lateral wavy walls and crack openings as well as glandular hairs. On the edges of guiding beams can be found multiple pointed basic cells with big diameter.

C. The test follows the chromatography method on the small layer (V.6.20.2) using a layer of silicate gel GF 34 R.

D. The identity test. Into reagent flask is placed 0.1 ml of reagent solution with a 2.5-ml solution containing 0.25 g dimethylaminobenzaldehyde in the solution of 5 ml 25% phosphoric acid, 30% acetic acid, and 40 ml water. The solution is heated for 2 min in water bath. After adding 5 ml petroleum ether, the solution is stirred up and the water phase is distinctly green-blue to blue in color.

Purity Test

External state means no more than 25% particles separated by mesh 710.

Foreign matters : (V.4.2). The drug has to correspond to the ash test (V.3.2.16), upper limit 13.0%.

Content Determination

The determination follows the determination of volatile oil content in the drug (V.4.5.8) using 30.0 g drug, a 1000-ml flask, 300 ml water as distillation liquid, and 0.5 ml xylol R as receiver during 4 h distillation with the distillation velocity of 3–4 ml/min.

Storage: Protect against light!

Tests of Purity

Crushed and overripe flower heads (undersized on the sieve IV): max. 20%

Flower heads with the stem longer than 20 mm on other parts of the mother plant (relics of stems, leaves, etc.): max. 8%

Flower heads with the stem longer than 40 mm: max. 1.0%

Undeveloped flower heads: max. 5.0%

Bare receptacles: max. 5%

Foreign organic impurities: max. 2%

Inorganic impurities: max. 5%

Loss due to drying: max. 14.0%

Ash: max. 11.0%

Ash insoluble in HCl: max. 3%

5.6.4 MECHANIZATION: PICKING TECHNIQUE AND PICKING MACHINES

5.6.4.1 Picking Technique

5.6.4.1.1 Manual Picking

This method of picking is sufficient for smaller areas in gardens. It is carried out by cutting, trimming, or combing using combs, which are similar to those used for picking of bilberries. This method of picking is very extensive and labor consuming, but it cannot be replaced by a mechanized method for picking areas determined for production of seed.

5.6.4.1.2 Mechanized Picking

Great attention has been paid to the mechanized picking of chamomile during the last 20 years. The question arises about the direct picking in several rows with the width of span from 200 cm to 610 cm according to the type of picking machine. The adapter is equipped with a comb dresser and cutting roller. The proper combing machine is represented by the finger comb with uniform radius of curvature along the entire length of the bar and with constant distances between individual fingers. Flower heads are released by the rotary brush. Flower heads are transported by worm conveyers or by inclined scraper conveyers, or pneumatically using underpressure [5].

5.6.4.2 Picking Machines

At the present time, the farms in Slovakia have various types of picking machines.

- SKM-2 R. Attached to the RS-09 implement carrier is a single-row picking machine with low capacity of 0.03 ha.h^{-1}, with a great ability to comb the crop without losses.
- SH 2 R: Attached to the RS-09 implement carrier, with the capacity of 0.4 ha per shift.
- ST-1-003 NESET: Has a higher capacity, it is designed for multirow picking with the working span of 2 m and capacity of 0.85 ha per shift. It is also mounted on the chassis of the RS-09 implement carrier, or it is suspended on the tractor of the Zetor type, the Horal system [14] (Figure 5.6.4).

The concentration of the chamomile-producing area depends directly on the technique of its picking and on the efficiency of the picking machine. Nowadays, the most specialized farms are of a large-scale nature as to the production and processing of chamomile. The VZR 4 large-area chamomile picking machine meets this standard (Figure 5.6.5).

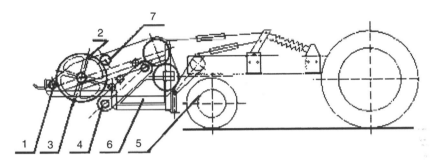

FIGURE 5.6.4 Frontally attached picking machine for chamomile. 1. cutter roller, 2. comb, 3. comb dresser, 4. conveyer, 5. implement carrier, 6. container for chamomile flowers, 7. rotary brush.

FIGURE 5.6.5 VZR machine for picking chamomile.

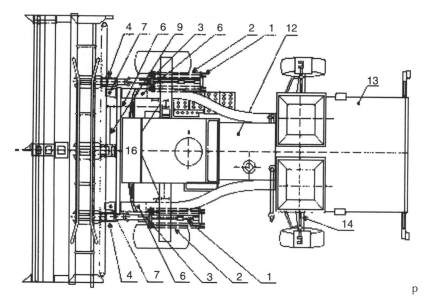

FIGURE 5.6.6 Ground plan pf the VZR 4 picking machine for chamomile. 1. lifting cylinders, 2. relieving springs, 3. lifting arms, 4. suspensions of bodyadapter, 5. pins, 6. pneumatic flexible hoses, 7. collecting mouth, 8. distribution system for the drive of picking machine adapter, 9. countershaft, 10. pivot of the upper arm, 11. adapter for picking of chamomile, 12. modified chassis of E-307, 13. bin, 14. wheel spacing for drive of fans, 16. modified chain transmission for travel.

The VZR 4 represents an adaptor for chamomile harvesting performed by an adapted unit of the harvesting windrower Fortschritt E-303. The whole unit is designed as E-307 and is a self-propelled chamomile collector. The undercarriage E-303 is composed of the following main parts: frame with wheels, motor, hydraulics, and plateform with cabin.

The picker has front-wheel drive and the back wheels have hydraulic control. They can be adapted to the spacing of the front wheels, which means an enlargement on both sides by 120 mm. This modification was implemented with the aim of allowing the back wheels to follow in the same path as the front wheels. It is designed for direct picking of chamomile flower heads. The separated flower heads are pneumatically transported to the sectional storage bin. After it fills, they are mechanically ploughed out to the vehicle body, or to the container, which is transported for further processing. The VZR 4 is a self-propelled chamomile picking machine that consists of the following main functional parts (Figure 5.6.6):

1. Adapter for picking of chamomile (11)
2. Modified chassis of the E-303 (12)
3. Two underpressure pneumatic conveyers (6)
4. Storage bin (13)

The adapter for picking of chamomile consists of the following main parts: frame (17), sectional combing dresser (18), sectional trimmer (20), sectional wipe roller (21), both left-hand and right-hand worm conveyer (22), distribution systems (23), and protective guards (25) (Figure 5.6.7).

The frame is made of a tube-welded structure, to which the sides are attached, and of the wall of gears for installation of shaft-bearing boxes for picking chamomile. The sectional comb dresser (18) is mounted in the front part of the frame. It consists of ten combs with oblique teeth. The sectional trimmer (20) is mounted in front of the comb dresser. It partially cuts and partially tears the chamomile stems protruding from combs, using the five cutters attached along the periphery of the drum. The cutter can be easily moved toward the comb dresser, so that the gap between the

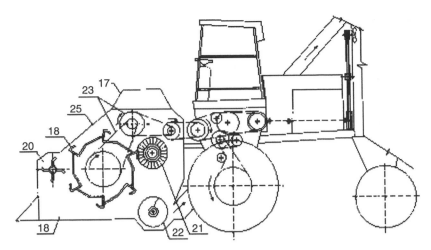

FIGURE 5.6.7 Scheme of the adapter for the chamomile picking machine. 17. frame, 18. comb dresser, 19. scraper, 20. stemming machine, 21. wiper, 22. right-hand and left-hand worm conveyer, 23. distribution systems, 24. divider, 25. guards.

cutter and the combs is as small as possible. The sectional wipe roller (21) is mounted in the upper rear section of the frame. It wipes the caught chamomile flower heads into the left-hand and right-hand worm conveyer (22). The wipe roller consists of plastic hairs, to avoid damaging flowers. The countershaft (9) is mounted in the rear section of the frame. It is driven by the chain transmission (8) from the engine. The distribution system (23) of the functional parts is located in the center of the adapter in its longitudinal axis. Three dividers (24) divide chamomile stems during operation of the adapter. Two dividers are near the edges, and one in the middle. The central divider is available only for design purposes, to divide the functional mechanisms of the picking machine into two parts due to a wide span. The protective guards (25) serve as protection for workers against the touch of rotating functional parts.

The storage bin (13) catches chamomile flower heads. It forms an independent part of the picking machine and is attached to the rear part of the E-303 chassis. Underpressure fans (15) for underpressure transport of the material are attached to the bin. The lower edge of the bin is 900 mm above the ground. The bin is symmetrical to the longitudinal axis of the machine symmetry plane. Two discharging holes are in the rear part of the bin, which must be airtight after their closing, so the fans do not suck air from outside.

5.6.4.2.1 Technical Data

Adapter for chamomile picking of the VZR type:

- Length: 4230 mm
- Width: 2100 mm
- Height: 1180 mm
- Weight: 1943.8 kg
- Number of revolutions of:
  - Comb dresser: 0.58 s^{-1}
  - Stemming machine: 12.25 s^{-1}
  - Wipe roller: 10.50 s^{-1}
  - Worm conveyer: 2.07 s^{-1}
  - Countershaft: 10.50 s^{-1}
- Capacity: $1.4 \text{ m}^2/\text{s}$

Engine: D-50 type:

- Type: four-stroke, compression ignition
- Number of cylinders: 4
- Volume of cylinders: 4750 cm³
- Power output: 41.25 kW
- Engine speed: 1700 RPM
- Alternator: 12 V, 500 W

Modified chassis: E-303 type:

- Length: 3960 mm
- Width: 3200 mm
- Driving wheel track: 2770 mm
- Control wheel track: 2400 mm
- Wheel base: 2400 mm
- Radius of curvature: 4200 mm
- Weight with the cabin: 3565 kg
- Brakes: service brake: hydraulic Duo-Duplex
  - Hand brake: mechanical
- Clutch of travel mechanism: single-plate, dry
- Fuel tank: 100 l
- Battery: 12 V, 180 Ah
- Tires: driving wheel: 16–20 PR A 19; p = 0.17 MPa
  - Rear wheel: 10–15 AM A 13; p = 0.20 MPa

Working speeds:

1st gear: 1.7–4.3 km/h
2nd gear: 4.2–10.7 km/h
Driving speeds: 3.4–8.6 km/h

Underpressure pneumatic conveyer:

- Diameter of piping: 160 mm
- Number of pipings: 2
- Fan speed: 2900 RPM
- Volume of sucked air: 0.7 m³/s

Storage bin:

- Length: 1920 mm
- Width: 1800 mm
- Height: 2050 mm
- Capacity: 3 m³

VZR 4 self-propelled chamomile picking machine:

- Length: 6280 mm
- Width: 4230 mm

- Height: 3900 mm
- Total weight: 5300 kg

The new machine for picking chamomile flowers is now in the prototype stage and there is a question about a single-purpose machine for picking chamomile flowers. It is designed to pick chamomile flowers of very high quality with maximum efficiency. The picked chamomile substance contains up to 80% chamomile flowers with maximum length of stem up to 20 mm.

The new picking machine can be installed on any carrier with the front three-point suspension and the front power take-off driving shaft. The transport of chamomile flowers from the adapter to the bin is pneumatic, which is simple and does not require any changes in the carrier design. The bin can be solved as a common tractor trailer, or in the form of an attached tilting container in the rear part of the carrier.

The technical data are:

- Total width of adapter: 3000 mm
- Working width: 2600 mm
- Weight: 660 kg
- Capacity: 0.32 ha/h

5.6.5 PLANT RAW MATERIAL AND PROCESSING IN SLOVAKIA

5.6.5.1 Picking Conditions and Picking Methods

The most suitable time for picking chamomile occurs when the peripheral white petals are fully opened in the horizontal plane, when the plane of the flower forms a right angle with the axis of the receptacle and of the stem. The first picking shall be made when one third of flower heads flower, by which the uniform crop with richer flowering and a greater number of pickings will be obtained. Further, it is important to pay attention to the intermediate period between individual pickings to prevent overripening, which could result in premature disintegration of flower heads and deterioration of drug quality.

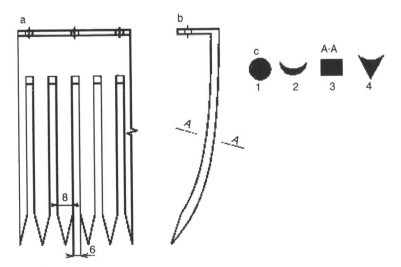

FIGURE 5.6.8 Shape of a comb and of a finger. a: comb; b: shape of finger; c: cross sections of fingers: 1. circular, 2. semicircular with a groove, 3. rectangular, 4. triangular with a groove.

The number of pickings depends on the date of sowing, cultivation, fertilization, and prevailing weather. On average, two to three pickings can be obtained from spring sowing, and five to eight maximum pickings can be obtained from autumn sowing. Time between individual pickings depends on weather, and is in the range of 10 to 20 days.

Due to a better regeneration ability of plants, it is necessary to carry out the picking earlier. Two methods or techniques can be used in picking:

- Machines for picking of chamomile by combing, e.g., based on the VZR 4 principle (Figure 5.6.8).
- By beating, using the steel fingers of a rotating drum. A disadvantage of this method of picking is later renewal of flowering, because the cutting surfaces are irregular and more stem damage takes place [7].

The sectional comb dresser is mounted on the front part of the frame. It consists of ten combs with oblique fingers. The fingers are arranged next to each other in the single plane. The distance between fingers should be from 4 to 6 mm. During later pickings (the second one up to the last one), the flower heads have smaller diameters; therefore, the distance between fingers must be smaller, so that the flower heads do not slip between the fingers. The shape and cross section are different. The circular cross section is the most abundant. Not only is its quality of work the best, but this finger is easy to manufacture at a bargain price. The finger with the rectangular cross-section works better than that of circular. However, it is more demanding to manufacture and is costly. The finger with semicircular cross section with a groove combs very well, but its penetration into the crop is worse. The finger with a triangular cross section with a groove is also demanding to manufacture, but it combs very well and penetrates into the crop. The edges of the combing finger function partially as cutters; therefore, it is suitable for the finger to be made of high-quality material and to have sharp edges [5].

The sectional sweeper is mounted in front of the comb dresser. It allows better catching of chamomile flowers to the comb dresser. The sectional stemming machine is mounted in front of the comb dresser. It partially cuts and partially tears chamomile stems protruding from combs, using eight cutters attached along the periphery of the drum. The cutters can be moved toward the comb dresser, so the gap between the cutter is as small as possible. The cutters can be attached alternatively to the combs, which helps regular cutting of flower heads.

5.6.5.2 Postharvest Processing

5.6.5.2.1 Sorting

Two types of sorting lines are used in Slovakia:

- Sorting line of the ST-1-005 type with a minimum capacity of 120 kg/h
- Presorter of the AST-034 type with a capacity of 1000–1200 kg/h (Figure 5.6.9)

After picking, the chamomile is sorted on sieves, or on the chamomile sorting line of the ST-1-005 type. The sorting line consists of the proper sorting machine and the presorter (Figure 5.6.10).

This equipment is stationary and is designed for the sorting of chamomile flowers into four qualitative classes [5].

The machine consists of the following parts: the frame, discharging conveyer (7), upper roller separator (9), lower collecting conveyer (11), lower sorting conveyer (12), and accessories. The preseparator consists of the frame, separating drum (2), lifting mechanism (3), drive (5), and hopper (6).

FIGURE 5.6.9 AST-034 presorter of chamomile.

During cleaning, the material is manually fed into the hopper, and from there, it moves to the separating drum (2). Small material, which is suitable for sorting, falls through openings in the separating drum into the hopper of the discharging conveyer (7). The long and coarse material falls out of the separating drum outside the machine.

The machine is driven by an electrical motor, and its daily capacity is minimum 1000 kg.

During picking of chamomile using the VZR 4 picking machine, undesired organic impurities, such as stalks, can get into the bin. In order to remove them, the chamomile presorter is used. The raw material gets rid of undesired organic impurities, stalks, and parts of chamomile, which corresponds to the third-quality class of classification (Figure 5.6.11).

Chamomile substance is transported manually or by the conveyer to the internal sieve. Both the internal sieve (1) and external sieve (2) are mounted on the wheels with the rubber rim, which are seated in the wheel holders with milled grooves. Using these grooves, it is possible to adjust the inclination of these sieves from 0 to 3°. The raw material thrown in the internal sieve is entrained along the periphery in the shape of helix. The undesired organic impurities and stalks are moved toward the end of the sieve. Chamomile substance, which has fallen through the apertures of the internal sieve, is sorted again on the external sieve in the same manner. Falling substance is on the roller conveyer. The preliminary sorted substance gets rid of oversizes. Chamomile substance is transported on rotating rollers for further processing.

The driving wheel of the internal sieve is guided in the groove, in order to be secured in the axial direction. The outer sieve is secured on one side by the locking roller, and by the semigroove on the opposite side, in which the wheel with the rubber rim travels.

5.6.5.2.2 Drying

Chamomile must be dried immediately after sorting. All qualitative classes are dried separately so the homogeneity of the dried material is ensured. Various types of driers are used, from simple ones as for drying hazels (heated by hot air in chambers, box driers, and continuous driers [3].

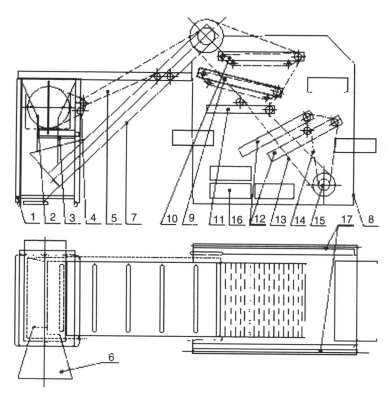

FIGURE 5.6.10 ST-1-005 sorting line for chamomile. 1. frame of the preseparator, 2. separating drum, 3. lifting device, 4. drive of the drum, 5. chain of the drive, 7. conveyer, 8. frame of the sorting line, 9. sorting rollers, 10. feeding conveyer, 11. collecting conveyer, 12. sorting conveyers, 13. chain for the drive, 11, 14. chain for the drive 12, 15. electric motor, 16. six bins, 17. side guards.

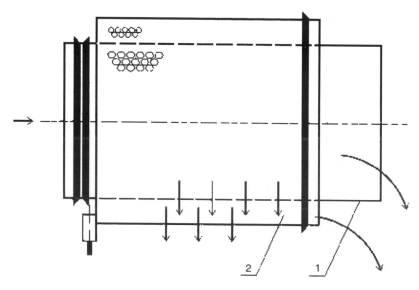

FIGURE 5.6.11 Internal and external sieves of the preseparator.

FIGURE 5.6.12 Stemming machine for chamomile.

The operating temperature must not exceed 40°C. At the higher temperature, a large escape of essential oil takes place, and drug is of brown color, which does not meet the criteria of the *Czechoslovak Pharmacopoeia* IV. After drying, chamomile drug must be left for 10 to 15 days in open packages to breathe.

The chamomile stemming machine is used for removing long stems. The chamomile stemming machine of the TR-6-002 type is a one-purpose stationary unit, which processes lower-quality classes of chamomile drug, by which a higher percentage of the first-quality class will be obtained (Figure 5.6.12).

5.6.5.2.3 Storage
Chamomile drug is stored in cartons or in paper bags with a net weight of 7–11 kg (paper bag) and 15–20 kg (carton box). [6].

5.6.6 PROCESSING IN SLOVAKIA

5.6.6.1 Essential Oils

Chamomile is most frequently used for production of essential oil, which takes place during its distillation with water steam. At the same time sterilization and assurance of microbiological safety takes place, with subsequent extraction of the obtained extract. The generally known method used for isolation of essential oil is distillation with water steam. This method is demanding for technological equipment. The quantity of essential oil obtained depends on a large number of factors, such as pH, drug homogeneity, and quality [16]. The technological equipment, which is used in Slovakia, is protected by the author's certificate of invention No. 245316 (Figure 5.6.13).

The distillation kettle of the apparatus for production of essential oils from medical plants consists of a closed conical vessel (1), with the discharging mechanism on its lower base (2). The steam supply is connected above them (3). The filling neck (4) and the outlet (5) are located on the upper base. The homogenizing rotor (6) driven by the driving mechanism (7) is inside of the closed vessel.

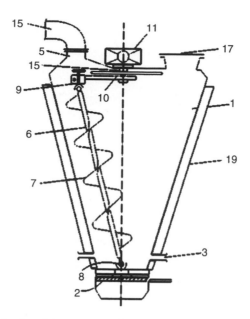

FIGURE 5.6.13 Distillation kettle of the apparatus for production of etheric oils from medical plants. 1. vessel of the kettle, 2. discharging mechanism, 3. steam supply, 4. filling neck, 5. discharging neck, 6. homogenizing rotor, 7. helix, 8. joint bearings, 10. eccentric, 11. drive, 15. transmission, 17. quick-closing device, 18. glass piping of the cooler, 19. thermal insulation; 12, 13, 14, 16: discharging mechanism.

The distillation kettle has the homogenizing rotor (6) consisting of the eccentric (10), transmission (15), and helix (7) located parallel with the wall of the closed vessel (1) in such a vat that its driven end is seated in the bearing (9) of the eccentric (10) and its free end is seated in the joint bearing (8), which is situated in the axis of the closed vessel (1).

The distillation kettle has the discharging mechanism (2) consisting of one fixed plate (12) and one rotary plate (13), which have at least two overlapping cutouts (14).

5.6.6.2 Extracts

Processing chamomile substance using the wet method has been performed since 1976. From 1995 it was calculated with essential changes in the production program via the introduction of new technologies for chamomile processing into the form of extracts, which can be used in cosmetics using the alcohol method and water-alcohol method and their granulation into instant tea. It was verified in the pilot plant in 1985 [15].

The preparation of extracts according to the developed technological procedures has the following three stages:

1. Separation of solid phase: the proper extraction
2. Fine filtration and concentration
3. Dehydration

The obtained water extract was concentrated in the classical manner in the vacuum circulation evaporators to the desired content of dry substance. Water-alcoholic extracts were prepared according to the required qualitative parameters by extraction and percolation of chamomile drug.

REFERENCES

1. Černá, K. (1988) *Systém pestovania liečivých rastlín*, VŠÚ Hurbanovo.
2. *Československý lékopis* IV (1984) Avicenum Praha.
3. *Československý liekopis* IV (1987) Avicenum Praha.
4. Dovjak, V. (1984) *Vplyv minerálnej výživy na dynamiku prijmu živín, na úrodu drogy a na tvorbu obsahových látok u rumančeka kamilkového.* Dizertačná práca, pp. 11–26. Nitra.
5. Hončariv, R. (1979) *Výskum pestovania rumančeka kamilkového na slaných pôdach.* Čiastková úloha ŠPTR, Košice.
6. Jech, J. (1984) Stroje pre RV II.; Príroda, Bratislava, pp. 354–359.
7. Oravec, V., (1985) *Štúdium biologických a agrotechnických vlastností vybraných liečivých rastlín*, Nitra.
8. Oravec, V. (1986) *Nové poznatky vel'koplošného pestovanie v podmienkach JRD "Rozkvet."* I. Celoštátna konferencia liečivých rastlín, Piešt'any, pp. 25–30.
9. Oravec, V., Varga, I., Repčák, M., Huliková, A. (1988) Author's certificate of the invention, Praha.
10. Oravec, V., Repčák, M., Černaj, P. (1993) Production technology of *Chamomilla recutita. Acta Horticulturae* **331**, 85–87.
11. Oravec, V., Seidler-Ložikovská, K., Holubař, J. (1996) *Genéza odrôd rumančeka kamilkového — Chamomilla recutita (L.) Rauschert) v Stredoeurópskom priestore.* III. Medzinárodná konferencia o liečivých rastlinách, Piešt'any, pp. 43–44.
12. Repčák, M. (1991) Šl'achtenie rumančeka a flavonoidy. Pestovanie, zber a spracovanie liečivých rastlín, *Medzinárodná konferencia o liečivých rastlinách*, Vysoké Tatry 1991, pp. 11–12.
13. Rcpčák, M., Halásová. J., Hončariv, R., Podhradský, D. (1980) *Biologické Plantarum* **22**, 183–191.
14. Tyl, M. (1984) Nové poznatky při pětování, sklizni a posklizňové úpravě heřmánku. Zborník: *Celostátní seminář o pětování heřmánku pravého.* JZD Úsovsko, Československo.
15. Tyl, M. (1986) *Metodika pětovaní heřmánku lékařského.*, MZVž, pp. 5–10.
16. Varga, I. (1985) *Súčasná úroveň spracovania liečivých rastlín na Slovensku.* Zborník referátov z vedeckého kolokvia, Vyžné Ružbachy, pp. 27–30.
17. Varga, I. (1990) Final report, *Výskum liečiv na báze liečivých rastlín a prírodných látok II*, Hlohovec, pp. 39–40.

5.7 GROWING VARIETIES OF CHAMOMILE IN THE CZECH REPUBLIC

JOZEF HOLUBÁŘ

Cultivation of chamomile (*Matricaria recutita* L.) started in the former Czechoslovakia about 1955 and the majority of the production up to this date came from the plucking of the wild-growing chamomile. Growing in the greater areas began in the 1960s. This happened thanks to the changing structure of agricultural production and the decrease of chamomile growth under natural conditions. The whole production was coming from increased growing from the 1970s up to the mid-1980s. The increase was more important in the Czech Republic than in Slovakia. The share of the total production in the Czech Republic was 40%; in Slovakia even in 1983 this share was not even 10%. At this time the total cultivated area in the Czech Republic was approximately 200 ha and the production 90 t, and in Slovakia about 80 ha and 10 t [2–4]. From the beginning of the 1990s the total crop area decreased rapidly and in the middle of the 1990s decreased about 50 ha. The development of the areas and chamomile production from the mid-1980s up to the mid-1990s is presented in Table 5.7.1.

The chamomile flower can be grown in the Czech Republic in all regions except the mountain regions. It tolerates lighter and heavier soils. Optimal annual rainfall is about 450 to 650 mm. It is classified just after unweeded winter crops or spring cereals; oliphérous plants and medicinal plants are cultivated because of their tops. Fertilized root crops are appropriate preceding plants as for nutritive elements in poorer soils.

TABLE 5.7.1
Development of Areas and Production in the Czech Republic from 1985 to 1995

Year	Area (ha)	Production (tons)
1985	240.9	142.1
1986	247.0	130.9
1987	251.0	145.6
1988	235.0	110.4
1989	277.0	152.4
1990	279.0	150.1
1991	137.0	65.8
1992	21.0	10.7
1993	3.5	2.0
1994	56.6[a]	22.8
1995	50.0[b]	c

Source: Leros Ltd Praha Zbraslav

[a] Investigation of MZLV Brno.
[b] Evaluation UKZUZ Brno.
[c] Production not found.

TABLE 5.7.2
Registered Chamomile Varieties in the Czech Republic

Variety	Country of Holding	License Year	Registration Prolongation
Bohemia	R	1952	1997
Bona	SR	1984	1997
Goral	SR	1990	1997
Lutea	SR	1995	1997
Novbona	SR	1995	1997

Optimal nutritive rates per 1 ha are recommended as 1.0 N/0.5 up to 0.7 P/5 up to 3.3 K, which means 20 up to 40 kg N/15 up to 20 kg P/66 up to 100 kg K [1]. The manure will be worked in during the middle-deep tillage and the lot will be prepared as usual. The chamomile seed needs light and stemmed soil to grow. Because of this it is seeded during calm weather on the soil surface and is then rolled down the whole area or in rows 10 to 12 cm wide. The sowing is 2 kg per 1 ha, and the distance between rows is 40 to 50 cm. The experience shows that about 50% of the programmed chamomile surface should be prepared during the autumn and 50% during the spring. The autumn sowing is done from the middle of August to the end of September. In the following year the growth will bloom about 3 weeks earlier than the growth coming from the spring sowing. The spring sowing should be done in the early spring. On smaller areas and with manual harvesting five to eight harvests can be achieved. Mechanical harvesting means three to four harvests.

In the Czech Republic five varieties of chamomile are actually registered; of those one is domestic and four are foreign varieties (see Table 5.7.2).

TABLE 5.7.3
Containing Substances of the Chamomile Varieties from SOZ Determined during 1993 and 1994

Variety	Essential Oil Content (%)	Content of Chamazulene (%)	Content of Bisabolol
Bohemia	1.2	21.0	a
Bona	1.0	18.9	46.1
Goral	1.2	21.4	23.0
Lutea	1.2	21.2	43.3
Novbona	0.9	18.0	46.1

[a] Content was not determined by technical means.

Bill No. 92/1996, in effect since July 1, 1996, determines that a variety registered in the National Variety Book before the above-mentioned bill was in effect can be considered as registered if the applicant applies within 1 year for the prolongation of the registration.

During 1993 and 1994 evaluations concerning the difference, uniformity, and steadiness (DUS) on the above-mentioned varieties were realized. One part of the evaluation was the determination of substances contained within a variety. See the average values for all harvestings in Table 5.7.3.

REFERENCES

1. Drozen, J., Kocourková, B. et al. (1995) *Léčivé, aromatické a kořeninové rostliny.* Situační a výhledová zpráva. Mze ČR, 8–9.
2. Tyl, M. (1984) Nové poznatky při pěstování, sklizni a posklizňové úpravý heřmánku. *Zborník: Celostátní seminař o pěstování heřmánku pravého.* JZD Úsovsko, Československo.
3. Tyl, M. (1986) *Agrotechnika velkovýrobní produkce heřmánku lékařského,* Léčivé rostliny, Zbraslav nad Vltavou.
4. Tyl, M. (1986) *Metodika pěstovaní heřmánku lékařského.,* MZVž, 5–10.

5.8 EXPERIENCES WITH THE CULTIVATION OF CHAMOMILE IN ARGENTINA

NORBERTO FOGOLA

5.8.1 CULTIVATION

5.8.1.1 Date of Sowing

In Argentina chamomile sowing is usually carried out in autumn, from March until June. In this way, time remains for the chamomile to establish well and to form good roots.

The advantages of the summer-autumn sowing over the spring sowing is not based on a higher azulene and essential oil content. The autumn sowing generates above all higher harvests, sometimes paralleling with a higher azulene content. Between the sowing in autumn and in spring there are great differences, not only the earlier flowering of the autumn chamomile. The autumn sowing offers the advantage of a more regular germination of the seeds.

The disadvantage is, however, that the chamomile sowed in autumn frequently becomes very weedy in spring. The autumn sowing requires more attention than the spring sowing. If the snow cover is missing, the chamomile plants could be killed by hoar. Deciding the date for sowing is frequently based on local climatic circumstances. Sometimes it is practical to carry out sowing at different dates in order to be able to harvest at different times. So it is sowed semiannually (in spring and summer), which results in two harvests (summer and autumn chamomile). Even an autumn sowing at two and a spring sowing at three (different) dates is recommended. According to the size of the fields, the sowing must occur in short intervals of 8–14 days. Sowing is accomplished in slightly raised lines. In order to reach a regular sowing, putting on the lines during calm weather is recommended. If the ground circumstances make it possible, it is reasonable to roll the soil surface. An old proverb tells us that where the chamomile is trod, it grows better.

5.8.1.2 Seed Amount

Seed material is mainly used, which is classified as Argentina's bisabolol oxide B-type. On a larger scale seeds from Germany were used, and a company-owned type that is derived from Spanish seeds has been used. Only recently has it been changed over to modern cultivation types.

The fine seed can be sowed without mixing with sand or semolina. For that a machine appropriate for the sowing into lines with lifted ploughshares or a modified seeding machine for fine seeds can be used. As chamomile seeds are plants that germinate in light, they must not be covered with earth. In any case the seeds are slightly pressed on the soil surface. The amount depends on the purity and the germination capacity of the seeds. The very small seeds (thousand kernel weigh 0.035–0.050 g) often are mixed with other parts of the blossoms, that can be separated by sieving due to their different specific weight.

The germination capacity is often insufficient; after 1 year of storage it is reduced by 7 to 49% and after 3 to 5 years after harvesting it is completely gone. Depending on the sowing machines available, one calculates 1 to 3 kg seeds/ha. If the germination capacity of the seeds is insufficient or the seeds are contaminated, a higher quantity has to be taken into consideration, i.e., up to 4 to 6 kg/ha. With a fluctuation range of 1.5 to 3.0 kg seeds/ha, no differences were found.

5.8.1.3 Row Distance

Carrying out broadcast sowing is not recommended, as it complicates the care and harvest. Concerning the optimal row distance, opinions diverge greatly. A maximal row distance of 70 cm is recommended, corresponding to a plant distance of 30 cm. The slightly lower yield per hectare is compensated for by better harvest conditions. The flower yield per hectare increases the degree by which the row distance gets smaller; the yield per plant, however, increases with the enlargement of the distance. An enlargement of the distance would reduce the harvest amount, whereas azulene and essential oil content are independent from the date of sowing, the row distance, and the quantity of seeds used.

It seems to be a fact that the plants have to stand so far each from another, so that a further harvest can be effected in the same year. The flower heads are always located at the end of the side shoots. The more shoots a plant has, the higher the quantity of flowers. Lack of light, a distance too close, and a soil too flat prevent the formation of side shoots, so that the plants grow with only one shoot.

Optionally an optimal row distance of 40, 45, 50, or 60 cm can be determined in the case of an automatically effected harvest. Finally, the optimal row distance is related to the frequency of rainfall. The drier the crop field is, the closer the row distance of the chamomile has to be; i.e., in dry zones a distance of 45 cm is recommended; in most other zones the distance should be 60 cm.

5.8.1.4 Germination Factors

The success with the sowing of the chamomile on the field depends to a great extent on the climatic conditions. In years with dry spring and autumn, the chamomile yield remains very small. Chamomile is a plant that germinates in light. Chamomile seeds depend on the humidity in the upper layer of earth in order to come to germination. If this is available, germination can take place. Sometimes, however, one single day of intensive sunlight can cause the seeds to die, as the humidity in the upper layer of earth evaporates. Under normal circumstances, germination takes place 14 days after sowing. With dry weather, it can be delayed.

5.8.1.5 Growth Factors

The overground parts of the chamomile can develop in very different forms according to their living conditions. The wild plants that grow unhindered develop a robust, straight, 55- to 60-cm-high stalk, while the plants that are found at unfavorable sites grow only up to 10 cm high. If the chamomile plants are planted not too closely, they can reach a height of 80 to 90 cm. The more the plant branches, the more capitulums it generates, which under favorable conditions can develop. If growth is constricted — even if the plants grow between other specimens slightly constricted — the development is hindered. In extreme cases, only plants with one shoot are produced or those that form only the final capitulum of the primary axis and the first upper offshoot.

5.8.1.6 Soil Properties

Generally the chamomile does not make demands; it grows on light and heavy ground, but favors sandy and loamy soil. It adapts itself to the ground conditions very well and grows with pleasure (as weed) in the areas that surround the crop fields. Nevertheless, the yield increases on black earth and meadow ground and decreases on brown, sandy earth without humus. Obviously the chamomile grows very well at a pH value between 7.3 and 8.1. So for example the alkaline and sodium monoxide grounds of the salt steppes of Hungary (SZIK grounds) are favorable. Chamomile can be grown successfully on grounds that due to their alkaline state (pH value 8.5 to 11) are not suitable for cereals any more.

Chamomile that is growing on saline grounds produces a more attractive appearance and greater purity; however, it does not contain high amounts of essential oils and azulene.

Although the drug yield is influenced by the ground quality, there is no relation between the basic factors of ground quality and active substance content of the chamomile.

5.8.1.7 Climate

The oil content is highly influenced by factors such as air temperature, rainfall, and solar radiation. The optimal temperature for the formation of essential oil during the flowering period would be 20 to 25°C. In an experiment with accentuated ecological differences it could be stated that the weather conditions do not have any influence on the amount and the composition of the essential oil of the chamomile. The increase of the content of essential oil and chamazulene depends mainly on genetic factors. There is no relation between the height of the plant and the content of ingredients, either.

5.8.1.8 Fertilizing

The fertilizing depends on the composition of the ground. For big sowing areas an analysis is recommended to consider the nutrient content and the pH value. The chamomile reacts positively to potash. However, it is very sensitive to excessive phosphoric acid. The fertilizing with N and P has a great influence on the growth, the flower size, and amount as well as the content of prochamazulene. When the N- and P- amounts are different, the result decreases. The best result is achieved with N + P in relation 1:1. The uptake of the active substances of the chamomile per hectare is as follows: 16 kg N, 4 kg P, 20 kg K, 4 kg Ca, 1 kg Mg < 1 g Na. The ground must contain sufficient phosphoric acid so that the stalk becomes strong enough and flowers with short stems can be cut. The same is valid for potash. In spite of that, the nitrogen gifts should be small. If the nitrogen gifts are too big, the plants are overgrown with weeds and have the tendency to burst. The flower number decreases. On the other hand, the unique fertilizing with nitrogen increases the flower number, as in the case of simultaneous fertilizing with phosphate and potash.

Fertilization with phosphate, potash, and calcium — alone or combined — does not bring any increase. A mineral fertilizing with a great amount of potash supports the flower formation. The chamomile should not be fertilized with stable manure. Because of its ability to release nitrogen slowly, it supports the vegetative development of the plants, which comes along with a drop of the flower number. Before sowing, a complete fertilizing (N, P, Ca, K, etc.) is necessary. After every harvest the ground is fertilized again, whereas the P amount has to be higher than during the first fertilizing. The usual fertilizer doses are as follows:

40–60 kg N: 200–300 kg ammonia and calcium-containing nitrate/ha.
36–45 kg P_2O_5: 200–250 kg super-phosphate/ha
80–120 kg K_2O: 200–300 kg potash (40%)/ha.

5.8.2 Weed and Pathogen Control

5.8.2.1 Herbicides

One must distinguish between the elimination of the chamomile in other cultures and the elimination of the weed within the chamomile fields. The elimination of the chamomile, for example, in cereal fields, is not very easy. Chamomile is resistant against almost all herbicides that are applied for cereals. Therefore, chamomile is often a weed in the following cultures.

In Argentina Triflan (commercial denotation) is used as herbicide before sowing; chemical nomenclature: alpha, alpha, alpha-trifluor-2,6-dinitro-N,N-di-propyl-p-toluidine. The chemical classification reads: Dinitroanilin; active substance: Trifluoralin. After this the chamomile plant is processed if it begins to sprout with the herbicide 2,4-D.

A weed-free chamomile field is the vital precondition for an impeccable quality of the drug. Particularly the application of the local harvest equipment is only possible in weed-free chamomile fields. It is advisable to plant chamomile on a maximally weed-free ground. The best precrops are sugar beet as well as potato, which create a good ground structure and a favorable ground state. Because of their high N-content, legumes (peas) are less suitable.

The chamomile tolerates also different cereals as precrop. Cornfields — weed-free or preprocessed with herbicides — are especially suitable.

The herbicides have to be applied at the appropriate time. If they come into direct contact with the seeds during germination, in most cases annihilation of the chamomile crop will result.

5.8.2.2 Insecticides

In Argentina the use of insecticides is not necessary.

5.8.3 Yield Formation

5.8.3.1 Stage of Development and Content of Nutrients

The flowering heads pass through different development stages until they open completely.

The step-by-step flowering of the tubular florets from the lower to the upper edge lasts approximately 3 to 4 weeks. When the ligulate florets bloom, they unfold themselves. While the ligulate floret continues growing, it lifts and lowers itself daily. The bigger the flowers are, the weaker the daily periodic movement of the ligulate florets. They do not move anymore when the last tubular florets have opened.

The dry matter per capitulum increases during the development on an average of 1.23 mg to 39.24 mg.

The capitulums of the closed flower have the highest oil content. The second highest level is reached with full blossom. While the weight increases during the flowering process, the oil content in the fully developed flowering heads drops below the content of the half-flourished ones. The content in a field reaches a maximum if 50% of the tubular florets of a flowering head are opened.

The increase of the essential oil content and the content of chamazulene entails an increase of the dry weight up to the full flower. Here, the increase of the essential oil content is a little smaller than the increase of the dry weight. Therefore the essential oil content decreases in relation to the dry weight. The azulene content of the oil increases during the expansion period of the flower. After the capitulums have reached their maximum size, both the oil and the azulene content decrease as they wilt. Thus, the production of essential oil and azulene seems to be ended when the flowering heads have fully sprouted.

The chamomile variety "Bodegold" reaches a maximum essential oil content when the flowering head has almost flourished and the lower parts of the flowering head are already hollow; a daily periodic growth movement, however, is still notable.

5.8.3.2 Distribution of the Active Substances

In the variety "Bodegold" the following distribution of the essential oil content (g/100 ml/ drug) was measured:

Flowering head:	0.95
Tubular florets:	0.82
Ligulate florets:	0.22
Lower part of the flowering head:	
With top:	1.18
Small fragments:	0.52

5.8.4 HARVEST AND PROCESSING

5.8.4.1 Time of Harvest

The chamomile harvest takes place in several gathering processes in the period October 10 to approximately November 30.

The chamomile harvest has to be started as quickly as possible so that the plants can form new flowering heads. The harvest starts when the capitulum of the main bud on the main shoot has flourished fully. Then it blooms for two weeks, and in this time many capitulums of the side shoots develop in the same way; in general they develop up to half and more. A suitable time for harvest is considered to be appropriate when the first three rows of the tubular florets are open; otherwise, the capitulums fall apart before they dry.

In Argentina, up to two harvests of the same plants can be achieved per annum at intervals of 10 to 14 days.

In contrast to the wild chamomile, the flowering of the frequently harvested chamomile stops with the appearance of frost.

TABLE 5.8.1
Chamomile Exports from Argentina in the Years 1987 to 2002

Year	Tons Dry Matter
1987	1800
1988	2321
1989	3281
1990	2413
1991	2843
1992	2900
1993	2800
1994	2000
1995	1800
1996	1500
1997	1500
1998	1300
1999	1200
2000	1100
2001	800
2002	600

Source: Office of the Authority for Agriculture, Stock Breeding and Fishing of the
Republic of Argentina).

The time of day the harvest is carried out is not exact. Because of the sun influence, in Argentina as a rule it is carried out from sunset until 10:00 the next morning, although about noon a maximum pro-azulene content is found.

Optimal harvest conditions are to be found when clouded sky, dry weather, and temperatures around 20°C prevail.

5.8.4.2 Yield

The flower yield strongly depends on the number of harvests. Argentina usually produces approximately 2000 to 3000 kg/ha fresh product (400 to 500 kg not-cleaned dry product). In the former Soviet Union a yield of approximately 300 to 500 kg/ha at four harvests per year was achieved. In Egypt approximately 3152 kg/ha (fresh) were harvested, in Germany up to 800 kg drug/ha, and in India approximately 600 kg/ha.

The drying relation is approximately 1:5 to 6.5. According to Heeger [1] the yield of dry flowers is 500 to 2000 kg/ha, in general 500 to 1200 kg/ha, and the yield of the whole dry plant is 2000 to 5000 kg/ha, in general from 2000 to 3000 kg/ha.

In the years 1987 to 1996, the total harvest in Argentina amounted to more than 2000 tons of drug per year with a drop during the last years of the period. The greater part of the drug crop is exported, above all to Europe (Table 5.8.1).

5.8.4.3 Harvest Methods

The collection or the harvest of the flowers is carried out as follows:

1. Manually: the flower head is torn off by shifting the stalk between the fingers and pressing the fingers briefly, as is still the case in Egypt today.
2. With a rake similar to the one that is used for blueberry and cranberry harvests. The following systems exist:

FIGURE 5.8.1 Manually pushed picking cart.

    a. Meyer: Tubes with a multipronged fork with a mobile cutting device (scissors) on top. No stalk-free product is produced.
    b. Sartorius: Sheet steel appliance with filed-off teeth. Long stalks are obtained.
    c. Central Germany: Rake with 20 prongs, 14 cm width, thorn body 65 mm in length at intervals of 6 mm
    d. Heeger [1]: Rake with 20 peaked thorns with 110 mm length each at intervals of 5 mm, 10 cm width. The device is completely open at the top.
    e. Checo: Quadratic sheet box that has a grip at the upper part. Below the aperture lie peaked teeth at intervals of 5 mm.
  3. With toothed shovels.
    In the salt steppes of Hungary a device is used in the form of a toothed shovel is used, similar to a potato harrow, that must be operated with both hands because of its size (see Figures 5.2.2 and 5.2.3 in Section 5.2).
  4. With special rakes, that are pushed by humans (Figure 5.8.1) or — the next larger size — pulled by horses.
  5. With mechanized harvesters.

At first, with simple devices 50 to 100 tons of chamomile drug were harvested per year. After 1950 the hand combs, common until then, were replaced by pushcarts (Figures 5.8.2 and 5.8.3). As of 1950 the first cultivation attempts were started. The quantities could then be increased rapidly. The functional pushcarts were pulled by horses or motor (Figure 5.8.4).

For the formerly common solar drying, approximately 35,000 racks were brought outside in the mornings and back to the storage area in the evenings. Temporarily more than 5,000 persons were working the harvest and processing during a season. A further increase of production, however, was only possible through the use of harvesters (Figure 5.8.5); the first prototype of modern harvesters was used starting in 1971 (Figure 5.8.6). This Argentina harvester was working effectively. By 1974 production already amounted to 2000 tons.

FIGURE 5.8.2 Manually pushed picking cart.

FIGURE 5.8.3 Manually pushed picking carts.

For this large-scale production, hand picking is not suitable, a mechanization of the chamomile harvest was established for economic reasons. The Linz harvester (variants I to III) developed in the former GDR proved workable on larger cultivation areas, but it is no longer manufactured. In Slovakia chamomile harvesters are used with good success, too.

Furthermore, it should be considered that the chamomile harvest lasts only a few weeks so that improvements can only be tested during the following season. The cultivation of chamomile is relatively specific in comparison to other agricultural crops. Therefore, the machine manufacturers do not show great interest because of low demand.

Depending on the collecting method, the yield varies. With a manual harvest it amounts to 0.75 kg fresh drug (0.15 kg dry drug) per hour, 4 to 5 kg fresh drug in 6 to 8 hours. With the Heeger device the yield can be increased to 2.0 kg fresh drug per hour (0.4 kg dry drug). The Slovakian machine was constructed to collect 800 kg fresh drug per day.

5.8.4.4 Seed Harvest

As the flowering period lasts for some months, the seeds as well mature at different times. If the seeds begin to fall from a plant of medium size, the main maturity has started and harvest can begin in the early mornings, when dew still lies on the plants. Harvesting is stopped when the plants are dry and the seeds fall out.

FIGURE 5.8.4 Mechanized picking cart.

FIGURE 5.8.5 Harvester for chamomile herb.

Large cloths are most suitable for the transport of the chamomile cut for seed extraction. The seed yield varies between 30 and 300 kg/ha; in general it amounts to 100 to 200 kg/ha.

5.8.4.5   Extraction of the Drug, Drying

The harvested flowers must not be pressed and should be transported in baskets (narrow on top, wide at the bottom) or in flat cartons. Then they should be spread out for immediate drying in the shade (an airy place at best) in a thin layer on drying racks or on the ground covered with paper. The water loss is 75 to 85%. If possible, the flowers should not be turned nor touched nor moved during the drying process. Fast drying is also guaranteed through natural air drying if the chamomile is distributed in thin layers and is well ventilated.

FIGURE 5.8.6 Argentinian chamomile harvester

Often drying in the shade at a normal temperature (under 35°C) is preferred because a speeded-up drying process results in a loss of active substance. The loss of azulene with solar drying amounts to more than 30%. The drug dried in the shade contains 20% more essential oil and bisabolol than the product dried in the sun.

Nevertheless, the natural drying requires a period of 3 to 4 days (with unfavorable climatic conditions up to 14 days), while the artificial drying at 30°C needs only 8 to 12 hours. Natural drying in the shadow cannot be used if the product is cultivated in big scale, particularly if the very lower loss of active substance is taken into consideration.

Today, during a season, 60 to more than 100 tons of fresh drug is processed per day. It continuously passes through the sorting machines to the belt drying machines and is sorted with the help of conveyor belts after drying and freed from weeds.

REFERENCES

1. Heeger, E.F. (1956) *Handbuch des Arznei-und Gewürzpflanzenanbaus. Drogengewinnung* [Handbook of the cultivation of medicinal and spice plants]. Deutscher Bauernverlag, Berlin, 775 pp.

5.9 CHAMOMILE IN CHILE: CULTIVATION AND INDUSTRIALIZATION

EDUARDO WELDT S.

5.9.1 BOTANICAL CONSIDERATIONS

Chamomile is a foreign species in the Chilean flora [9], first introduced in the *Catalog of the Cultivated Species* in Santiago's Botanical Garden [12].

The name *chamomile* in Chile is applied to several species of the *Asteraceae* family [10], but Common chamomile is the most commonly used [15]. Several authors have referred to this plant using a Latin term [4], widely accepted in Europe as *Matricaria recutita* L. (*Chamomilla recutita* L. [Rauschert]) [2, 3], also in America in the taxonomy of commercial species [7] as well as in Chile [5]. In Europe, it is well known as Common chamomile in England, *Echte Kamille* and

Gemeine Kamille in Germany, *Camomille vulgaire* in France, *Camomilla* in Italy, and *Manzanilla común* in Spain [13]. The synonym *Matricaria chamomilla* L. is accepted [10, 14]. Some authors in Chile have called it *Chamomilla recutita* L. (Rauschert), according to the denomination for Chile in the catalogs of R.A. Philippi [12] and F. Philippi [11].

5.9.2 Use

In Chile, the chamomile shows the highest consumption among the medical herbs, in folk medicine as well as for use in the herbal tea and infusions industry.

The common chamomile was first mentioned in Chile in 1881. There is a great deal of information about its use since the arrival of the Spaniards in South America [5, 6]. Later, during the German colonization in southern Chile, from the year 1854, the common chamomile is mentioned as being used for medicinal purposes in the chronicles of that time.

The first industrial crops were harvested in 1977, and they were developed by Puelche S.A. to export flowers [1]. These are the first records of this activity in the country, using German seed. Later, in 1980, some rural rustic crops for obtaining dry flowers used in the popular pharmacopoeia were found in villages near the town of Traiguén. This material was used in comparison trials with the material from the experimental crops using German and Argentinean seeds.

The results of these trials showed that the naturally selected and acclimatized seeds in Chile produce a sweet flower, whose quality can be compared to the German product, and proved much better than the Argentinean product, which presents a more bitter taste. This Argentinean chamomile represent mostly a bisabolol oxide B-type.

Puelche S.A. started working on selecting the seeds, isolating individuals of special flowering precocity and homogeneity. From this, Puelche S.A. was able to develop a high-purity line, which has been cultivated until today. Since 1984, this variety of sweet chamomile has been grown to be packed for herbal tea infusions and commercialized as *Matricaria recutita* L., type Manzanilla Primavera Puelche.

According the analysis, the Manzanilla Primavera Puelche represents a bisabolol oxide A-type and its composition is as follows:

Essential oil:	761.81 mg/100 g
Chamazulene:	12.47 mg/100 g
Bisabolol oxide A:	105.97 mg/100 g
Bisabolol oxide B:	26.75 mg/100 g
α-Bisabolol:	8.81 mg/100 g
cis-Dicicloether:	56.36 mg/100 g
trans-Dicicloether:	4.01 mg/100 g

The industrial use of the chamomile for infusion in individual teabags started increasing in 1980 in Chile. Before this happened, the offer was provided by rural suppliers, which were quickly replaced by a consistent offer. This allowed an interesting development of the infusion industry in Chile, though there were sporadic imports from Argentina. The offer of an excellent quality product allowed a sustained development of the Chilean Packing Companies until the present day.

5.9.3 Cultivation

According to López [8], chamomile's habitat is found in Mediterranean climate, between 0 and 1000 meters above sea level (m.a.s.l.). It is easily adaptable to different types of soils, preferably the silica-clayish type, deep and fresh, avoiding excess moisture.

Selecting clean and fresh soils, and using artificial irrigation, helps achieve an optimum phytosanitary condition.

The Puelche chamomile cultivation program considers handling 100% tech-irrigation surfaces, using Side-Roll.

The seed time goes from May to August every year, depending on the seed land, type of irrigation, and crop rotation periods.

The irrigation is performed using the Side-Roll system, using clean weed-free water, which helps to obtain weed-freer soils year after year in the yearly rotation of crops. Roll sprinkling irrigation is also used in new soils being added to the cultivation area. Other growers prefer organic cultivation.

5.9.3.1 Sowing

Cover sowing is performed using selected seed of Manzanilla Primavera Puelche or seeds from Europe. The dose is normally 6–8 kg/ha. Before sowing, the seeds are mixed with the fertilizer, the dose varying, depending on the soil analysis.

Average of soil characteristics:

pH:	7.1
Organic matter:	4.2%
N:	6.3 ppm
P:	11.0 ppm
K:	190 ppm

Average fertilization per hectare

50–80 kg:	Nitrogen
100–120 kg:	Phosphorus
50–70 kg:	Potassium
15–20 kg:	Sulphur
10–15 kg:	Magnesium
30–40 kg:	Calcium

5.9.3.2 Weed Control

In Puelche a weed control plan was developed, determined by the rotation of chamomile crops.

FIGURE 5.9.1 Handpicking with a special comb.

The rotation considers an intermediate cleaning crop once a year, using oats sown in December–January every year. These seed are sown over the stubble formed by the chamomile that has been immediately incorporated after the harvest.

This rotation of crops allows a weed control; the main species are:

Convolvulus arvensis L.
Polygonum persicaria L.
Veronica persica Poiret
Raphanus raphanistrum L.
Capsella bursa-pastoris (L.) Medikus
Galega officinalis L.
Spergula arvensis L.
Plantago lanceolata L.
Echium vulgare L.
Silene galica L.
Leucanthemum vulgare Lam.
Anthemis cotula L.
Chenopodium album L.
Taraxacum officinale G. Weber ex Wigg.

5.9.3.3 Harvest

The chamomile harvest starts in November using machines that cut and select the floral capitulums (Figures 5.9.1., 5.9.2, 5.9.3).

The drying process is performed in tunnels at a controlled temperature of 45°C.

Associated with harvest we can find insects that naturally live in the farm under chamomile crops, showing healthy living conditions for the crops.

FIGURE 5.9.2 Tractor-drawn picking cart.

FIGURE 5.9.3 Chamomile harvester (Argentine type).

These insects are[1]:

Schistocerca sp. (Orthoptera, Acridiidae)
Megalometis cacicus Kuschel (Coleoptera, Curculionidae)
Naupactus xanthographus (Germar) (Coleoptera, Curculionidae)
Chrysolina gemellata (Rossi) (Coleoptera, Chrysomelidae)
Pycnosiphorus costatus Beuesh (Coleoptera, Lucanidae)
Astylus gayi Solier (Coleoptera, Dasytidae)
Hylamorpha elegans (Burmeister) (Coleoptera, Scarabaeidae)

---

[1] We thank Professor Dr. Andrés O. Angulo (Biologist/Entomologist, University of Concepción) for the exact determination of insects.

Praocis curta Solier (Coleoptera, Tenebrionidae)
Colletes cognatus Smith (Hymenoptera, Colletidae)
Apis mellifera L. (Hymenoptera, Apidae)
Synhgrapha gammoides (Blanchard) (Lepidoptera, Noctidae)

5.9.4 FINAL PRODUCTS

The harvest process, drying process, and industrial processing are destined to obtain capitulums as the main product, which is obtained by threshing the pollen (the tubular flowers of the capitulum), which measures 0.25–1.25 mm. Of the remnant, an industrial-quality product is made.

Yields averages:

Dry flowers:	200–250 kg/ha
Pollen:	300–350 kg/ha
Industrial:	600–800 kg/ha
Seeds:	80–100 kg/ha

REFERENCES

1. Banco Central de Chile (1978) *Registros de Exportación*. Santiago, Chile.
2. *European Pharmacopoeia* 3rd Edition (1997) Council of Europe, Strasbourg, France.
3. Font Quer, P. (1982) *Plantas Medicinales*. El Dioscórides Renovado. Barcelona, Spain.
4. Foster, S. (1999) Chamomile, *Matricaria recutita* and *Chamaemelum nobile. Botanical Series* No. 307. American Botanical Council.
5. Hoffmann, A. (1992) *Plantas Medicinales de uso común en Chile*. Ed. Fundación Claudio Gay. Santiago, Chile.
6. Laval, E. (1953) *Botica de los Jesuitas de Santiago* (1767). Biblioteca de la Historia de la Medicina en Chile, Tomo II. Santiago, Chile.
7. Liberty Hyde Bailey Hortorium (1976) *Hortus* 3rd Edition. New York: Macmillan.
8. López, C. (1988) *Plantas Medicinales: Cultivo y Perspectivas*. El Campesino, Santiago, Chile, August/September.
9. Matthei, O. (1995) *Manual de las Malezas que crecen en Chile*. Santiago, Chile. Alfabeta Impresores.
10. Muñoz, O. M., Montes, M., Wilkomirsky, T. (2001) *Plantas medicinales de uso en Chile, Química y farmacología*. Andros Imp. Ltda., Santiago, Chile.
11. Philippi, F. (1884) *Memoria y Catálogo de las plantas cultivadas en el Jardin Botánico hasta el 1° de Mayo de 1884*. Imprenta Nacional. Santiago, Chile.
12. Philippi, R. A. (1881) *Catálogo de las plantas cultivadas para el Jardín Botánico de Santiago hasta el 1° de Mayo de 1881*. Imprenta Nacional, Santiago, Chile.
13. Steinmetz, E. F. (1957) *Codex Vegetabilis*. Amsterdam, Netherlands.
14. Tucker, A. O. (1986) Botanical Nomenclature of Culinary Herbs and Potherb. In Craker, L. E., Simon, J. E. (Eds.), *Herbs, Spices and Medicinal Plants: Recent Advances in Botany, Horticulture and Pharmacology*. Vol. 1. Phoenix, AZ: Oryx Press.
15. Zin, J., Weiss, C. (1980) *La salud por medio de las Plantas Medicinales*. Santiago, Chile, Ed. Salesiano-Chile.

5.10 CULTIVATION EXPERIENCES IN EGYPT

TAMER FAHMI

5.10.1 CULTIVATION REGIONS

Chamomile (*Matricaria recutita / Asteraceae*) has been cultivated in Egypt since 1960 in two areas:

  a. Oasis Fayoum, approximately 80 km southwest of Cairo, where approximately 2000 feddan (1 feddan = 0.42 ha) are cultivated (900 ha). Because of its quality, this product is mostly used for industrial purposes.
  b. Beni Suef, approximately 130 km south of Cairo, where approximately 1000 feddan of the fertile Nile soil are used for cultivation (450 ha).
  c. Small amounts of chamomile are cultivated in Belbes (northeast of Cairo), in the Nile Delta, and also in Assiut (375 km south of Cairo).

Chamomile is densely cultivated in Beni Suef, Fayoum, and Sharqya Governorates. The best is cultivated in Al-Menya and Assiut.

The entire cultivation area in Egypt adds up to approximately 1500 ha, depending on the market situation.

5.10.2 CULTIVATION AREAS

In former times in Egypt, the land was "one big plot of land cultivated with one crop." Then it was divided between many woneers in the course of the agriculture reform, and this affected the actual growing method and the cultivation and the harvesting.

Due to the agricultural reform in Egypt only small areas are cultivated (approximately 0.5 to 15 feddan = 0.225 to 6.75 ha). This results in advantages but also disadvantages:

  • Due to the small surfaces, mechanical harvest is rather impractical.
  • On the other hand, this results in individual families harvesting from their own surfaces. Some of these families are specialized in box material with good quality.

During the last years the surface of ecologically grown chamomile has increased. For the time being ecological cultivation proceeds mainly in the Oasis Fayoum, partially also in the Nile Delta and in Beni Suef: approximately 850 feddan (approximately 350 ha). The Egyptian organization ECOA maintains quality control in tight cooperation with internationally active control institutions. Evaluations of accommodating the entire Oasis Fayoum into ecological cultivation are made.

5.10.3 CULTIVATION PROCEDURE

Cultivation in Egypt is mostly done by growing the plants in nurseries followed by planting in prepared soils.

  • Soil
       The plants can bear severe cold but not hot weather. Chamomile is better cultivated in yellow soil and light alluvial soil with good drainage and airing, as well as in new and reclaimed sandy soil where a drop-irrigation system is used, and in fertile soil.
       The plants bear a proportion of salinity (12,000 parts per million), and are better cultivated in light neutralized alkaline and in somewhat acidic soil.

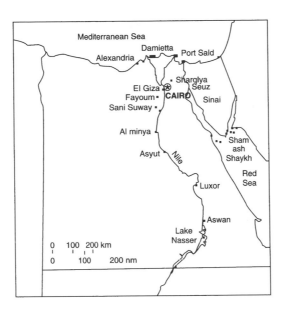

FIGURE 5.10.1 Map of Egypt.

- Growing of plants

  During August 15 until September 15 plants are prepared in nurseries. 1 Irate (176 m²) of land is prepared with 2 m³ compost followed by full irrigation until the water covers the compost. After dryness it should be well plowed, softened, and divided into equal 1.5 × 3 m² area basins, so we can control the sowing operation of plant nursery seeds in the plant nursery. The plant nursery should be in a dark place and during growing not be directly exposed to the sun so that buds do not fly away. Then the land should be plowed and the seeds must be covered.

  Then 1 kg of good seed should be mixed with 3 kg of sand (3 times as much sand as seeds) to facilitate the distribution of seeds. The land then must be plowed and seeds must be covered one and one half times their weight with soil or sand, then water sprayed to fix the seeds, then slowly and densely irrigated so that seeds do not compile. The plant nursery should be continuously irrigated and cleaned of weeds so that water does not remain.

- Preparing the land for cultivation

  20 m³ of compost + 200 kg of rocky phosphates + 2 m³ furnace dust are distributed with vibration. Then the land must be arranged in lines with a range of 12 lines per 2 Egyptian poles or divided into a 1-m-wide (mastaba) bench, then to segments and basins to control irrigation water.

- Planting

  At the end of September/start of October the soils are prepared and the young plants are planted into the soil. The soil is divided into rows (distance between is 0.75 m) and fertilized with superphosphate. The soil is watered immediately before planting. Manual watering is done as furrow irrigation.

  Per feddan approximately 20,000–25,000 seedlings are planted (45,000–55,000/ha). This constitutes into a seeding amount of 200–250 g/feddan (ca. 500–600 g/ha).

  The plant nursery should not be irrigated 10 days before uprooting, which is done with an axe from beneath the roots.

  The nursery plant should be planted in the upper 1/3 portion of the line and must be well covered with mud. Flowers should be picked if found. The distance between one plant and the next must be from 30 to 40 cm, and cultivation must be at the side facing the sun. During cultivation, the soil is usually sprayed with Kroon 500 fertilizer at a rate of 200 g for each 20 liters of water, then stirred in both directions. It is better to cultivate the nursery plant on the same day.

- Water supply
  The land is irrigated 3–5 days after the cultivation irrigation, then resown. The land is irrigated 10–12 days after that, or according to a flower gathering program. Intervals between irrigation must be less in sandy land and drop-irrigation land. Thirst decreases the blossoms' crop and generally it needs to be irrigated approximately 14 times during the growing season. Irrigation is accomplished mostly as furrow irrigation in intervals of 20 days.
- Care and weed control
  A light first hoeing is done to fill up chinks and to clean up weeds after success with the nursery plant, leaving only one or two plants in the hole. Then a second hoeing is done. After that supporting plants are done and they are left without irrigation until roots get deeper. Flowers must be picked if found. A third hoeing is done, adding 4 m^2 compost and covering it by hoeing, and irrigation is done after 1 month at least.
- Fertilization
  After approximately 3 weeks the small plants are fertilized. Fertilization is repeated after 1 month. The same fertilizer is used during the preparation of land for cultivation. With the third hoeing 4 m^3 compost is added, and at flowering, quartz fertilizer is sprayed before sunrise at a rate of 2 g for each 20 liters of water and stirred in both directions.

5.10.4 INSECTS THAT INFECT THE PLANT AND THE METHODS OF FIGHTING THEM

1. White fly (*Bemisia tabasi* GENNADIUS):
   This insect feeds on plant juice, causes plant weakness, transfers leaf-wrinkling disease, and transfers a virus that decreases flowering.
   - Resistance: Use "Biofly" compound at a rate of 100 cm per 100 liters of water.
2. Honey-dew (*Aphis gossypii* GLOVER):
   An insect that sucks the plants' juices with avidity. This insect reproduces in great numbers and secretes a sweet material called "Honey Dew," which attracts "Black Rot" fungus that leads to the deterioration of the crop.
   - Resistance:
     A. Eradication and burning of the plants carrying the virus.
     B. Spraying of Potash soap at a rate of 1.5 liters for each 100 liters of water.
3. Cotton worm (*Spodoptera litoralis* BOISDUVAL):
   Infects the plants in the plant nursery and damages it.
   - Resistance:
     A. Setting traps at a rate of four traps per feddan (=10 traps per ha)
     B. Stop irrigation of trefoil after 10 days
     C. Spray "Bio-Ibcotic" compound of a rate of 200 g per 400 liters of water for each acre (500 g per 1000 liters per ha)
     D. *Bacillus thuringiensis* bt*2 (double) at a rate of 200 g per 400 liters of water for each acre (feddan)
4. Rodent worm and the digger (*Pectinophora* BUSCK ssp.):
   Rodent worm infects the plants in the plant nursery and in the permanent land. It causes weakness of the crop and feeds on the area that connects the root and the stem, so the root separates from the stem. The buds die while the worm and the digger are under the surface of the soil.

- Resistance:
  Using poisoned bait that consists of 15 kg bran (in the case of rodent worm) or crushed corn (in the case of digger) + 2 kg molasses + 100 g of leavened bread + 150 g of crushed alum + 3 kg green material, left for 3 hours then thrown (in the lines in the case of rodent worm) and thrown after irrigation and before sunset (in the case of digger).

5. Tiny whiteness (*Empoasca* WALSH ssp.):
   Infects the plants severely. It is resisted with the "Bentonite" mixture (powder of rocks from Sinai Mountain) (1 kg bentonite + 1 kg miconic sulfur + 1 kg slaked lime). This must be well mixed and sprayed at a rate of 2 kg in the early morning or from 3 to 5 kg for each 600 liters of water. The best spraying condition occurs when the weather is sunny and a high pressure machine is used.

6. Red spider (*Tetranychus* ssp.):
   This is a dangerous pest that causes the leaves to fade, dry, and fall. The pest is under the surface of the leaves so it looks dull; its color turns brown, with dust stuck to spider webs.
   - Resistance:
     Eradication of weeds, close irrigation to the plant, and use of micronic sulfur at a rate of 250 g per 100 liters of water.

7. Strips (*Sitophilus granarius* L.)
   It appears as a silver stain on the surface of the leaves that become black, dry, and die in cases of severe infection.
   - Resistance:
     A. Agricultural sulfur is added as spray at a rate of 20 kg per feddan (50 kg per ha).
     B. Spraying of potash soap at a rate of 1.5 liters per 100 liters of water.

5.10.5 CULTIVATED VARIETIES/TYPES

Specific sorts are not cultivated. The Egyptian chamomile belongs to the low matricin/chamazulene type.

For the time being, there are no breeding activities in Egypt. Seeds are harvested from existing chamomile cultivations, cleaned, and afterward used for new cultivation areas.

For the future it is important to use selected seeds that yield flowers with a high percentage of azulene.

5.10.6 HARVESTING, DRYING, AND PREPARATION

Harvesting starts in mid-December. Plants are picked every 18 to 20 days, five times until the end of April. Immediately after picking the soil is watered again. When chamomile is cultivated there must be assurance that there is enough manpower trained to collect, because only one picking machine is used in Egypt (Linz III).

5.10.6.1 Harvest Date

Starting in the middle of December, flowering begins; then collecting takes place, when radial flowers or white petals are in a horizontal position or parallel to the ground. This is the suitable phase for crop ripeness.

Flower collecting occurs in December/January, performed by a workforce trained to gather flowers with horizontal, parallel-to-the-ground radial petals, because this is the suitable phase for collecting. If the petals are leaning upward then they are not ripe, and if they are leaning downward they are in the late phase when the flowers scatter, as shown in Figure 5.10.2.

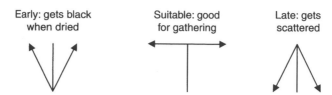

Early: gets black   Suitable: good   Late: gets
   when dried   for gathering   scattered

FIGURE 5.10.2 Positioning of flowers/petals in relation to harvest dates.

FIGURE 5.10.3 Handpicking harvest.

FIGURE 5.10.4 Drying in palm leaf hurdles and cleaning.

5.10.6.2 Harvest

The flower neck, especially the remaining part of the stem, must not exceed 0.5 cm. Collecting is done in baskets that are categorized immediately after collection to discard flowers with long stems, fallen petals, and small flowers, while good flowers are sent to the drying shelf. Collecting is done every 10–15 days at maximum according to the nature of the land and the flowers; irrigation is done after collecting ends in April and the beginning of May. The good feddan gives up to about 2000 kg of good fresh flowers with a ratio of fresh to dry 5 to 1 or 400 to 500 kg of dry flowers (ca. 900–1200 kg/ha).

5.10.6.3 Treatment after Harvest

Flowers are sent to the drying shelf. They are stored on clean drying shelves that are made of wood or palm leaves lined with snack cloth; 1–1.5 kg of good flowers are put in each cage after they are riddled while being fresh. The layer of the flowers on the drying shelf must not exceed 2 cm, so they can dry quickly.

The drying shelf must be in the shade with good airing and away from the stables and compost piles. The drying shelves can be put over each other in opposite directions to let in air. They should be left in the sun on the first day and covered with an upside-down drying shelf so they will not be directly exposed to the sun; this helps in the drying process. At the end of the day drying shelves must be put onto the big shelf covered with a ceiling so that they are not exposed to the dew.

Material must not be stirred on the drying shelves. Drying should be in the shade, except for the first day when the shelves are exposed to sunlight in order to lose humidity. The drying process takes from 6 to 7 days, oil percentage 0.45 or 0.9–1.1% relative to dry weight.

Depending on the weather conditions, air drying takes from 1 to 4 weeks.

Cleaning and sorting of the dried products is done manually. There are only a few existing machine units.

5.10.6.4 Packaging, Storage, and Shipping

Blossoms are packed in 12.5- or 25-kg boxes; pollen and industrial quality are packed in 20-kg plastic sacks.

Export is mainly handled through shipping from Alexandria, Damietta, or Port Said to Europe (Italy, Germany, France) as well as the United States and North and South America.

5.10.6.5 Production Quantity, Export Quantity, and Usage

Annual production is estimated to be approximately 1600 to 1800 tons. This results in a crop yield of approximately 420 kg/feddan (1000 kg/ha). Production quantity of ecological goods is approximately 300 tons. These amounts are exported. Domestic use within Egypt is very rare.

Packaging in filter bags is done only to a small extent. These filter bags are mainly exported to Arab countries. For the time being, evaluations for establishing oil distillations are done.

- Boiled flowers are used as a drink for stomach pains. It also activates digestion.
- Used in most medications to decrease fever, and in creams to cure eyelid swelling.
- Oil is used as a hair tincture as it contains azulene. It is used as well to activate blood circulation, especially in children; it is also used as a flu preventative.

5.11 CULTIVATION IN GERMANY

ROLF FRANKE AND HANS-JÜRGEN HANNIG

5.11.1 INTRODUCTION

In Germany between 1930 and 1945, though chamomile flowers were collected, only about 6 ha were cultivated; the drug requirement was about 1000 tons [5]. In 1955 the main regions of origin of the drug *Chamomillae flos* for Germany were Germany (mainly Saxony and Franconia), Hungary, the Balkan countries, the USSR, the CSR, Yugoslavia, Belgium, France, and Spain [2, 3]. Varieties or origins used up to the 1980s were "Holsteiner Marschenkamille" (Holstein Marsh Chamomile), "Quedlinburger Großblütige Kamille" (Quedlinburg large-flowered chamomile), and "Erfurter Kleinblütige Kamille" (Erfurt small-flowered chamomile) [1]. Only the tetraploid variety "Bode-gold" brought the breakthrough for the use of cultivated forms in Germany in 1962.

In the meantime a clear shift took place in the main cultivation areas. Today, the main suppliers are Argentina, Egypt, Hungary, Poland, and the Balkan countries; the major quantity comes from Argentina (with decreasing quantities since 1995) (Table 5.11.1). Meanwhile, a big part is imported from Egypt as well.

TABLE 5.11.1
Chamomile Import to Germany in the Years 1980–1987 (in tons) [6]

	1980	1981	1982	1983	1984	1985	1986	1987
Argentina	2162	1730	1849	2334	1673	1807	962*	2020
Egypt	605	594	723	741	773	760	1537	989
Others	369	424	522	850	924	623	447	1045

* Decrease in production in 1986 was due to climatic influences.

FIGURE 5.11.1 Chamomile harvest in Thuringia.

In Germany as well, cultivation was extended after 1975. Today exclusively cultivated varieties are grown. The biggest part of these are tetraploid varieties with high (–)α-bisabolol content.

Meanwhile, the main culture of the medicinal plants and spices cultivated in Germany is True chamomile. The annual cultivation area amounts to more than 800 ha, more than 700 ha thereof are to be found in Thuringia. Other German cultivation areas are Saxony (about 40 ha organically grown chamomile) and Hesse (approximately 100 ha). The total area of organically grown chamomile has increased continuously during recent years and comprises about 100 ha at present. A comprehensive investigation as well as an evaluation of procedures and equipment of the production of chamomile in Saxony and Thuringia is discussed by Herold et al. [4] and Seitz [7].

The cultivation of chamomile in Germany is still increasing slightly and is limited in principle by the availability of harvesting equipment (in 2003 and 2004, over 800 ha alone in Thuringia).

There are two reasons for the increasing cultivation of chamomile in Germany:

- Cultivation of high-quality, protected chamomile varieties with a special profile of ingredients for the production of pharmaceutical products
- The increasing requirement of product safety concerning undesired residues of pesticides and heavy metals

5.11.2 CULTIVATION

Chamomile cultivation in Germany and similar climatic regions is normally effected through direct seeding in late summer (September) and spring (March/April). The splitting of the seeding time in autumn and spring reduces the risk of emergence that is inevitable due to the low thousand seed mass of the chamomile seeds and the characteristic properties of a plant that germinates in light. Furthermore, the splitting of the seeding time leads to different maturity dates, so that existing picking techniques and drying capacity can be used reasonably over a longer period of time.

The seeding is effected with special drilling machines for fine seeds on a weed-free recompacted and well-rolled soil.

Dependent on the available soil humidity and temperature, germination takes places within 1–2 weeks. In general, row distances of about 25 cm are preferred, and the seed density is approximately 2.0–2.5 kg per ha.

The nutrient need of chamomile is not very high and it also flourishes on moderately supplied soils.

Increased quantities of nitrogen fertilizer lead to undesired additional herb growth and a delayed formation of flowers. For the fertilizer need, 40 kg N, 50 kg P_2O_5, and 100 kg K_2O can be considered as reference points.

In case of well-supplied soils, an additional fertilizing can be completely set aside.

Weed regulation is effected by soil herbicides that on the one hand are worked into the soil during preparation of the seed bed, and on the other hand via machine hoeing after emergence.

TABLE 5.11.2
Cultivation of Chamomile in Germany 1992–2003 (in ha)

	1992	1993	1994	1995	1996	1997	1998	1999	2000	2001	2002	2003
Total	260		445	590	650	703	796	761	723	824*	882*	936*
Among them in Thuringia					600	653	673	620	633	674	723	815

*Among them approx. 100 ha organic.

FIGURE 5.11.2 Chamomile harvest with the Linz III picking machine.

FIGURE 5.11.3 Complete picking of the flowering horizon with the Linz III harvester.

Post-emergence herbicides are available only on a very limited scale and are legally regulated in the individual cultivating countries.

Chamomile has a slow development at the beginning and forms an opulently dense population in a late stadium, that is capable of effectively suppressing weeds.

Besides downy mildew (*Plasmopara leptosperma* [de Bary] Skalicky), in central European cultivation areas there are almost no other diseases or parasites that could cause economically relevant damage.

Harvest is mostly effected with special picking machines.

Harvest time is indicated by the beginning of full blossom. In the upper flower horizon, at least 80% of the flowers should have fully blossomed out, i.e., the white ligulate flowers should stand horizontally or bend slightly downward.

The harvested material is highly endangered by fermentation. Therefore, the period of time between harvest and the beginning of drying should be limited to a maximum of 2 to 3 hours. Any rising pressure due to high storage levels is to be prevented.

Immediately before drying, a sorting of herb parts that have entered the product due to the mechanical picking often takes place. This is frequently achieved by double-sided countercurrent drum sieves.

Drying is realized with belt-drying plants or grating dryers at a maximum product temperature of 40°C.

Depending on the number of possible picking procedures, a harvest of 350–600 kg drug per ha is obtained.

5.11.3 SEED PRODUCTION

Special varieties with a high content in (–)α-bisabolol are used, especially in the German chamomile cultivation. These varieties in many cases are company property and the corresponding seeds cannot be bought commercially. Therefore, seed production is mainly realized by the companies themselves.

For this separate propagation, special seed-propagation areas are sown and cultivated until full ripeness of the flowers. The distance between the individual plants is often very wide and the cultivation is treated analogously to a root crop in order to assure that seeds are only obtained from one special plant.

With the selection of the propagation areas it is important to observe that natural weed pressure of wild chamomile can be excluded.

The fully flourishing chamomile herb is mostly harvested by cutting and drying on grating driers with cold air. This leads to an after ripening of immature seeds, and a good yield of chamomile seed can be obtained through posterior threshing.

Cleaning is effected through the classical methods of seed preparation. The results are seed qualities with a germinating power of 80–85% with a yield of 100–200 kg seeds per ha.

REFERENCES

1. Ebert, K. (1982) *Arznei-und Gewürzpflanzen. Ein Leitfaden für Anbau und Sammlung.* 2nd ed., Wiss. Verlagsgesell., Stuttgart, Germany, 221 pp.
2. Freudenberg, G. and Caesar, R. (1954) *Arzneipflanzen. Anbau und Verwertung.* Parey, Berlin, Hamburg, 204 pp.
3. Heeger, E.F. (1956) *Handbuch des Arznei-und Gewürzpflanzenanbaus. Drogengewinnung.* Deutscher Bauernverlag, Berlin, 775 pp.
4. Herold, M., Pank, F., Menzel, E., Kaltofen, H., Loogk, E., Rust, H. (1989) Verfahrens technische Entwicklungen zum Anbau von *Chamomille recutita* (L.) Rauschert und *Calendula officinalis* L. für die Gewinnung von Blütendrogen. *Drogenreport*, **2**, 2, 43–62.
5. Jaretzky, R. (1948) *Taschenbuch für den Heilpflanzenanbau.* Verlag Dr. Roland Schmiedel, Stuttgart, Germany, 33 pp.
6. Kirsch, C. (1990) Kamillenanbau in Argentinien. *Dragoco Report*, **2**, 67–75.
7. Seitz, P. (1987) Arznei-und Gewürzpflanzen in der DDR. *Deutsch. Gartenbau* **51**, 3040–3046.

6 Abiotic and Biotic Stress Affecting the Common Chamomile (*Matricaria recutita* L.) and the Roman Chamomile (*Chamaemelum nobile* L. syn. *Anthemis nobilis* L.)

Andreas Plescher

CONTENTS

Numerous abiotic and biotic stress factors can affect plant growth and survival, and the quantity and quality of drug yield. This chapter presents some of the most common diseases and stress agents of chamomile plants. Additional diseases may affect the two chamomile species, and the frequency and type of disease will vary with the local climate.

6.1 ABIOTIC DAMAGE

6.1.1 Hail

Leaves, buds, and flower heads can get scratched or knocked off by hailstones. At first the damaged tissue turns lighter in color. Later on, necrosis is observed. Shoot tissue above the point of damage will start wilting. Shoot tips may collapse.

6.1.2 Excessive Soil Moisture, Waterlogging

High soil moisture caused by intense rainfall or poor drainage may restrict chamomile growth. Especially young plants, but also old ones, will turn yellow, wilt, collapse, and die. There is also an increased risk of root rots (cf. Section 6.4).

6.1.3 Nutrition

Although both chamomile species do prefer certain soils, especially those with high calcium concentrations, stress caused by deficiencies or an excess of nutrients has not been described.

6.1.4 Low Temperature

In general, very tiny chamomile plants are sensitive to extended periods of freezing temperatures in the spring. Older plants of the common chamomile, at a stage of having six to eight leaves, are already remarkably resistant to frost periods, whereas the Roman chamomile remains sensitive to freezing temperatures in springtime or early in the autumn. Flowers may be affected and the ray florets will turn brown.

6.1.5 Drought

Periods of drought may lead to the loss of the second and the following harvests. Depending on the soil structure-related water supply, plants are likely to get scorched after the first harvest and may die.

6.1.6 Herbicides

Herbicide treatment can result in abnormal or restricted growth associated with irregular tissue bleaching and brown necrotic patches on the leaves. Bent, crooked, or twisted shoots with increased internodal growth are indicative of hormone-containing weed killers. However, similar symptoms are caused by mollicute infection or bug (*Heteroptera*) damage.

6.2 VIRUSES

Virus infection has so far not been reported to affect the quantity or quality of chamomile yield. However, the common chamomile is one of the host plants for the lettuce big vein virus (LBV-V) and the cabbage black ring virus (CBR-V), but without showing any visible symptoms. Common chamomile thereby acts as an important reservoir for these viruses, which are transmitted by aphids.

6.3 MOLLICUTES

Infection of preflowering plants by mollicutes may cause reduced elongation of the shoot tip internodes. Flower buds will show the same symptoms in combination with abnormal greening. Flower malformations ("double and triple flowers") and fasciations may occur. Secondary shoots or numerous small leaves develop at these points. Similar symptoms can result from herbicide treatment or bug (*Heteroptera*) feeding.

6.4 FUNGI

6.4.1 *FUSARIUM* SPP.

Fusarium infection is a frequent reason for the inhibition of plant growth. Plants get stunted, become chlorotic, lose turgidity, and turn yellow. The base of the stems turns dark brown to black in color and sometimes appears to be girdled. In addition, longitudinal cracks may appear at the base. The roots turn dark and decay. Chamomile wilts and basal stem rots are primarily caused by *Fusarium culmorum* (W. G. SM.) SACC. Similar symptoms are observed in conditions of excessive soil moisture and afterdamage by stem-feeding herbivores.

6.4.2 POWDERY MILDEW (*ERYSIPHE CICHORACEARUM* D.C. EX MERAT AND *E. POLYPHAGA* HAMM.)

Infection is characterized by the appearance of white powdery patches of fungal growth. Very soon, the entire plant is covered by the powdery mildew growth. Newly emerging flowers are dwarfed. Leaves fall dry, starting from their tips. In the older areas of infection, tiny pinhead-sized yellow-brown cleistothecia become visible, which later turn into black spots.

6.4.3 DOWNY MILDEW (*PERONOSPORA RADII* DE BY., SYN. *PERONOSPORA DANICA* GÄUM.; *PLASMOPARA LEPTOSPERMA* [DE BY.] SKAL.; SYN. *PERONOSPORA LEPTOSPERMA* [DE BY.] GÄUM.)

Infection first results in bleached patches on the leaves, from where the fungus starts to grow as a white lawn with pinhead-like structures on both sides of the leaves. Leaves first turn yellow, then brown, and die.

6.4.4 CHAMOMILE RUST (*PUCCINIA MATRICARIAE* SYD., SYN. *PUCCINIA TANACETI* D.C. A. L.)

Plants cultivated in temperate to cold climates may get infected by the chamomile rust. Pale-brown rusty pustules (uredia) form on the leaves and stems, whereas the black powdery telia are only found on the stems.

6.4.5 White Rust (*Albugo tragopogonis* [Pers.] Schroet.)

White rust infection occurs on leaves, shoots, and buds. It is characterized by pustules that change their color from pale yellow into white, and burst, releasing their lime-like contents. White rust infection has so far only been observed on Roman chamomile.

6.4.6 Leaf Spot Disease (*Stemphylium botryosum* Wallr.)

Stemphylium infection is likely to occur after long periods of wet weather and results in spherical light brown to grey or dark brown to black spots on leaves and shoots. The midribs collapse.

6.5 PLANT PARASITES

In southern Europe, broom rape (*Orobanche* sp.) has been recorded to parasitize common chamomile.

6.6 HERBIVORY

6.6.1 Chewing Herbivores on Roots and Stem Bases

Insects feed on the roots, the stem base, and the leaves close to the ground. Plant growth is restricted. Plants wilt and die prematurely. Some of the plants get completely detached from their roots and are easily pulled from the soil.

Yellowish-white beetle larvae, up to 6 cm long, with dark-colored abdominal segments (May bug larvae, *Melolontha* spp.; *Phyllopertha* spp., *Rhizotrogus* sp., etc.), 3–4 cm long, brown-grey legless fly larvae with fleshy abdominal segments (cranefly larvae; *Pales* spp., *Tipula* spp.), or about 2.5 cm long, thin beetle larvae having a rigid cuticle (wire worms; *Agriotes* spp., *Athous niger* L., *Melanotus brunnipes* GERM.) can be found in the vicinity of the plants that have been affected.

Mainly at night, earth-colored, grey, or greenish-grey lepidopteran larvae (*Scotia* [*Agrotis*] spp.) feed on the tissues close to the ground. They spend the day curled up in the soil. Occasionally, the small, yellowish white larvae of the root fly (*Delia* [*Phorbia*] spp.) may cause the plant to die. In warmer climates, the mole cricket (*Gryllotalpa vulgaris* LATR. = *Gryllotalpa gryllotalpa* L.) and several millipede species (*Blaniulus guttulatus* [BOSC.], *Cylindroiulus teutonicus* [POCOCK], etc.) can seriously damage the roots. In humus-rich soil with a high proportion of decaying plant tissues (e.g., in gardener's substrates for the propagation of Roman chamomile), the larvae of the St. Mark's fly (*Bibio* spp.) may attack the plants.

6.6.2 Gall Formation on Roots

Chamomile plants are attacked by the northern root-knot nematode *Meloidogyne hapla* CHIT-WOOD. At the point of infection roots swell and develop spherical or spindle-shaped galls in which the females (up to 1 mm long and 0.5 mm wide, pear-shaped) can be found. Plant growth is inhibited, and infested plants are more sensitive to drought than healthy plants.

6.6.3 Chewing Herbivores on Leaves and Shoots

Various insects and their larvae are known to cause more or less severe damage to chamomile plants by feeding on the leaves and shoots. Some lepidopteran larvae (such as the owlet moth; *Cucullia tanaceti* SCHIFF, but also other *Cucullia* species) damage leaves by skeletal feeding. *Phalonia implicata* WCK. larvae (10 mm long, pale-yellow bodies with brown heads and a yellow dorsal neck plate) mainly feed on the upper plant parts and spin their webs around them.

In areas of high humidity, snails (*Helix* spp., *Arianta* spp.) and various types of slugs (*Arion* spp., *Deroceras reticulatum* MÜLL.) can cause considerable damage to the plant tissues near the ground as evidenced by their slime tracks.

6.6.4 SAP SUCKING ON LEAVES AND SHOOTS

Various aphid species live on the leaves and shoots or at the tips of the plants. The leaf blades curl up from the tip or the sides to their base, turn yellow, and finally brown.

Whereas attack by the green aphid *Cerosipha gossypii* HB. has mainly been recorded in warm climates, the black aphid *Aphis fabae* SCOP. and the green aphids *Myzus persicae* SULZ. and *Brachycaudus* spp. are found on chamomile species all over the world.

Sucking by cicadas (*Cicadinae*, for example, *Eupteryx atropunctata* GOEZE, *Empoasca pteridis* DAHLB., *E. flavescens* F., and *Chlorita viridula* FALL.) can initially be recognized from white spots on the leaves. Severe attack results in leaf fading and death. Patches of tissue first turn dark green or brown and then die. Larvae usually stay on the underside of the leaves.

Leaf and stem deformations and growth abnormalities result from sucking by various types of bugs (*Heteroptera*, for example, *Lygus lucorum* MEY. D., *L. pubescens* REUT., *Exolygus pratensis* L., *Plagioganthus chrysanthemi* WOLFF., *Adelphocoris lineolatus* GOEZE, and *Calocoris norvegicus* GMEL.). If plants are cultivated in a greenhouse, the white fly, *Trialeurodes vaporariorum* WESTW., and its oval-shaped larvae, living on the underside of chamomile leaves, cause considerable damage.

6.6.5 LEAF AND STEM MINING

Shoot weevil *Ceutorhynchus rugulosus* HERBST larvae (legless with brown head capsule and whitish-yellow body) mine the central stem pith of chamomile plants. As a consequence, the lower stem parts first turn red and then brown, while the leaves turn yellow and wilt. Flowers are degenerate, and flowers and stems easily collapse. The adult shoot weevil is 2.0 to 2.5 mm long and grey to chocolate-brown in color. The tiny larvae of the shrew weevil *Apion confluens* KHY. similarly mine the stem, but also feed small holes into chamomile leaves.

Hardly detectable are the mining structures of the grey-brown larvae from the leaf miner species *Phytomyza atricornis* MEIG., *Phytomyza matricariae* HAND., *Liromyza strigata* MEIG., and *Typetha zoe* MEIG.

6.6.6 HERBIVORES MINING AND CHEWING ON FLOWERS

Beetles of the genus *Meligethes* (1.5 to 2.7 mm in length, metallic green, shiny blue-grey to blue-purple, or nonshiny black in color) generally feed only on mature pollen already released by the anthers and rarely damage single florets and their anthers.

Severely reduced drug quality is caused by different insect larvae mining within the receptacle. Larval feeding tunnels form more or less circular horizontal patterns in the tissue. The above growing disc florets are the first to wither. Later on the entire flower head turns brown and larvae move further down into the receptacle.

The head and abdominal end of *Olibrus aenaeus* FABR. larvae are dark in color. Larvae have their thoracic legs developed. Beetles emerge from the early summer on. They are 1.8 to 2.5 mm long and black, sometimes with a shining metallic-green appearance. The same type of damage is caused by the legless dark-headed larvae of the weevil *Pseudostyphlus pilumnus* GYLL. Adults are brown to brown-black with grey-white scales and 2.5 to 3.3 mm in length. In rare cases, larvae of the weevil *Ceutorhynchus rugulosus* HERBST (see Section 6.6.5) also mine the receptacle. Finally, larvae of the blossom boring fly *Trypanea stellata* FRUNSLEY cause a very similar type of damage to chamomile flowers. Larvae are legless with a light-colored head capsule.

Flower mining does not significantly reduce allover yield, but affects flower color and integrity and thereby drug quality.

6.6.7 SAP SUCKING ON FLOWERS

A large number of thrips species feed on chamomile flower heads. They are whitish to yellow-brown, slender insects (from 0.5 mm to a maximum length of 1.5 mm) with short legs, and sometimes fringed wings. Different species of grass thrips suck between the tubular florets and thereby impair flower head integrity. Single tubular flowers wither and turn brown. *Thrips physapus* L. and *Thrips tabaci* LIND. are frequently observed on chamomile. Again, this type of damage mainly reduces drug quality not yield.

6.6.8 OTHER QUALITY DETERIORATIONS CAUSED BY INSECTS

Various ladybug (*Coleoptera*: *Coccinellidae*) species colonize aphid-infested chamomile plants. Since their brightly colored wing covers are difficult to remove from the harvested plant material, chamomile marked value can be significantly reduced.

6.6.9 STOCK PESTS

A wide range of pests feeds on the dried chamomile flowers, depending on the climate of the country of origin and the storage conditions. In Central and South America, the tobacco beetle *Lasioderma serricorne* F. (*Coleoptera*) has been recorded. In South America the hay or cocoa moth *Ephestia elutella* HB (*Microlepidoptera*) infests the chamomile flowers. In central Europe the dry fruit moth *Ploida interpunctella* HB (*Microlepidoptera*) and the carpet moth *Anthrenus verbasci* L. (*Coleoptera*) are known to feed on the stocks.

7 Raw Plant Material and Postharvest Technology

Horst Böttcher and Ingeborg Günther

CONTENTS

7.1 POSTHARVEST PHYSIOLOGICAL RESPONSE

Freshly harvested crops of chamomile are live plant products or parts of them. They are characterized by a water content of about 80% and also by high metabolism. Therefore, the postharvest period is of decisive importance for the maintenance of excellent external and internal quality traits; it is the key for ensuring a stable and reliable quality of raw and processed chamomile products with a high standard of therapeutic and medical effects. However, our knowledge about what is happening between harvest (i.e., mowing and gathering in the field) and subsequent drying or further processing is still very poor. The physiological processes occurring in fresh horticultural crops, during the postharvest period in general, were described recently by References 9, 10, and 11 and more profoundly by Reference 8, but for medicinal and aromatic plants, and for chamomile herbs and flowers in particular, little information can be found. Recently, References 4 and 6 reported first results on the respiration activity of chamomile flowers.

The main loss factors responsible for the rapid decline of the biological quality of medicinal herb crops are characterized in Table 7.1 by Böttcher and Günther [4].

In the postharvest period respiration, senescence, transpiration, ripening, and changes in the biochemical constituents caused by the secondary metabolism were found to take place in the live product. The relationships between these processes are demonstrated in Figure 7.1.

The design of the technical lines reflects the importance of the single processes with regard to external and medical value. The external quality, important for use as tea components or powder, is largely influenced by transpirating, senescing, and ripening and by the development of harmful microorganisms. Medical values, however, are marked mainly by senescence and ripening. It is typical for medicinal crops that the loss factor "respiration" will not in all cases react directly, but mainly via transpiration, wilting, and secondary metabolism responses (Figure 7.1).

TABLE 7.1
Physiological Processes in Freshly Harvested Medicinal Plants and Their Effects [3, 4]

Respiration
→ Heating of the stacked crop
→ Extreme heating up to spontaneous combustion
→ Fermentation

Senescence processes
→ Chlorophyll degradation
→ Leaf siccation
→ Changes in the quantity of special constituents

Transpiration
→ Wilting and shrivelling
→ Changes in quantity of constituents

Microbiological contamination and spoilage
→ Lesions and bruises
→ Partial heating
→ Rotting losses
→ Influence of special constituents

Injuries and damages
→ Mechanical abrasion
→ Disintegration of flowers and buds
→ Quality shifts

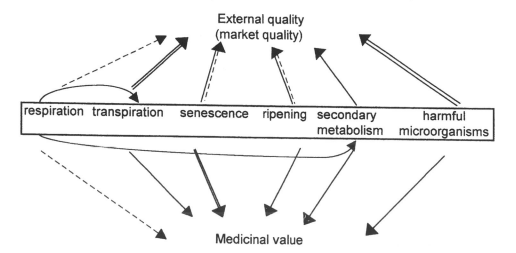

FIGURE 7.1   Relationship between physiological loss factors and quality of the chamomile product [3].

The reactions of the flowers are of special interest, because they are used often and are a valuable phytopharmaca and because they contain a high level of essential ingredients. So chamomile flowers are well known in general for quickly becoming perishable.

7.1.1 RESPIRATION: THE PHYSIOLOGICAL FOCUS

Respiration is a very stringent process in living cells of freshly harvested chamomile crops. It mediates the release of chemically bound energy through the breakdown of carbon components and the formation of carbon skeletons necessary for maintenance and synthetic reactions after harvest. A secondary result of respiration is the release of energy as **respiration heat** expressed in terms of energy in $W \ t^{-1}$ [watt. ton^{-1}]. Respiration is very important under postharvest conditions; therefore, we have to check it and abduct the heat from the stacks of the stored product.

On the other hand, the rate of respiration is also an indicator of the total rate of metabolism in plants, plant parts like flowers, or herbs.

Crops of freshly gathered chamomile flowers, variety "Bodegold," picked in the full-flowering stage in the field using the gathering harvester LINZ 3, have at free-flow measuring conditions of 10°C an unexpected high mean value of respiration rate of $999 \pm 134 \ W \ t^{-1}$ [7]. In all trials the plants grew under representative agrotechnical and growing conditions in the field. Their activity was much higher than that of horticultural plants with well-known high values [2, 3]. In fresh chamomile flowers the high mean value exceeded parsley by 4.3-fold ($235 \ W \ t^{-1}$) and broccoli by 2.9-fold ($350 \ W \ t^{-1}$). It is also much higher than that of other medicinal and aromatic herbs harvested at the beginning or in the half-blooming stage: 1.6-fold for marjoram ($632 \ W \ t^{-1}$), 1.4-fold for savory ($696 \ W \ t^{-1}$), or 1.6-fold for sage ($615 \ W \ t^{-1}$) [5, 6].

The high activity may be explained by the fact that harvest occurs in the full-flowering stage at the end of the generative phase of plant development, which takes place in June, when sunny and warm weather prevails in Germany.

Although each year the crops grow under equal agrotechnical and site conditions, a definite significant influence by the annual weather and growing situation can be observed ($p = 0.00357$) (Figure 7.2): 1994: $\bar{x} = 1141 \ W \ t^{-1}$; 1995: $\bar{x} = 878 \ W \ t^{-1}$; 1996: $\bar{x} = 979 \ W \ t^{-1}$. There was measured a difference in the rate between autumn (a) and spring (s) sown crops (a: $\bar{x} = 958$; s: $\bar{x} = 1040 \ W \ t^{-1}$). But no difference could be detected between respiration and the harvest date as well as climatic situation at the harvest date.

With **rising product temperature**, the respiration rate in each test series with chamomile increased considerably ($p < 0.001$) (Figure 7.2). In single cases, up to 4500 to 5000 $W \ t^{-1}$ were reached at the 30°C level. The following mean values in the above-mentioned trials of Böttcher, Günther, and Franke [7] were determined for 20°C $2438 \pm 289 \ W \ t^{-1}$; for 30°C $4552 \pm 570 \ W \ t^{-1}$.

The extent of the **temperature increase** in steps of 10 K (Q_{10}) was ascertained for:

10 to 20°C: $Q_{10} = 2.47$ (mean value of 6 series)

20 to 30°C: $Q_{10} = 2.00$ (mean value of 6 series).

Despite the primarily high respiration rate of chamomile flowers these parameters between 10 and 20°C were in *full correspondence with van't Hoff's rule* and confirm in this range also its validity for freshly harvested chamomile. A heat-caused depression at the level of 30°C was definitely obvious. [7].

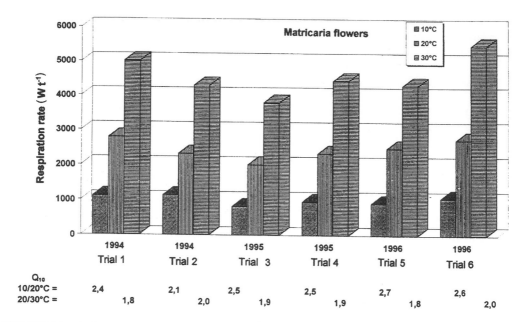

FIGURE 7.2 Influence of temperature on the respiration rate of freshly harvested chamomile flowers [7]. (Reprinted from *Postharvest Biology and Technology* **22**, Böttcher, H., Günther, I., Franke, R., Warnstorff, K., Physiological postharvest responses of Matricaria flowers (*Matricaria recutita* L.), pp. 39–51, Copyright (2001), with permission from Elsevier.)

So far, no measured values are known for the respiration rate of chamomile herbs in total and herbs with flowers.

Because of this very high respiration activity measured for chamomile flowers, it is urgent to take this into account, especially for the postharvest technological treatments like ventilating, cooling, or drying.

7.1.2 INFLUENCE OF SENESCENCE ON RESPIRATION RATE

The high respiration intensity of chamomile flowers during postharvest storage was unexpectedly stable, irrespectively of the actual storage temperature, when calculated in W t^{-1} (Figures 7.3, 7.4, 7.5). Its decline varied in each series, but its course clearly followed the regression function:

$$\hat{y} = \alpha_1 + \beta_1 \cdot e^{-c\,(x-z)}. \ (c < 0) \qquad \text{(Figures 7.3, 7.4, 7.5)}$$

where $\hat{y}$ = respiration rate in W t^{-1}

   x = postharvest time in hours

   α_1 = remaining respiration rate under senescence conditions

   β_1 = variable respiration rate during postharvest senescence

   c = parameter for the intensity of degression during the postharvest period

   z = time from harvest to the state of equilibrium in the respiration-measuring equipment.

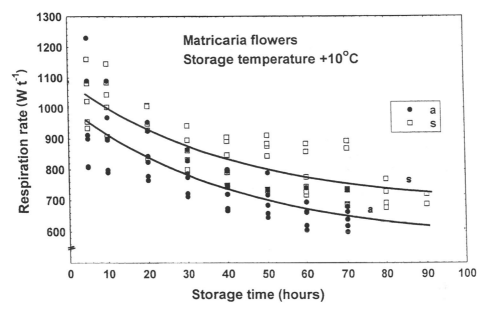

FIGURE 7.3 Influence of senescence on the respiration rate of chamomile flowers at postharvest conditions of 10°C [7]. (Reprinted from *Postharvest Biology and Technology* **22**, Böttcher, H., Günther, I., Franke, R., Warnstorff, K., Physiological postharvest responses of Matricaria flowers (*Matricaria recutita* L.), pp. 39–51, Copyright (2001), with permission from Elsevier.)

Regression: a (autumn sown): $\hat{y} = 566.54 + 390.49 \cdot e^{-0.023570 \cdot (x-5)}$ $R^2 = 79.6\%$

s (spring sown): $\hat{y} = 682.80 + 360.00 \cdot e^{-0.024707 (x-5)}$ $R^2 = 82.4\%$

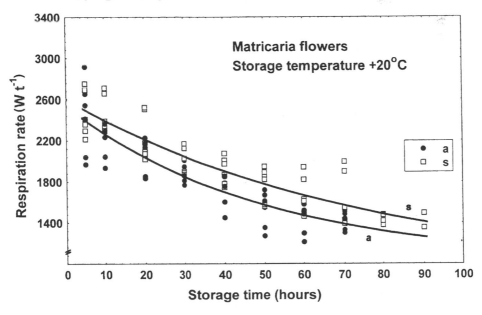

FIGURE 7.4 Influence of senescence on the respiration rate of chamomile flowers at postharvest conditions of 20°C [7]. (Reprinted from *Postharvest Biology and Technology* **22**, Böttcher, H., Günther, I., Franke, R., Warnstorff, K., Physiological postharvest responses of Matricaria flowers (*Matricaria recutita* L.), pp. 39–51. Copyright (2001), with permission from Elsevier.)

Regression: a (autumn sown): $\hat{y} = 1003.20 + 1391.11\ e^{-0.019800 (x-5)}$ $R^2 = 89.0\%$

s (spring sown): $\hat{y} = 857.68 + 1638.07 \cdot e^{-0.012814 (x-5)}$ $R^2 = 89.5\%$

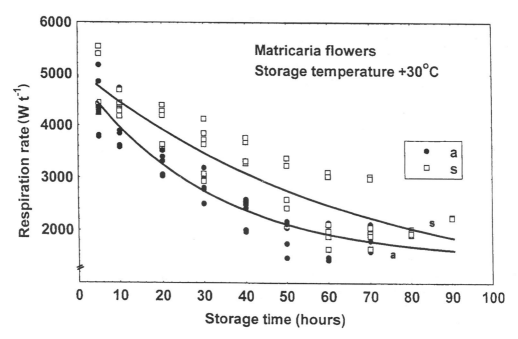

FIGURE 7.5 Influence of senescence on the respiration rate of chamomile flowers at postharvest conditions of 30°C [7]. (Reprinted from *Postharvest Biology and Technology* **22**, Böttcher, H., Günther, I., Franke, R., Warnstorff, K., Physiological postharvest responses of Matricaria flowers (*Matricaria recutita* L.), pp. 39–51, Copyright (2001), with permission from Elsevier.)

Regression: a (autumn sown) $\hat{y} = 1452.63 + 2973.04 \cdot e^{-0.033322\,(x-5)}$ $R^2 = 94.2\%$

s (spring sown) $\hat{y} = 874.04 + 3884.37 \cdot e^{-0.016079(x-5)}$ $R^2 = 89.9\%$

In the chosen postharvest period of 72 to 80 hours in the trials (in practice normally not so extended) Böttcher, Günther, and Franke [7] recorded a mean decrease in respiration for the different postharvest conditions:

10 ± 0.2°C, ~ 98% relative air humidity:

> *delayed aging (senescence) due to low temperatures*

20 ± 0.1°C, ~ 95% relative air humidity:

> *normal process of aging (senescence)*

30 ± 0.2°C ~ 98 ...92% relative air humidity:

> *accelerated aging due to increased physiological temperatures.*

The extent of senescence-related decrease during the postharvest period of 80 hours amounted in these trials:

at 10°C to absolutely   313.7 W · t⁻¹ or relatively 31.4% of the initial value

at 20°C to absolutely   1,043.8 W · t⁻¹ or relatively 42.7% of the initial value

at 30°C to absolutely   2,725.1 W · t⁻¹ or relatively 59.3% of the initial value

It turned out that chamomile flowers have a different response to the *sowing date*. At all tested temperature levels, spring-sown crops showed a significantly higher rate at harvest than autumn-sown crops: at 10°C +85.8 W t^{-1} (+8.96%), at 20°C +101.5 W t^{-1} (+4.24%), at 30°C +332.7 W t^{-1} (+7.52%). This results from the fact that flowers of spring-sown chamomile were harvested 11 to 25 days later in June than autumn-sown stands. This means that they were exposed to the warm summer conditions in June for 11 to 25 additional days during the last part of the growth and development period in comparison to the autumn-sown stands.

This difference, which was statistically significant (p < 0.000), was also maintained during postharvest storage. Therefore, it was necessary to make separate statistic estimations and graphics for autumn- and spring-sown crops in Figures 7.2, 7.3, 7.4, and 7.5.

On the other hand, the sowing date had a different influence on the course of respiration, especially the radius of curvature, characterized by factor c in the equation, also at the different temperature levels. At conditions of 10°C this factor was nearly equal (autumn sown (a) = –0.02357; spring sown (s) = –0.024707). However, at temperature steps of 20 and 30°C an increasing faster senescence-caused respiration decline in autumn-sown crop was observed: at 20°C a = –0.01980; s = –0.012814. At 30°C it dropped even twice as fast: a = –0.033322; s = –0.016079 [7]. This proves that at the optimal harvest date autumn-sown chamomile flowers are inevitably physiologically older and react with a stronger decline of the respiration rate than spring-sown plants.

The complex relationship between storage temperature, postharvest storage time, and respiration course is shown in Figure 7.6.

7.1.3 Changes in Quality Parameters

Besides an essential respiration rate, freshly gathered chamomile flowers also show a trend to high transpiration. The same applies to marjoram and sage [5, 6]. The mean *fresh matter losses* amounted to 1.83% during 24 h+ (1.2 to 1.9% in 24 h) under favorable postharvest conditions (10°C, φ ≈

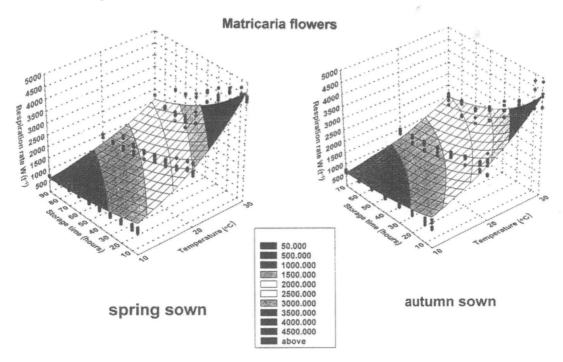

FIGURE 7.6 Complex influence of storage temperature and storage time on the respiration rate of spring- and autumn-sown chamomile flowers [7].

98%) and to 6.04%/24 h[+] under normal conditions (20˚C, $\varphi \approx 95\%$). When the temperature was increased to 30˚C ($\varphi \approx 92$ to 98%) the losses rose to $\bar{x} = 13.4\%/24$ h[+], resulting from higher release of transpiration energy by the ascertained high respiration rate. A depression of the product temperature by 0.6 to 1.5 K was also recorded [5, 7].

During 80 to 90 hours of postharvest time the very high released respiration heat led to pronounced *dry matter losses* in the product, dependent on the temperature, calculated on the base of dry matter at harvest: 10˚C 7.5%[+]; 20˚C 12.3%[+]; and 30˚C 15.8%[+] (mean values of six trials) (Figure 7.7) [7].

The external quality traits of the chamomile crops are very important for a great many purposes. They drop clearly in dependence on the storage temperature [7] and are marked by wilting and

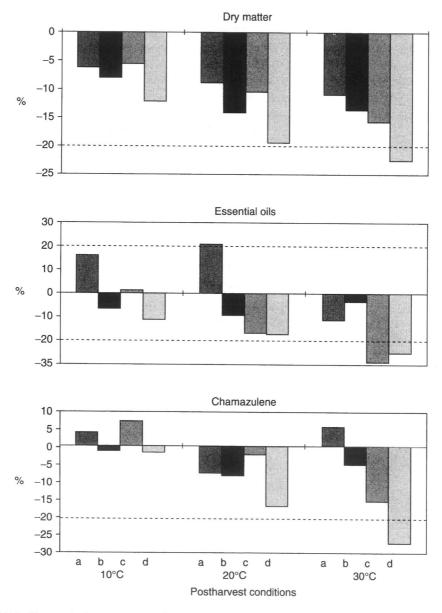

FIGURE 7.7 Changes in dry matter, essential oil, and chamazulene of chamomile flowers during a post-harvest period of 80 hours at different temperatures calculated of the base of balance sheets (a = 1995 autumn sown; b = 1995 spring sown; c = 1996 autumn sown; d = 1966 spring sown) [7]. (Reprinted from *Postharvest Biology and Technology* **22**, Böttcher, H., Günther, I., Franke, R., Warnstorff, K., Physiological postharvest responses of Matricaria flowers (*Matricaria recutita* L.), pp. 39–51, Copyright (2001), with permission from Elsevier.)

shrivelling flowers, by losing the bright white and yellow color impression of their blossoms, by degreening the fresh green parts of the crops (small stems, leaves, etc.). The stored crop made an increasingly faded impression, especially at 30°C, and to a smaller extent at 20°C. This is correlated with the continuing generative development of the inflorescences and can be recognized very distinctly by rising *disintegration of the blossoms* and the development of seeds in the inflorescences, particularly at 30°C.

A temperature of 10°C guaranteed crop material of fresh and bright quality even up to 70 hours after harvest. Acceptable chamomile quality after gathering could be supported at 20°C for 25 to 30 hours, but at 30°C for 15 to 20 hours only.

The *main active constituents* of chamomile flowers show different changes in their contents in the drug herb (mg/100 g) during the postharvest period, in most cases not very pronounced or significant. But the high respiration and transpiration rates of this crop material led to distinct losses in dry matter, so that changes of the constituents could only be discovered by calculations of balance sheets [1]. On the other hand, the plant material responded not uniformly in all trials, so that it is not advisable to consider the trials in total.

For some constituents, there are clear differences in the extent of the reactions in dependence to the prevailing microclimatical conditions during the growing and storing season.

Thus, the quantity of constituents, in general, turned out to have been relatively stable during the postharvest period, but there were some unfavorable reactions [7].

In three trials the *essential oils* showed at 10 and 20°C on the base of balance sheets small decreases up to 20%, and only in one trial did they rise to the same extent. At 30°C decreases were only on the level of up to 30% ($\bar{x}$ = –17.0%) (Figure 7.7). At 20°C a mean value of only –5.3% loss was estimated. So natural samples of chamomile flowers contained after a postharvest storage period of 80 days a quantity of +46 ml essential oils/100 g dried drug at 10°C and of +40 ml/100 g dried drug at 20°C in comparison to the quantity at harvest date. However, after conditions of 30°C there was a decline by 9.75 ml/100 g dried drug [7]. *Chamazulene* was marked by small decreases, rising with higher temperatures $\bar{x}$ at 10°C –1.95%; at 20°C -8.8%, and at 30°C –10.7% (Figure 7.7).

The valuable constituent (–)-α-*bisabolol* and its oxidation forms showed, on the contrary, clear reactions; however, this only at cooler and microclimatic conditions of 10°C: 60–70% of the amount of (–)-α-bisabolol as well as bisabololoxid A and B from the 1995 grown chamomile crops got lost (Figure 7.8).

Autumn- and spring-sown crops have the same reaction. This fact is considered not as a result of oxidation of (–)-α-bisabolol, but as a result of temperature-involved changes in the secondary metabolism in dependence on the annual growing and developing conditions. In the other year and at temperatures of 20 and 30°C only small changes up to 10 to 20% in the bisabolol and -oxid quantities occurred (Figure 7.8). The same reaction was shown by *cis-EN-IN-dicyloether* at temperatures of 10°C and to a smaller extent at 20 °C ($\bar{x}$ at 10°C = –27.5%, at 20°C = –18.2%), but at 30°C the changes came only to $\bar{x}$ = –3.4% (Figure 7.9).

The flavonoid quantities of *apigenin-7-glycoside* on the base of balance sheets are characterized by small decreases of 10 to 35% in the single samples, apart from the reaction in the autumn-sown trial 1995. Altogether, the decreases were $\bar{x}$ at 10°C –7.7%, at 20°C –16.1%, and at 30°C –19.6% in relation to the quantity at the moment of gathering [7] (Figure 7.9).

Whether the determined changes of the single constituents may cause a shift on the therapeutic value of the drug cannot be decided yet. Temperatures of 20°C led to the smallest changes in the crops. Best external traits were obtained at 10°C.

7.1.4 CHARACTERIZATION OF RESPIRATORY ACTIVITY

The characteristic behavior of the respiration activity of freshly gathered chamomile flower crops was demonstrated in a nomogram (Figure 7.10). It is recommended for calculating ventilation, drying, and manufacturing processes where chamomile is involved or for the assessment of physiological activities.

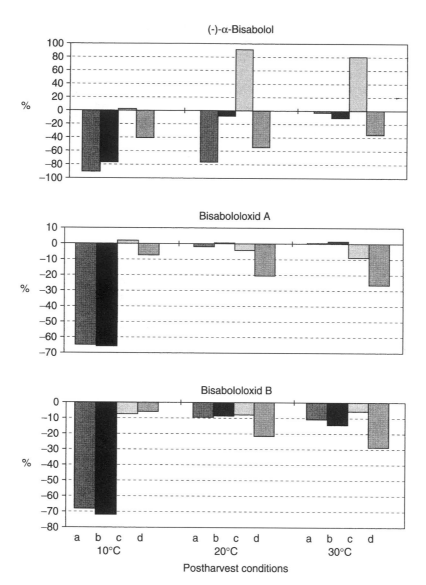

FIGURE 7.8 Changes in (-)-a-bisabolol and its oxide A and B of chamomile flowers during a postharvest period of 80 hours at different temperatures calculated on the base of balance sheets (characterization of a–d see Figure 7.7 [7]. (Reprinted from *Postharvest Biology and Technology* **22**, Böttcher, H., Günther, I., Franke, R., Warnstorff, K., Physiological postharvest responses of Matricaria flowers (*Matricaria recutita* L.), pp. 39–51, Copyright (2001), with permission from Elsevier.)

7.2 POSTHARVEST TECHNOLOGICAL TREATMENTS

The gathered flowers should be immediately transported from the field to a processing site in a shaded place. At first, equipment for postharvest **cleaning** was used. Fresh chamomile flower crops gathered by using high-capacity chamomile combines or by hand contain different quantities of leaves, stem parts of the herbs as well as of weed plants, and other *impurities*. These worthless contaminations should be removed very carefully as soon as possible, because they increase the energy needed for drying or steam distillation of the crops, prolong the drying time, and raise the

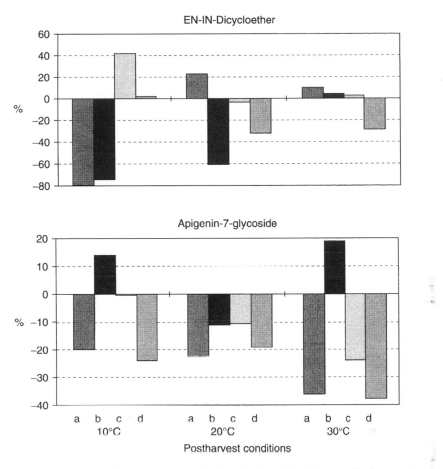

FIGURE 7.9 Changes in EN-IN-dicycloether and apigenine-7-glycoside of chamomile flowers during a post-harvest period of 80 hours at different temperatures calculated on the base of balance sheets (characterization of a–d see Figure 7.7) [7]. (Reprinted from *Postharvest Biology and Technology* **22**, Böttcher, H., Günther, I., Franke, R., Warnstorff, K., Physiological postharvest responses of Matricaria flowers (*Matricaria recutita* L.), pp. 39–51, Copyright (2001), with permission from Elsevier.)

temperatures in the stacks (younger parts of plants are mostly characterized by higher water content and higher respiration rates). All these processes are managed by the *actual product temperature* in the summer months, additionally. Therefore, the fresh flowers will go brown within a few hours, especially if they are stored in sacks or box pallets with insufficient ventilation. Besides, their microbiological contamination will rise drastically. The removal of the coarse impurities is an important contribution toward a better and more aesthetic quality of the drug.

Large, slowly rotating drums separate the bigger waste parts that are passing the drum, while the chamomile flowers fall through small, 5- to 12-mm mesh split openings of the cleaning drum cylinder. The latter works like a screening system. In the second step the flowers pass highly rotating small rubber rolls to pull all small particles like leaves and other inorganic impurities out of the crop mass. If it is not possible to start the cleaning immediately after gathering, it is essential, based on the above-demonstrated extremely high respiration intensity especially during high summer temperatures, to **ventilate** the crops **with outdoor air** to abduct the released respiration heat out of the stack. It is also necessary to take away the condensed water of the intensive transpiration

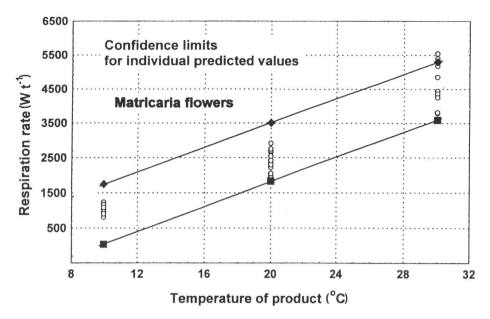

FIGURE 7.10 Respiration rate of mechanically gathered fresh chamomile flowers expected at different product temperatures during postharvest time

from the storing crop, to limit the microbiological developments, to cut off the losses of essential oils, and to prevent quality decreases. It is favorable to stack the crop only up to a height of 30 cm.

Chamomile herb crops do not require an exclusively intensive preparation before drying or steam distillation.

During all technological steps good agricultural practice, as recommended by the European Herbae Infusions Association (EHIA) in 1993, should be observed [12].

REFERENCES

1. Böttcher, H., 1986. Zur Problematik des Erfassens von Qualitätsveränderungen während der Lagerung von Gemüse. *Nahrung*, **30** S, pp. 723–728.
2. Böttcher, H. (Ed.), 1996. *Frischhaltung und Lagerung von Gemüse*. Ulmer-Verlag, Stuttgart, Germany, 252 pp.
3. Böttcher, H., 1998. Freshness of vegetables — a decisive precondition for sales prospect. *Zahradnictvi, Horticultural Science*. Praha. **25**, pp. 67–73.
4. Böttcher, H., Günther, I., 1995. Nachernteverhalten und Nacherntephysiologie von Arznei-und Gewürzpflanzen. *Herba Germanica* **3**, pp. 47–66.
5. Böttcher, H., Günther, I., Bauermann, U., 1999a. Physiological postharvest responses of marjoram (*Majorana hortensis Moench*). *Postharvest Biology and Technology*. **15,** pp. 41–52.
6. Böttcher, H., Günther, I., Warnstorff, K., 1999. Nicht-destruktive Bestimmung des Gasstoffwechsels zum Erfassen des Seneszenzverlaufes während der Nacherntezeit von Arznei-und Gewürzpflanzen. Deutsche Gesellschaft für Qualitätsforschung DGQ, XXXIV. Vortragstagung Zerstörungsfreie Qualitätsanalyse. 22–23 March 1999, Freising Weihenstephan. S, pp. 107–118.
7. Böttcher, H., Günther, I., Franke, R., Warnstorff, K., 2001. Physiological postharvest responses of Matricaria flowers (*Matricaria recutita L.*). *Postharvest Biology and Technology* **22**, pp. 39–51.
8. Cantwell, M.J., Reid, M.S., 1993. Postharvest physiology and handling of fresh culinary herbs. *J. Herbs, Spices, Med. Plants*. **1**, pp. 83–127.
9. Kays, St. J. (Ed.), 1991. *Postharvest Physiology of Perishable Products*. Van Nostrand-Reinhold, New York, 532 pp.

10. Lieberman, M. (Ed.), 1983. *Postharvest Physiology and Crop Preservation*. Plenum Press, New York and London, 572 pp.
11. Wills, R.B.H., McGlasson, W.B., Graham, D., Lee, T.H., Hall, E.G. (Eds.), 1989. *Postharvest* 3rd Edition. BSP Professional Books, Oxford, 174 pp.
12. Richtlinien für die gute landwirtschaftliche Praxis von Arznei- und Gewürzpflanzen, 1997. *Zeitschr. Arznei-Gewürzpfl.* **4**, pp. 202–206.

8 Processing of Raw Material

Horst Böttcher and Ingeborg Günther with cooperation of Reinhold Carle and Albert Heindl

CONTENTS

8.1   BASIC PHYSIOLOGICAL AND TECHNOLOGICAL PROPERTIES OF THE PRODUCT

After harvesting, chamomile is a very perishable product during storage. This applies equally for the harvested whole herbs and, to an even greater extent, for the flowers. The main deterioration phenomena are:

- Dropping of the external quality traits, especially the natural green color of the leaves and stems and the light colors of the flowers
- Loss of particularly valuable ingredients due to material conversions, wilting due to transpiration, and physically induced escape

- Increased formation of grit through the decay of the flowers and separation of the leaves from the herb, both triggered by aging (senescence) processes
- A rise in the microbiological contamination of the product to be dried through warming and heating, especially in combination with mechanical damages and pressure points
- A marked shortening of the subsequent storability of the dried product

The very high **respiration rate** of harvested chamomile, which reaches a mean value of 999 W t^{-1} for flowers at 10°C (Section 7.1.1), accelerates these changes and leads, without ventilation, to an extremely fast rise in the stack temperature. On the other hand, chamomile reacts very easily to various external factors. For this reason, chamomile flowers and herbs should be regarded as very **drying-sensitive** crops.

Rapid, sufficient lowering of the water content in the harvested crops is the most important and — besides extraction — most frequently used option for preservation to avoid these changes. However, the drying conditions in question also determine the quality and stability of the dry product to a large extent. These are characterized by:

- The maximal occurring product temperature during the drying process
- The dwell time of the crops in the hot air section
- The saturation deficit of water vapor in the drying medium (hot air), and also by the movement of air in the drying facility
- The coverage density (pile height and density) of the harvested product

The aim of drying must be to remove both the water that is physiologically bound in the harvested product and also the external moisture (precipitation, dew) in the shortest possible time, in order to *reduce the water content of the dry product to 8–10%.*

This will then cut off the ongoing respiration processes, the fermentative breakdown, and conversion reactions and the physical changes, since these can easily cause changes to the natural plant colorants (component of the external quality traits) and lead to a considerable loss of valuable active substances. The water content quoted of 8–10% for chamomile dried to a high quality is thus below the physiological water activity $a_w < 0.60$, which prevents the dried product from being affected even by the most xerophilic types of harmful microorganisms, the *Aspergillus* and *Penicillium* molds, during subsequent storage.

A sufficiently high drying temperature and a fast drying process have a decisive effect on the quality of the dry product. These factors are difficult to achieve with drying in the outdoor air (drying sheds, drying shelves, etc.), even if the chamomile is picked primarily in the months with the most favorable weather (June and July). It is thus necessary to improve the efficiency of air as the drying medium. This is done in practice either by adding thermal energy to the drying air (*heated air-drying*) or by lowering the moisture level in the drying medium through the use of a chilling machine such as a dehumidifier. The latter option, however, is less commercially viable in terms of both energy consumption and economics.

The **permitted temperature of the product being dried** is physically very limited, since increased evaporation of various components in the essential oils during the drying process can lead to an increased loss of the active substances or shifts in the spectrum of active substances. For chamomile flowers with particular, but also for the chamomile herb, which both naturally have a low level of resistance to transpiration with respect to water vaporization, this is a particularly high risk, requiring particular caution. The boiling points of the main components in the essential oils are not that low (chamazulene 160°C, bisabolol 121°C) [1], but in combination with the large quantities of water present in the freshly harvested plants, a further reduction in the partial steam pressure occurs in the resultant oil and water mixture, which thus lowers the vaporization to temperatures below 100°C, as Rinder and Bomme [4] proved for the conditions for water vapor distillation for plants containing essential oils. In addition, the particular thermolability of the covering membranes of the secretion containers for the essential oils are greater for chamomile than for other medicinal plants. This can cause considerable evaporation losses during drying, even at lower temperatures.

Thus Schilcher [5] attributes the rising losses that may occur particularly at high drying temperatures and a high air humidity (>60%) to "a type of micro water vapor distillation of the essential oil from the glandular scales," which primarily affects the low-boiling fractions (including farnesene).

Constant checks must be carried out to ensure that the **flower base** of the chamomile is **completely dried**, since it has a fairly high resistance to transpiration and thus takes longer to dry.

Basically, the aim should be to achieve short drying times in order to minimize any changes in color, odor, and tissue structure and reduce the increasing microbiological contamination and the loss of important active substances during drying. This applies particularly to outdoor air drying.

Smaller quantities of the harvested product can also be preserved using a well-designed **outdoor air-drying system**, making use of suitable outside air conditions. A good forced-air ventilation system through the chamomile flowers that are spread flat on hurdles is important to achieve sufficiently fast drying within 5 to 6 days.

But the **physiological characteristics** of the harvested chamomile also affect drying and quality [3]. One element that is extremely important is the crushing of the flowers (gritting, or crushing of the flower heads, which leads to the separation of the different parts of the heads) during drying. This crushing process is of most relevance for large flowers in which more than three quarters of all the tubular blossoms are open [2]. For tetraploid genotypes, values between 75 and 88% were measured, with values of 64% for diploid genotypes. If the flowers are plucked at the "medium mature" stage (second circle of tubular blossoms opened), the tendency to decay for tetraploids is only 14 to 27% and for diploids 11%. If they are picked even earlier, the small flowers (buds opened to first ring), there is hardly any decay (< 0.8%). In addition to the favorable influence of the ambient temperature, the tetraploid genotypes are another positive factor [2].

Harvested chamomile should be taken to drying *basically without any preliminary wilting*, as fresh as possible in order to guarantee that the dried medicinal product is of the highest possible quality. Drying has to be started within 2 hours after harvest, unless the stack is not ventilated (Section 7.2).

The *cleaning* and *maintenance* of the drying equipment should be such that microbiological contamination and pollution are avoided. During drying, the *Guidelines for a Good Agricultural Practice (GAP) of Medicinal and Aromatic Plants* [6] should be complied with at all stages.

REFERENCES

1. Gildemeister, E., Hoffmann, F. (1960) Die ätherischen Öle. Band IIIa. Akademie-Verlag, Berlin.
2. Letchamo, W. (1991) Vergleichende Untersuchngen über die nacherntetechnisch bedingten Einflüsse auf die Wirkstoffgehalte in der Droge bei Kamille-Genotypen. Drogenreport. Sonderausgabe zur Fachtagung in Erfurt, pp. 129–134.
3. Marquard, R., Kroth, E. (2001) *Anbau und Qualitätsanforderungen ausgewählter Arzneipflanzen*. Agrimedia Verlag, Bergen/Dumme.
4. Rinder, R., Bomme, U (1998) Wasserdampfdestillation ätherischer Öle aus frischen und angewelkten Pflanzen. Bayerische Landesanstalt für Bodenkultur und Pflanzenbau, Freising-München, pp. 1–12.
5. Schilcher, H. (1987) *Die Kamille — Handbuch für Ärzte, Apotheker und andere Naturwissenschaftler*. Wissenschaftl. Verlagsgesellschaft, Stuttgart, Germany.
6. O.V. (1998) *Guidelines for Good Agricultural Practice (GAP) of Medicinal and Aromatic Plants*. Z. Arznei-und Gewürzpflanzen, **3**, S. 166–174.

8.2 DRYING OF CHAMOMILE FLOWERS (*MATRICARIA RECUTITA* L.)

ALBERT HEINDL

8.2.1 PROCESSING OF HARVESTED PRODUCT BEFORE DRYING

Before drying, the mixture of herbs and flowers is fed to a double-drum sieve to separate stems and other undesired matter (like stones or weed plants) from flowers. The inner sieve drum, made of perforated plate, has a hole diameter of 25 mm, the outer sieve drum a hole diameter of 20 mm [15]. The flowers and small stems fall down to a roller course with clear span between rollers of 3–4 mm. Flowers are discharged in conveying direction, whereas small stems fall down to the waste. Figure 8.2.1 shows a sieving machine with an input capacity of approximately 1200 kg/h [15].

8.2.2 BASICS OF DRYING

Chamomile flowers have an initial water content of around 80% (wet weight basis, or w.w.b.) and are dried to a final water content of 10–11% (w.w.b.). The weight relation of raw chamomile flowers to dried can be calculated to 4.5:1. So to produce 1 kg of dried flowers approximately 3.5 kg of water has to be evaporated from 4.5 kg of raw chamomile flowers. If a herb portion of 50% is considered, this value will increase to 9:1 (weight relation of harvested chamomile herb and flowers to dried flowers).

Drying time of chamomile flowers is influenced by air temperature, air velocity, height of layer, and relative humidity of drying air. Müller [13] examined the influence of these parameters for a thin layer of flowers (approximately 1 cm of height, drying area charged initially with 1.75 kg/m²). Doubling drying temperature from 30 to 60°C reduced drying time by 97%, dramatically showing the influence of temperature. Increasing air velocity from 0.1 m/s to 0.4 m/s at a drying temperature of 60°C resulted in a decrease of drying time of only 30%. At an air temperature of 60°C drying

FIGURE 8.2.1 Chamomile sieving machine [15].

time was significantly increased by a relative humidity (r.H.) of air exceeding 50%. At 45°C an increase of drying time can be stated at relative humidities exceeding 20%.

During the first stage of drying it is recommended to use higher air velocities. With progressive drying, air flow should be reduced. Thus the air flow going through the product layer is adapted to the water quantity, which has to be evaporated out of the product layer. Figure 8.2.2 shows the drying of chamomile flowers in a layer of 40 cm in height at an air temperature of 60°C and with a constant and a staged air velocity [2]. A higher air velocity at the beginning of drying process leads to a quick removal of the high evaporated water quantity in the first 5 hours and to a faster drying. Thus recondensing of water vapor in the top layer is avoided. This has a positive effect on the essential oil content of the flowers in the top layers, and the energy consumption per kg of dried product will be reduced. These effects underline the account of time-staged control of air velocity for static dryers.

For low-temperature drying with dehumidified air the so-called sorption isotherm in Figure 8.2.3 is important [2]. When drying is carried out at a low temperature of 25°C, the relative humidity of the drying air has to be below 55% to reach a final water content of 11% w.w.b. For reasonable drying times (below two days) the dehumidification unit has to remove sufficient water from the air to get a relative humidity of below 40–45%.

For chamomile flowers a drying time of 24 hours was achieved in a special drying box working with dehumidified air at a layer height of 30 cm. Drying temperatures ranged from 18 to 29°C [9].

8.2.3 INFLUENCES OF DRYING PARAMETERS ON QUALITY AND ENERGY CONSUMPTION

8.2.3.1 Quality

Convective drying of chamomile flowers causes losses of essential oils of approximately 25% in a wide range of drying parameters [13].

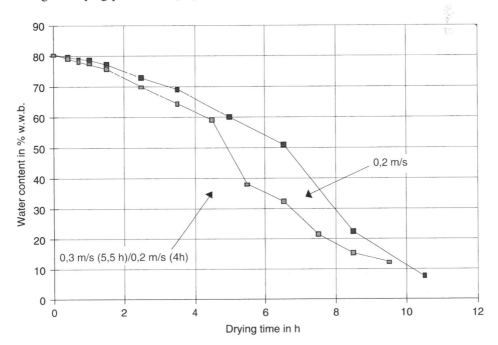

FIGURE 8.2.2 Drying of chamomile flowers in a high layer (air temperature 60°C, height of layer 40 cm) [2].

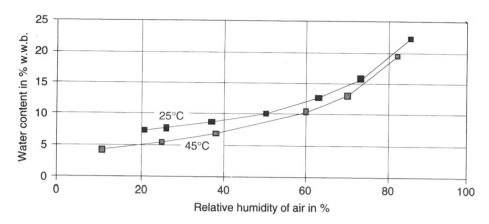

FIGURE 8.2.3 Sorption isotherms of chamomile flowers [2].

8.2.3.1.1   Influence of Air Temperature

Older publications on drying of chamomile recommended a maximum drying temperature of 45°C to limit the losses of essential oils. By 1969 Buschbeck [2] assumed, due to his experiments, that a higher air temperature of 60°C could be applied without having to accept decisively higher quality losses. Müller [13] confirmed this statement by intensive examinations. Both made their tests at an air velocity of 0.2 m/s. Drying at temperatures above this limit causes remarkably higher losses in essential oils according to Reference 2, as shown in Figure 8.2.4.

Müller [13] could not find a sharp increase in losses even up to 90°C. No clear link between drying temperature and the composition and losses of the four main components of the chamomile oil (chamazulene, bisabolol, bisabololoxide A, bisabololoxide B) for this temperature range could be discovered [13].

In Hungary quality examinations were carried out on samples from industrial drying units in 1966. In the case of a band dryer the content of essential oils decreased by 27% at temperatures of 80–90°C in comparison to a sample dried on a loft [18]. There were no remarkable differences in essential oil contents at air temperatures between 40–50°C and 60–70°C. The chamazulene content even increased at 40–50°C by 6% and at 60–70°C by 21%. Only the high-temperature drying with 80°C resulted in a loss of 6%.

Drying of chamomile flowers in a modern five-band dryer can be carried out with a staged temperature profile. The temperature under the top band can be adjusted to 60–70°C, the temperature under the second and third band to 50–55°C, and the temperature for the fourth and fifth band to 45–50°C. Thus the drying capacity of the band dryer is increased without additional deterioration

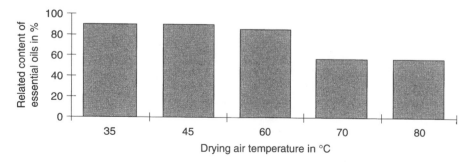

FIGURE 8.2.4 Related content of essential oils of chamomile flowers in dependence of drying air temperature (basis sample naturally dried in a shaded place), air velocity was 0.2 m/s for convective drying experiments [2].

of product quality. Product temperature in the layers on bands one to three does not reach air temperatures due to the cooling effect of the water evaporating from the surfaces.

Examinations carried out by Schmitt [17] show the influence of an uneven drying within a lot on the product quality. Samples were taken from fast-drying spots of an industrial static dryer, which were overdried and exposed over a longer period of time to a higher temperature nearby the air temperature of 55–60°C (up to 8 hours); the samples had an essential oil content of 0.5% instead of 0.8%. This value corresponds to a related additional loss of 35% [17]. On the other hand, the microbial counts and the fungus contamination were lower than the values of other normally dried spots (factor 30 for microbial counts and factor 6 for fungus contamination).

8.2.3.1.2 *Influence of Relative Humidity of Air*

Samples of chamomile flowers dried at 60°C and 0.2 m/s showed at r.H. of drying air above 50% higher losses in total content of essential oils. Losses increased from 25% in the region of r.H. below 50% to 30% at a r.H. of 60% and to 50% at 70% r.H. [13].

The losses of single components of essential oil reach their maximum in the case of chamazulene at a relative humidity of 40%, for bisabolol at 30%, for bisabololoxide A at 50%, and for bisabololoxide B at 50% (air temperature 60°C, air velocity 0.2 m/s). Therefore it can be concluded that at relative air humidities above 30% increased losses for the four main components of the essential oil occur. It is recommended that the portion of recirculated air flow has to be limited so that the relative humidity of the drying air does not exceed 30% at a temperature of 60°C and an air velocity of 0.2 m/s [13].

8.2.3.1.3 *Influence of Air Velocity and Load of Dryer*

Low air velocities and high dryer loads (in kg raw material per m² of drying area) result in a high air humidity of drying air and even condensing of water vapor especially in the top layers of a drying bed. Water vapor uptake capacity of drying air is limited, and air is cooled when coming into contact with product in cool top layers. High air humidity or even condensing of water vapor leads to a micro water vapor distillation of essential oils out of the gland chambers [16]. This reduces the total content of essential oils in the dried product and deteriorates the oil components as stated in Section 8.2.3.1.2. Therefore, the air flow per drying area or air velocity through a bed has to be coordinated with load of bed and applied air temperature. Drying chamomile flowers in big layers and at low air velocities even limits the admissible value of the relative humidity to 10% [14].

8.2.3.2 Energy Consumption and Energy Costs

The specific energy demand per kg of dried chamomile flowers depends on:

- Maximum drying temperature and resulting drying time
- Applied air flow in m³/h per m² of drying area
- Number of product turning during drying period
- Initial and final water content
- Height of product layer or load per drying area
- Kind of raw material (sieved flowers or mixed with herbs)
- Portion or recirculated air flow (e.g., relative humidity of air max. 30% at 60°C and 0.2 m/s)

The energy consumption of a static dryer for chamomile drying with partial recirculation of air flow can be estimated according to the assumptions stated below. Changing conditions will change the energy consumption, too.

Height of layer is 30 cm, bulk density is approximately 200 kg/m³, area charging with raw flowers is approximately 60 kg/m², the initial water content is 80% (w.w.b.), final water content is

11%, and drying time is approximately 20 hours. Air temperature is raised by 35°C by the oil-fired air heater with an efficiency of 90%. The air velocity and the specific air volume per hour are 0.15 m/s and 540 m³/(m²*h).

The following specific values can be calculated: The thermal energy input is approximately 7.14 kWh/(m²*h) corresponding to a fuel oil light consumption of approximately 0.71 l/(m²*h) (fuel oil light, ASTM No. 2: 1 liter = 9.96 kWh). The specific production of dried chamomile flowers is 0.67 kg/(m²*h) and results in a specific fuel oil consumption of approximately 1.05 liter fuel oil per kg of dried chamomile. The electric energy consumption can be calculated to approximately 0.15 kWh/(m²*h) and to 0.14 kWh/kg dried chamomile. The costs for the energy consumption are € 0.43 for oil (assumed price of € 0.41/liter) and € 0.02 for the electric current (assumed price of € 0.13/kWh) for 1 kg of dried chamomile flowers.

Increasing the portion of recirculated air flow raises relative humidity of the drying air and decreases energy consumption per hour for heating the drying air. On the other hand, the drying air has a lower capacity for water uptake, leading to a longer drying time. Thus, the energy saving by partial recirculating exhaust air might be compensated by a longer drying time caused by a higher humidity of drying air.

Müller stated a minimal energy consumption for a drying temperature of 45°C at a r.H. of 40%, for a drying temperature of 60°C at 60% r.H. But exceeding 30% r.H. at 60°C leads to a quality deterioration of essential oils as already mentioned [13].

According to Table 8.2.1 the following related energy savings could be stated for a partial recirculating of air operation at an air velocity of 0.2 m/s in comparison to a pure fresh air operation (60°C/10% r.H., 45°C/15% r.H.) [13].

An absolute limit of relative humidity is 70% for 45 and for 60°C. Aside from the quality deterioration of the essential oil, the drying time is extremely prolonged, so that fungus growth and a general deterioration of the chamomile flowers can be assumed.

Concerning the energy consumption per kg of evaporated moisture Müller [13] found a minimum for an air temperature of 60°C and for an air velocity of 0.2 m/s.

Practical measurements of an industrial five-band dryer equipped with a cross-stream heat exchanger for heat recovery showed a consumption of fuel oil light of 0.6 liter/kg of dried chamomile flowers despite applied low temperatures of 46/43/41°C under the first/third/fifth band and despite a low final water content of 7–9% [1]. The energy consumption of a five-band dryer with partial recirculating of air flow and without heat recovery equipment can be estimated to 0.78 liter of fuel oil light per kg of dried product for a final water content of 11% and for temperatures of 60/55/46°C under the first/third/fifth band [5].

8.2.4 MICROWAVE-ASSISTED WARM AIR DRYING

By applying microwave energy, drying of chamomile flowers can be accelerated. Hereby the energy input per kg of dried product and the time of application are decisive. Figure 8.2.5 shows the drying

TABLE 8.2.1
Possible Energy Saving through Partial Recirculating of Air Flow [13]

Drying temperature	Relative humidity due to partial recirculating of air flow	Prolongation of drying time/reduction of dryer capacity	Energy saving in comparison to pure fresh air operation
45°C	30%	20%	50%
45°C	40%	35%	47%
60°C	30%	10%	44%
60°C	60%	100%	50%

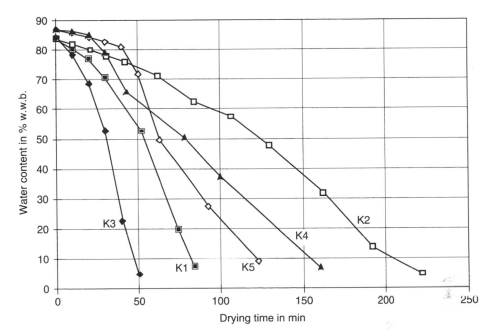

FIGURE 8.2.5 Microwave-assisted warm air drying of chamomile flowers [8], influence of microwave energy application on drying time.

curves for different microwave energy concentrations, calculated in kWh per kg of dried product with a final water content of 10%. All experiments were carried out at an air temperature of 60°C, a relative humidity of 10–14%, and an air velocity of 0.2 m/s. A high microwave energy input of 10.8 kWh/kg results in a reduction of drying time of 60% in comparison with nonmicrowave warm air drying [8].

Focusing the interest on the results of the experiments of K4 and K5, a shorter drying time in the case of K5, it can be stated despite equal microwave energy concentration. In K4 microwave energy is applied in the water content range of 85 to 65% w.w.b. (5.67 kg/kg to 1.86 kg/kg bone dry basis, or b.d.b.), whereas in the case of K5 the water content ranges from 81 to 50% w.w.b. (4.26 kg/kg to 1.00 kg/kg b.d.b.). Obviously there is a higher efficiency in microwave energy absorption in this lower water content range. The microwave energy input in these experiments is rather high; further trials are necessary to optimize the quantity and the time of application of microwave energy.

The analysis in Table 8.2.2 shows higher total contents of the essential oil for the microwave-treated samples and supports the application of microwave energy. K0 is a reference sample naturally dried in a shaded place at ambient temperature [8].

Microwave energy application also affects the optical quality of dried chamomile flowers. A better color preservation and bigger flower heads were stated for microwave-treated samples in comparison with air-dried products [10].

8.2.5 DRYING EQUIPMENT

The following systems and dryers can be applied for the drying of chamomile flowers or mixtures of herbs and flowers: drying on a covered floor, drying in trays, discontinuous static dryers with or without solar preheating of air, half-continuous basculating tier dryers (kilns), and continuous-band dryers. Table 8.2.3 gives a summary on dryers for chamomile.

TABLE 8.2.2

Total Content of Essential Oil for Microwave-Assisted Warm Air Drying of Chamomile Flowers [8]

Experiment	Microwave energy concentration (basis product of 10% water content)	Total content of essential oil in ml/100 g of dried matter (b.d.b.)
K0	0 kWh/kg	0.84
K1	5.0 kWh/kg	0.83
K2	0 kWh/kg	0.66
K3	5.8 kWh/kg	0.91
K4	3.3 kWh/kg	0.78
K5	3.3 kWh/kg	0.90

TABLE 8.2.3

Comparison of Drying of Chamomile on Static Dryers, Basculating Tier Dryers (Kilns), and Band Dryers

Kind of Dryer	Drying Temperature	Height of Feed Layer	Drying Time	Max. Air Velocity or Specific Air Volume/h	Number of Product Turnings	Portion of Recirculated Air Flow
Static solar greenhouse dryer	Up to 45°C (lower in the evening/night)	10 cm	70 h	0.1 m/s or 360 m³/(m²*h)	No or one manually	Low to medium
Static dryer	Max. 60°C, often 45°C	30–40 cm	16–24 h	Up to 0.2 m/s or 720 m³/(m²*h)	No or one manually or with grab	Low to medium
Basculating tier dryer	Max. 60°C	15 cm	10–14 h	Up to 0.4 m/s or 1440 m³/(m*h)	Two to three by gravity basculating	Low to medium
Five band dryer	60–70/50–55/45 −50°C under 1/3/5 band	10–15 cm 1 band	7–10 h	Up to 0.8 m/s or 2900 m³/(m²*h)	Four	Medium to high

8.2.5.1 Advantages and Disadvantages of Dryers

8.2.5.1.1 Static Dryer (Figures 8.2.6, 8.2.7)

Advantages: flexible operation; corresponds to the structure of a working day in small, medium, and large farms or agricultural enterprises; simple operation; low investment costs; suitable for low, medium, and large throughputs; mainly suitable for uncut plants in large layers.

Disadvantages: higher specific energy consumption, danger of uneven drying across the height of layer, higher labor demand for filling and discharging of dryer or high investment costs for automation of product handling, staged drying can be carried out only with high expense in time control of relative humidity of waste air.

8.2.5.1.2 Basculating-Tier Dryer (Kiln) (Figures 8.2.8, 8.2.9)

Advantages: flexible operation, corresponds to structure of working day in small and medium farms and agricultural enterprises, staged drying possible by additional air conduits with low expense, automatic product turning through gravity dumping leads to an even drying, suitable for cut material that does not stick to the tiers, high drying area combined with low ground area demand, low labor demand for filling and discharging.

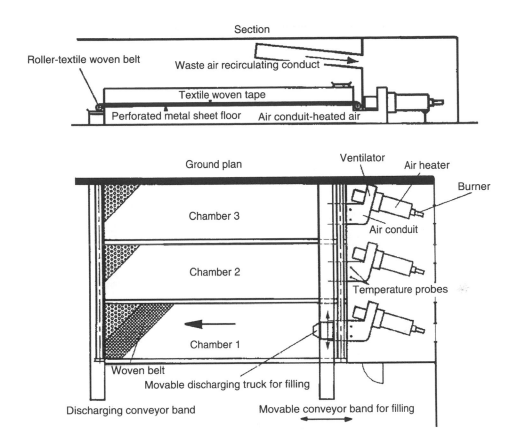

FIGURE 8.2.6 Static dryer with filling and discharging conveyor bands and air conduit for partial recirculating of air flow [17].

FIGURE 8.2.7 Static dryer for chamomile in Slovakian Republic.

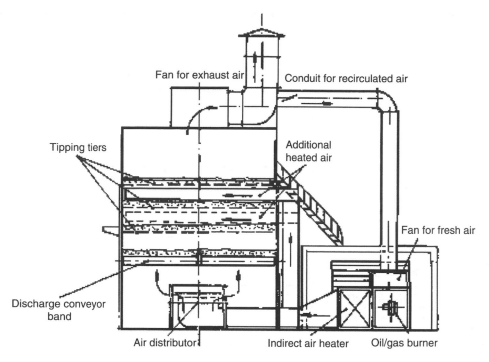

FIGURE 8.2.8 Basculating tier dryer (kiln) [3, 4].

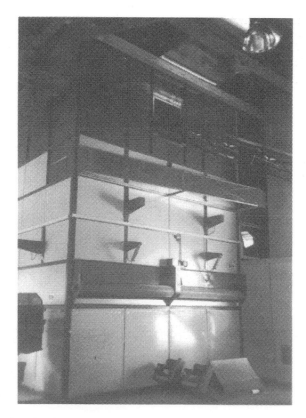

FIGURE 8.2.9 Basculating tier dryer (kiln) [3, 4].

Disadvantages: medium investment costs; not suitable for whole, uncut plants or material sticking to the tiers.

8.2.5.1.3 Band Dryer (Figures 8.2.10, 8.2.11)

Advantages: high throughput per drying area, suitable for a small range of different products, staged temperatures and air velocities easy to adapt to the drying curve of all products, even drying through several product turning by gravity dumping, lower energy consumption, corresponds to the structure of a working day in medium and large farms and agricultural enterprises, low labor demand for filling and discharging.

Disadvantages: no correspondence with structure of a working day in small farms and small agricultural enterprises, as a continuous feeding and operation in three shifts per day is necessary for an economic run with good product quality; high investment costs.

The investments for basculating-tier dryers and band dryers can only be justified if medium or higher throughputs in the range above 100–200 kg of raw material per hour will be reached and the period of operation will last up to 6 months per year. That means that other medicinal plants and spices and even vegetables or fruits have to be processed additionally.

In general the investment costs rise with increased degree of automation, causing lower specific energy costs and lower specific labor costs, too.

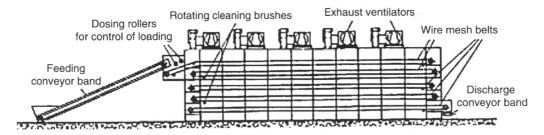

FIGURE 8.2.10 Five-band dryer, cross section [7].

FIGURE 8.2.11 Five-band dryer [6].

FIGURE 8.2.12 Inclined harp machine for cleaning of dried chamomile flowers [15].

8.2.6 PROCESSING OF THE DRIED PRODUCT

The final processing of dried plant material results in a first-class, nearly pure flower product. The dried flowers are cleaned in a so-called inclined harp machine. Long stems are separated from the flowers. Figure 8.2.12 shows such a machine with a throughput of up to 120 kg/h, which was developed in the Slovak Republic [15].

8.2.7 SUPPLEMENT: RECOMMENDATIONS FOR DRYING

The following are recommendations for drying of chamomile flowers in static and band dryers.

8.2.7.1 Static Dryers

- Quick drying after harvest, storage of harvested material not in high layers and not under the sun, maximum storage time 2 hours, for longer periods cooling with ambient air is necessary.
- Drying temperatures up to max. 55–60°C.
- Air velocities approximately 0.15–0.20 m/s for static dryers corresponding to a specific air flow of approximately 540–720 $m^3/(m^2*h)$, for solar-assisted static dryers values are approximately 0.1 m/s or 360 $m^3/(m^2*h)$.
- Maximum relative humidity of drying air should be below 30% at 60°C for partial recirculating of waste air, for lower air velocities or drying in deep beds, 10% r.H. should not be exceeded, affording a pure fresh air operation without recirculating [14].
- In case of limited heating capacity and a maximum drying temperature of 45°C, fresh air operation is recommended until water content of lot drops under 60% w.w.b.
- Feeding height of layer max. 30 cm corresponding to a raw material charging of approximately 60 kg/m^2 (air pressure drop of up to 170 Pa at air velocity of 0.2 m [11]), for solar-assisted dryers with lower air velocity max. 15 cm corresponding to 30 kg/m^2 (air pressure drop up to 85 Pa at air velocity of 0.2 m/s) of raw material load.
- Even spreading of material on dryer (bed height, density).

- Drying time at least 16–20 hours at 60°C, at 45°C up to 30 hours, for solar-assisted dryers up to 70 hours.
- Final water content under 10–11% (w.w.b.).
- Even drying by careful turning of product (manually), for static dryer turning is recommended after 8–10 hours, thus overdrying of bottom layers and recondensing of water vapor in the top layers is avoided and even drying throughout the lot is promoted.
- Reduction of air temperature and air velocity after advanced drying time recommended due to energetic and quality reasons (measuring of relative humidity of waste air as guideline).

8.2.7.2 Band Dryers

- Quick drying after harvest, storage of harvested material not in high layers and not under the sun, maximum storage time 2 hours, for longer periods cooling with ambient air is advisable.
- Drying temperatures staged 60–70°C/50–55°C/45–50°C under 1/3/5 band in a five-band dryer.
- Max. air velocity approximately 0.70–0.80 m/s for five-band dryers under the top band corresponding to a specific air flow of 2500 to 2900 $m^3/(m^2*h)$.
- Max. relative humidity of 30% at 60°C in the case of partial recirculating of waste air, portion of recirculated waste air max. approximately 40–50%.
- Feeding layer height of raw material max. approximately 10 cm for pure chamomile flowers (15 cm for higher herb portion) corresponding to a load of raw material of approximately 20 kg/m² (30 kg/m²).
- Drying time at least 7–10 hours depending on drying temperatures, initial and final water content, and herb portion.
- Final water content under 10–11% (w.w.b.).
- Even drying by automatic gravity product turning from one band to the following band, additionally with product-turning device above the middle of the first band or through dividing of first band in two bands arranged in a line with gravity turning at handing over position of product.
- Application of a heat-recovery device (e.g., cross-stream heat exchanger).

REFERENCES

1. (2001) Personal communication from practice.
2. Buschbeck, E. (1969) Forschungsbericht Arzneipflanzentrocknung. Technische Universität Dresden. 1969.
3. Heindl, A. (1997) Brochure of Heindl GmbH, D-84048 Mainburg.
4. Heindl, A. (1998) Brochure of Heindl GmbH, D-84048 Mainburg.
5. Heindl, A. (2000) *Datensammlung für das Kuratorium für Technik und Bauwesen in der Landwirtschaft e.V.*, Darmstadt.
6. Heindl, A. (2001) Brochure of Heindl GmbH, D-84048 Mainburg.
7. Heindl, A., Müller, J. (1997) Trocknung von Arznei- und Gewürzpflanzen. *Z. Arzn. Gew.pfl.* **2**, 90–97.
8. Heindl, A., Müller, J. (2001) Microwave assisted warm air drying of medicinal herbs and spices. *World Conference on Medicinal and Aromatic Plants.* Budapest, Hungary, July 8–11, 2001.
9. Herold, M., Förster, C., Mickan, P., Röhl, W. (1991) Kleintechnischer Boxentrockner für Arznei-und Gewürzpflanzen auf der Grundlage der Luftentfeuchtung. *Drogenreport* **4**, Nr. 6, 94–103.
10. Kartnig, Th., Lücke, W., Lassnig, Ch. (1994) Der Einsatz von Mikrowellenenergie zur Aufbereitung von Arzneidrogen. 1. Mitteilung. *Pharmazie* **49**, Nr. 8, 610–613.

11. Maltry, W., Pötke, E., Schneider, B. (1975) *Landwirtschaftliche Trocknungstechnik*. Verlag Technik, Berlin.
12. Marquard, R., Kroth, E. (2001) *Anbau und Qualitätsanforderungen ausgewählter Arzneipflanzen*. Agrimedia Verlag, Bergen/Dumme.
13. Müller, J. (1992) *Trocknung von Arzneipflanzen mit Solarenergie*. Diss., University Hohenheim, Stuttgart, Germany.
14. Müller, J., Köll-Weber, M., Kraus, W., Mühlbauer, W. (1996) Trocknungsverhalten von Kamille (Chamomilla recutita (L.) Rauschert. *Z. Arzn. Gew. pfl.* **3**, 104–110.
15. Polnohospodarske Drustvo ROZKVET (1997) Brochures and offer 11.3.97. Slovakian Republic.
16. Schilcher, H. (1987) *Die Kamille — Handbuch für Ärzte, Apotheker und andere Naturwissenschaftler*. Wissenschaftl. Verlagsgesellschaft, Stuttgart, Germany.
17. Schmitt, E. (2000) *Einfluss der Trocknungstechnik auf die Qualität von Arzneipflanzen am Beispiel von Echter Kamille und Ringelblume*. Dipl. Thesis, University Giessen, Germany.
18. Svab, J. (1966) Trocknungsversuche mit ungarischer Handelskamille. *Herba Hungarica* **5**, Nr. 1, 31–36.

8.3 DISTILLATION OF ESSENTIAL OIL

REINHOLD CARLE

8.3.1 INTRODUCTION

Essential oils are steam-volatile mixtures of complex natural substances of plant origin. In some cases, over a hundred different chemical compounds can be detected in an essential oil. Presently, a total of 3000 defined compounds are known [15, 16, 19]. The wide range of application of volatiles is due to the existence of an extensive diversity of compounds in essential oils. A market survey (Table 8.3.1) shows that essential oils are mainly used in the food industry and in the perfume industry. However, some essential oils, e.g., chamomile oil, are especially important in the pharmaceutical industry.

The increase in industrially processed foods has constantly increased the demand for flavoring agents over the last ten years. The world production of natural essential oils is estimated at around 45,000 tons per annum. The price of essential oils ranges between 1.25 Euro/kg for orange oil and 60,000 Euro/kg for genuine Melissa oil (Table 8.3.2). Therefore, genuine essential oils are often replaced by mixtures of synthetic compounds.

Advances in the field of chemical synthesis have been such as to suggest that natural essential oils would eventually disappear. However, in reality, these products have maintained their place on the market in the face of competition from synthetics.

The establishment of intensive methods and the simplification of production systems in the agriculture of developed countries — with fewer crop species grown over bigger areas with mechanization — led to a nearly complete abandonment of volatile oil crops in industrialized countries. Consequently, process engineering with respect to distillation of essential oils has been neglected.

Whoever is engaged in food technology or pharmaceutical technology would realize that process development in the production of essential oils has been stagnant since the 1950s. Even today, the production of essential oils is mainly performed by the traditional method of water distillation in rather primitive field distillation stills (Figure 8.3.1).

The plant material is put into a vessel and completely covered with an appropriate volume of water and finally closed. Steam is produced by directly heating the vessel. Technical requirements for the separate production of steam needed for the milder steam distillation are generally not available. Especially disadvantageous is the decomposition of hydrolysis-sensitive components as well as the batch processing method. The alternating feeding and emptying of the vessel is associated with high operational and energy costs.

TABLE 8.3.1
Market Importance of Essential Oils [14]

Aroma extract in the food industry	50%
Odoriferous agent for perfumes/cosmetics	20%
Raw material in the aroma industry for the isolation of substances	15%
Active component in pharmaceuticals	5%
So-called "natural products"	5%

TABLE 8.3.2
Important Essential Oils According to Prices [14]

Oil	Price per kg
Melissa	6300
Iris root	6300
Violet leaves (absol.)	4300
Rose (Bulgarian)	4000
Orange flower	2000
Jasmine flower (absol.)	1500
Angelica roots	1250
Chamomile (blue)	1250
Labdanum	880
Galbanum	400
Sandalwood (East India)	200
Bergamot orange	125
Cassia (Chinese)	75
Lavender 40/42%	60
Peppermint (*Mentha piperita*)	30
Spearmint	25

In order to improve the efficiency of the distillation process, a continuous method was developed as an alternative to the conventional batch and quasi-continuous distillation methods. Using the production of chamomile oil as example, the various methods will be presented. For all investigations, chamomile of the variety "Degumille," which is characterized by high contents of chamazulene and (–)-α-bisabolol, was used [10]. The quality of the chamomile oil as influenced by the method of distillation and pretreatment, especially drying and grinding of the plant material, will be investigated.

8.3.2 PRODUCTION OF CHAMOMILE OIL

As of now, with chamomile oil, there is uncertainty with regard to the nature and quality of the raw material to be used. While the *Food Chemicals Codex* (1981) [7] permits the use of flowering tops, that is the chopped flowering tops, pharmacopoeias (e.g., *Ergänzungsbuch zum Deutschen Arzneibuch* 1953 [6], *Österreichisches Arzneibuch* 1981 [11], and *Pharmakopöea Helvetica* 1971 [12]) specify the use of pure flower heads, that is, the picked material with a pedicel of maximum 2 cm. Due to economic reasons, the justification of this requisition is always questionable [8]. Up until now dried chamomile flowers have been used for the production of chamomile oil. This practice seems inappropriate since considerable losses of volatile components of the essential oil are incurred during drying [3, 4].

FIGURE 8.3.1 Field distillation still [14].

On the basis of comparable investigations, dried and freshly harvested chamomile with various stalk lengths in both ground and whole forms were used. With the aim of improving profitability, the conventional batch process of steam distillation is compared with the newly developed quasi-continuous and continuous distillation processes.

8.3.3 METHODS OF PRODUCTION

8.3.3.1 Batch Method

The principle of the batch steam distillation achieved with the technical apparatus is shown in Figure 8.3.2.

The plant material is brought and evenly distributed in a supporting basket of the still. After tightly screwing on the cover, steam is introduced. The distillation is carried out at atmospheric pressure. To prevent water distillation, condensed water is drained off through an outlet provided for this purpose. The specific lighter chamomile oil is separated in a "Florentine bottle." Because of the water-soluble nature of the chamomile oil, a series of several Florentine bottles is necessary. An almost complete separation of the oil is achieved through a coalescence filter at the end of the series [3]. The alternating feeding and emptying of the vessel results in process interruptions, which necessitate high operational and energy costs.

8.3.3.2 Quasi-Continuous Method

For the production of highly volatile peppermint oil, mobile distillation equipment has been developed [9, 17]. The plant material from the field is chopped into a mobile container, which can be directly used as a still by its connection to a stationary steam generator and a condenser with an attached oil separator. A so-called mint cooker (Newhouse, Redmond, WA) was designed to achieve a quasi-continuous operation through the alternating usage of two or more containers. However, energy losses cannot be avoided during heating and cooling cycles of the still. This technology has proven its effectiveness for the production of highly volatile monoterpene-containing oils. It should be examined whether this method is also suitable for the production of low volatile sesquiterpene-containing chamomile oil.

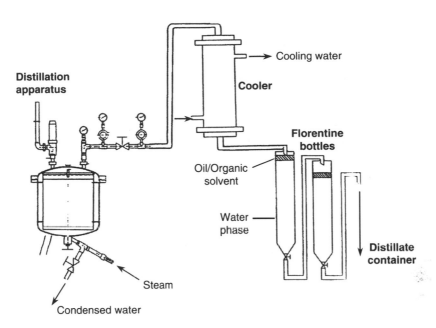

FIGURE 8.3.2 Batch steam distillation [1].

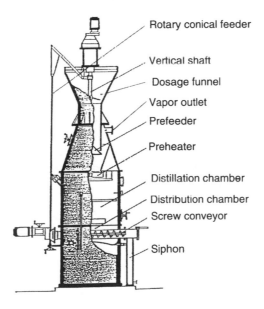

FIGURE 8.3.3 Continuous distillation apparatus [2].

8.3.3.3 Continuous Method

A still originally developed for the production of ethanol derived from marc enables a continuous operation (Figure 8.3.3).

The plant material is introduced into the distillation chamber through a conveyor belt. The steam flows against the plant material through a distributor plate. The steam saturated with oil leaves the still through the vapor outlet. To prevent the formation of canals, a stirrer is provided to ensure a uniform throughput and distribution of steam. The distillation waste is finally ejected by

TABLE 8.3.3
Distillation: Production Balance [2]

	Oil Yield		
	Fresh Weight		Dry Weight
	[kg/ha]	[‰]	[‰]
Continuous distillation			
Chamomile flowers			
- Picked	1.03	0.51	2.22
- Chopped	4.28	0.44	1.75
- Dried (drug)	1.04	—	3.18
Batch distillation			
Chamomile flowers			
- Picked	0.43	0.61	2.78
- Chopped	2.00	0.63	2.50

means of a screw conveyor. If necessary, the waste can possibly be redistilled. At the end of a distillation cycle, part of the waste is conveyed into the dosage funnel to prevent steam leakage. To prevent water distillation conditions, the condensate formed is drained off through a siphon.

8.3.4 EVALUATION OF THE METHODS

Comparing the production balance (Table 8.3.3), it becomes clear that the use of picked material in both fresh and dried forms is not economical.

Because of the high harvest, drying, and distillation costs the use of dried material is especially unprofitable. The best yields are provided by employing chopped flowering tops. If a complete separation of stalks is avoided, a large amount of biomass must indeed be treated; however, a considerably better yield is obtained. This result is in agreement with previous observations made in Reference 8, which recommended a mixture of chopped straw or chamomile weeds to prevent the flower heads from over baking.

With both methods, the technical yields of oil on a dry weight basis do not reach — even by redistillation — the amount of 0.4% specified in the pharmacopoeia. At any rate the lower yields of the continuous method are compensated by the higher capacity and the distinctly lower energy consumption.

Freshly harvested chamomile flowers give better oil quality than the dried chamomile drug (Table 8.3.4).

Because of drying losses, the resulting oil has low chamazulene and bisabolol contents. Chamomile oil from flowery sprouts is comparable to that produced from chamomile drug. Due to the lengthy heating periods in the batch method, the oil produced contains high azulene but low bisabolol contents. Due to their content of chamazulene, high-quality chamomile oils are dark blue in color (*Food Chemicals Codex* 1981 [7], *Ergänzungsbuch zum Deutschen Arzneibuch* 1953 [6], *Österreichisches Arzneibuch* 1981 [11], *Pharmakopöea Helvetica* 1971 [12], *Pharmacopoea Hungarica* 1986 [13]). Chamazulene is of course not a genuine component of chamomile flowers. It is formed during steam distillation from its colorless precursor matricine through dehydration, deacylation, and decarboxylation [18]. The heat-sensitive labile spiroethers are completely decomposed.

The physical and chemical properties differentiate the chamomile oils produced with the various methods (Table 8.3.5).

Chamomile oil produced by the batch method has a higher specific gravity and a higher refractive index. The high wax content of such oils reflects their significantly high saponification

TABLE 8.3.4
Influence of the Distillation Method and the Raw Material on the Quality of Chamomile Oil [2]

	Chamazulene [%]	Bisabolol [%]	cis-Spiroether [%]
Continuous distillation			
Chamomile flowers			
- Picked	7.8	33.3	2.5
- Chopped	5.9	24.0	6.1
- Dried (drug)	2.1	27.9	5.6
Batch distillation			
Chamomile flowers			
- Picked	11.4	21.1	n.d.
- Chopped	7.8	19.4	n.d.

n.d.: not detectable

TABLE 8.3.5
Physical and Chemical Properties of Chamomile Oil [2]

	Batch Distillation	Continuous Distillation	
	Chamomile Oil from Drug	Chamomile Oil from Fresh Plant	Chamomile Oil from Drug
Specific gravity [g/ml]	0.9334	0.8949	0.9077
Refractive value $\left[n_D^{20} \right]$	1.5198	1.5042	1.5045
Water content [%]	0.76	0.81	0.55
Saponification number	25.6	8.3	8.0
Acid value	13.1	5.4	5.6
Ester value	12.5	2.9	2.7
Ester value after acetylation	90.5	62.0	62.7
Optical rotation α_D^{20} [*]	n.d.	-3.0	-2.5

n.d.: not determined

numbers and acid and ester values. A comparison between chamomile oil from fresh and dried chamomile shows that these differences are attributable to the method of production and not to the raw material used.

According to the *DAB 6 Supplementary Book* (*Ergänzungsbuch zum Deutschen Arzneibuch* 1953 [6]), chamomile oils should congeal during cooling to a consistency similar to butter. Our investigations reveal that only wax-rich chamomile oil meet this requisition. High-quality low-wax chamomile oils are accordingly ruled out, so that this requisition presently seems unreasonable.

In accordance with the practice, therefore, the *DAB 10* Monograph, "Chamomile oil" (Deutsches Arzneibuch 1997 [5]) considered the present findings and registered chamomile oil produced from fresh or dried flower heads or flowery sprouts through water distillation. High-quality chamomile oils are no longer excluded from the list of key physical data.

Based on this finding, recommendations for a chamomile oil specification were established. In Table 8.3.6 data according to the *Food Chemicals Codex* are compared to those recommended.

TABLE 8.3.6
Recommendation for Chamomile Oil Specification [2]

	Recommended Data	Data According to FCC III (1981)
Content [mg/100 g]		
Bisabolol	15–30	n.m.
Chamazulene	≥3	n.m.
Farnesol	≤45	n.m.
Physical-chemical data		
Specific gravity $\rho 25$ [g/ml]	0.87–0.94	0.91–0.95
Optical rotation α_D^{20} (EtOH; c = 1,0)	–5 to 0	n.m.
Refractive value n_D^{20} (EtOH; c = 1,0)	13,640–15,250	n.m.
Saponification number	≤40	n.m.
Acid value	≤25	5–50
Ester value	≤15	≤40
Ester value after acetylation	≤100	65–155
Heavy metals [ppm]	≤2	n.m.
Water content [%]	<1	n.m.

n.m.: not mentioned

Especially limits of saponification number, acid, and ester value (with and without acetylation) were reduced in consideration of high-quality low-wax chamomile oils that were hitherto ruled out, i.e., by setting minimum ester values.

With regard to production technology the continuous distillation equipment proved to be superior to conventional techniques in yield and quality. Finally, further studies with Melissa and Mentha distillation confirmed a broad applicability of the continuous distillation technology.

REFERENCES

1. Carle, R. (1992) *Zur pharmazeutischen Qualität biogener Wirkstoffe — Untersuchungen an Kamillenöl (Chamomilla Aetheroleum)*. Habilitationsschrift, Regensburg.
2. Carle, R. and Fiedler, G. (1990) Über ein kontinuierliches Verfahren zur Gewinnung ätherischer Öle. *Pharm. Ind.*, **52**, 1142–1146.
3. Carle, R. and Gomaa, K. (1992) Technologische Einflüsse auf die Qualität von Kamillenblüten und Kamillenöl. *Pharm. Ztg. Wiss.*, **137**, 71–77.
4. Carle, R., Dölle, B., and Reinhard, E. (1989) A new approach to the production of chamomile extracts. *Planta Medica*, **55**, 540–543.
5. *Deutsches Arzneibuch* (1997), 10th Ed., Kamillenöl, Deutscher Apotheker Verlag, Stuttgart, Germany.
6. *Ergänzungsbuch zum Deutschen Arzneibuch* (1953), 6th Ed., Deutscher Apotheker Verlag, Stuttgart, Germany, p. 364.
7. *Food Chemicals Codex* (1981), 3rd Ed., National Academy Press, Washington, DC, pp. 81–82.
8. Guenther, E. (1952) *The Essential Oils*, Van Nostrand, Princeton, NJ.
9. Hannig, H.J., Herold, M., and Röhl, W. (1988) Resultate der Gewinnung etherischer Öle nach der Containertechnologie. *Drogenreport*, **1**, 73–87.
10. Isaac, O. (1974) German Patent Application No. 37 04 519.
11. *Österreichisches Arzneibuch* (1981) Verlag der Österr. Staatsdruckerei, Vienna, pp. 202–203.
12. *Pharmakopöea Helvetica* (1971), 6th Ed., Eidgen. Drucksachen- u. Materialzentrale, Bern, pp. 1031–1032.
13. *Pharmacopoea Hungarica* (1986), 7th Ed., Academia Verlag, Budapest, Hungary, pp. 1503, 1589.

14. Protzen, K.-D. (1993) Produktion und Marktbedeutung ätherischer Öle. In Carle, R. (Ed.), *Ätherische Öle — Anspruch und Wirklichkeit*, Wiss. Verlagsgesellschaft, Stuttgart, Germany, pp. 23–32.

15. Schilcher, H. (1986) Pharmakologie und Toxikologie ätherischer Öle. Anwendungshinweise für die ärztliche Praxis. *Therapiewoche*, **36**, 1100–1112.

16. Schilcher, H. (1987) *Die Kamille — Handbuch für Ärzte, Apotheker und andere Naturwissenschaftler*, Stuttgart: Wissenschaftliche Verlagsgesellschaft.

17. Small, B.E.J. (1982) Agfacts, Dept. Agriculture, New South Wales, Australia, pp. 1–4.

18. Stahl, E. (1954) Über das Chamazulen und dessen Vorstufe, 3. Mitt.: Zur Konstitution der Chamazulencarbonsäure. *Chem. Ber.*, **87**, 202, 205, 1626–1628.

19. Verlet, N. (1993) Commercial aspects. In R.K.M. Hay and P.G. Waterman (Eds.), *Volatile Oil Crops: Their Biology, Biochemistry and Production*. Longman Scientific and Technical, Harlow, pp. 137–174.

9 Storage of the Dry Drug

Horst Böttcher and Ingeborg Günther

CONTENTS

9.1 QUALITY DECLINE IN THE DRIED RAW PRODUCT

Freshly dried chamomile flowers and herbs are very *storage-sensitive products*, even if complete drying of the receptacles has been achieved after a few days. The most important causes of deterioration in quality after drying are the following conditions, which must generally also be considered in terms of the product physiology:

9.1.1 WATER ABSORPTION OF THE DRY PRODUCT (MOISTENING) IN UNFAVORABLE CLIMATIC CONDITIONS

Because the dried plant chamomile organs contain a large proportion of hydrophilic constituents (sugars, flavonoids, mucilages, phenyl carbonic acids, amino acids, choline, salts), chamomile flowers in particular, but also the chamomile herbs are **very hygroscopic products**. Their moisture content can therefore adopt the surrounding microclimatic conditions very quickly, absorbing moisture from the air in the stack or room very fast. This means that the water content of the dry product very soon exceeds the limit of the *physiological water activity* of $\varphi = > 0.60$, which is responsible for microbiological deterioration. This causes a wide range of reactions: Purely biochemical transformations occur more frequently, leading to discoloration, especially of parts of the plant that were previously damaged by pressure, heat, or a deficiency of oxygen. In the green leaves and stems, there is a breakdown of the chlorophyll that is responsible for the green color impression of the plants, and phaeophytin forms at the same time. In the tongue and tube blossoms, the white and yellow colors fade, through undesirable reactions occurring with the colorings in question.

What is particularly serious is the increase in the bacterial flora that often occur in large quantities already at the time of harvest.

Microbiological deterioration caused by fungal agents can also occur within a short time. Thus, at the marginal conditions of the dry product, the most xerophilic species, molds of the species *Aspergillus* and *Penicillium* form first. The metabolism of bacteria and fungal agents releases more and more moisture for more demanding microorganisms, such as *Fusarium* and *Rhizopus,* so that the attack continues to develop in a kind of cascade effect. The metabolic excretions from the microbiological agents also make the stored product smell musty or damp, which is rated very negatively in terms of quality. In addition, there is a risk that the stored product will be contaminated with **mycotoxins**, which are a health hazard.

9.1.2 Effect of the Residual Water Content in the Dried Product

Even the most perfectly dried drug still contains 4–6% reactable water in the plant tissues, which can lead, over longer periods of time, to the formation of undesirable flavor components and also to increased formation of phaeophytin (loss of green color).

9.1.3 Formation of Coarse Powder in the Raw Material

During the time it is in storage, the chamomile flowers in the drugs will be increasingly destroyed mechanically. This is very marked in large, fully opened flowers and at room temperatures. In cold stores, on the other hand, this generally does not occur.

9.1.4 Unfavorable Physical and Biochemical Reactions that can Lead to a Loss of Active Ingredients

These involve various reactions. For example, during storage, the steam pressure of the oil-water vapor mixture in the drug, created through the combination of the remaining water content in the drug with the essential oils, is lowered considerably in comparison with the essential oil, so that at lower temperatures there is a **constant evaporation of essential oils** during storage, especially if the product is not packed. Certain components of the essential oil can also lose their structure and effect as a result of *resinification*. Similarly, *auto-oxidations* (bisabolol/bisabolol oxide) or enzymatic *splitting and rearrangement* may occur, which are registered as a loss of active constituents. The role that individual value-giving components play in the processes is not yet clear in detail. All these destructive reactions are largely dependent on the storage temperature and duration. They run much faster at higher temperatures.

The extent of the specific quality changes in relation to the typical active ingredients in chamomile is described in Section 7.1.

Herbs that have been finely chopped and unpacked storage goods always show higher reductions than unprocessed crops.

9.1.5 Attacks by Insects and Pests (Stock Pests)

The dried product is also a favorite habitat for certain insects. Larvae and beetles generally damage the stored product by eating away at it and pollute it with excreta and webs. This considerably reduces the quality and can lead to total deterioration in a short time. The evaporating essential oils and other odor components attract the insects intensively over long distances. It is therefore essential to ensure that they cannot penetrate the store, since the universal use of organochloric insecticides is not possible, as it would have an adverse effect on the smell and flavor.

The main stock pests that affect drugs are (according to References 14 and 15):

- Copper-red Indian-meal moth (*Plodia interpunctella* Hb.)
- Dark-brown thief beetle (*Ptinus latro* F.)
- Yellow-brown thief beetle (*Ptinus testaceus* OLIV.)
- Smooth spider beetle (*Gibbium psylloides* GZEMP.)

9.1.6 SHRINKAGE LOSSES CAUSED BY CHANGING OF CONSTITUENTS THAT ARE NOT ESSENTIAL OILS

The changes during long-term storage in this important group of substances that determine quality to a considerable extent (and are generally also directly related to the storage process) have not yet been studied for chamomile.

9.2 REQUIRED STORAGE CONDITIONS

The safe long-term storage of chamomile flowers and herbs for several months requires the following storage conditions:

Moisture content of stored product: 8–10%
Relative humidity of ambient air: <50%
Room temperature: 4–15°C
Lighting: Largely in a darkened room
Stock pests: Free from living larvae and insects

Given these conditions, almost all the quality deteriorations described in the previous section could be prevented or reduced. Particular efforts are required to control the occurrence of harmful insects. Basically, the entry of such pests should be prevented by permanently covering all window, wall, and ventilation openings with fine-mesh wire netting (1–2 mm mesh width). All doors and gates should be kept automatically closed. The continuous operation of electrical UV insect traps in the storage room is even more effective; these will have the prophylactic effect of attracting all insects flying in and killing them with the current. Pheromone traps installed in storage rooms are good indicators of the presence of pests. If the occurrence of pests is detected in the store, *disinfestation* must be carried out immediately. All the dry crops stored in the room must be subjected to a CO_2 compressed gas disinfestation, since this is the most effective and most suitable method. In this process (e.g., Carvex and Pex), the drugs affected are first subjected in an autoclave to evacuation to 200 mbar (reducing the oxygen content) and then, through the addition of CO_2 gas, an atmosphere of 18–40 bar is created. Within a period of 150 min., complete destruction of all development stages of the pests is achieved [8]. The process has the advantage that the drugs can be left in their original packaging, the treatment can be carried out at normal outside temperatures, and, in contrast to chemical processes, there are no problematic residues.

Pyrethrum preparations have proved successful for the disinfestation of storage areas.

One of the basic problems with storing drugs is the *considerable space requirement* due to the low bulk density of the stored product. Nonetheless, a clear, permanently accessible layout of the individual batches in the storage room must be guaranteed so that quality can be constantly monitored [6, 13].

The stored product must be *regularly inspected* to determine:

- Its water content (percentage of water in the product)
- Its external quality traits
- Whether it is infested with stock pests
- The level of value-determining active ingredients by chemical analyses

These checks including the analyses must be carried out or repeated every 4 to 6 months. Only in this way can the drugs' minimum content requirements for the proposed usage be complied with.

9.3 STORAGE ROOM

The characteristics of the storage room also play a major part in maintaining quality and in successful storage. These characteristics should be as follows [7, 13]:

- Room cleaned and disinfected before products are taken into storage.
- Airy and cool.
- Dry room climate and constant curve of temperature.
- Building openings sealed with wire netting to prevent entry of harmful insects, pests, birds, and pets.
- Fire prevention measures have been taken.
- Avoid undesirable effects on odor caused by storing other drugs or products at the same time.
- Do not store together with toxic drugs.
- Only carry out disinfestation if there is an acute occurrence of pests.
- Keep dust down and prevent the transfer of the rubbed residues of other types of drug onto the chamomile in order to exclude the possibility of cross-contamination.
- Create the possibility of forced-air ventilation for dry crops that have just been taken into storage to balance moisture levels or for cooling.
- Select stack order for the sensitive chamomile drugs in such a way that the products are not placed on the ground or subjected to mechanical pressure (avoid stacking sacks on top of each other over 1 meter high).
- Larger storage rooms should have mechanical stacking devices.

9.4 CHANGES IN THE ESSENTIAL INGREDIENTS DURING STORAGE

Since demand for the whole year must be covered, the **long-term storage** of the dried drugs is also essential. In the case of the storage-sensitive chamomile flowers, the main problem is a loss of **essential oils** in the dry drug, which also contain a considerable proportion of the active ingredients. The main reason for the losses is the low steam pressure of the oil-water vapor mixture that forms, which encourages the evaporation of essential ingredients. The most important factors influencing this are the storage temperature, the length of storage, the relative air humidity, and a number of physiological-technical peculiarities of chamomile.

For loose chamomile flowers or those that are packed in simple paper bags, the representative analytical values available [5, 7, 11] show an almost linear increase in losses with the length of storage, which is shown in Figure 9.1.

At room temperature, it reaches 5.0%/month of storage and 2.0%/month of storage for cold storage at 0 to +2°C. This also proves the major advantage of lowering the temperature in order to maintain quality during the postdrying period. Only after the 11th month of storage at room temperature was a drop in the increase recorded. With extremely long periods of storage, this loss over the total storage period becomes even clearer: 24 months: 62.5% loss at room temperature, 24% loss in cold storage [4]; 31 months: 45.1% at room temperature, 25% loss in cold storage [11]. The greater deviation range recorded for losses at room temperature (Figure 9.1) is due to the values for the level and fluctuations in room temperature and relative humidity, which are not precisely known and vary considerably for this storage process.

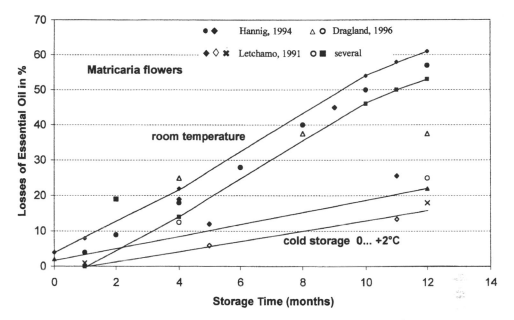

FIGURE 9.1 Losses of essential oil in chamomile flower drugs during storage and the prediction interval for losses at room temperature and cold storage at 0 to +2°C [4, 5, 11, 12].

The flower development stage of the chamomile at the time of harvesting also leads to major differences in losses over a storage period of 31 months, with smaller flowers showing greater storage stability at room temperature (16°C, $\varphi = 60\%$) [11, 12]:

- Large flowers (opening of >75% of the tube blossoms): 46.1%
- Medium flowers (half part of tube blossoms opened): 30.9%
- Small flowers (tube blossoms opened to first ring): only 29.8% loss

In cold storage (+2°C $\varphi = 85\%$), the losses were 27.9, 16.2, and 15.6%, respectively, during the same period.

It was also proved repeatedly that a disintegration of flowers as far as the formation of coarse powder obviously increases the amount of evaporation, so that the loss of essential oil was 14 to 18% [3] or 5 to 19% [7] higher than in undamaged flowers. In the chamomile powder, a greater loss in the first 4 months of storage at room temperature is typical (9.0%/month of storage), which subsequently drops in a smaller extent (3.2%/month of storage) [7].

A strong mechanical treatment of the chamomile flowers (processing as "finely chopped plants in filter bags") does not necessarily lead to greater losses of essential oils (25.6% loss in 18 months, with a level of 0.32% essential oil remaining in the stored drug [7]).

The active ingredient groups **matricin** and **chamazulene** in the chamomile flowers are subject to similar losses to the essential oils, but this process shows a negatively exponentially increasing loss as a function of storage duration (Figure 9.2), as seen in the values for representative losses determined in experiments [7, 9].

Thus, for the first 2 months of storage, a loss of 8.0%/month of storage was determined; after 10 months, the loss was only 3.0%/month of storage, so that after 12 months at room temperature, the losses were 48% [7] or 47.4 to 52.8% after 14 months [9]. As the mean of 33 trade samples, Lachhammer [10] determined after 6 months a loss of chamazulene of 37.2%. In chamomile powder, the losses of matricin are always 15% higher than in whole blossoms [7]. Blazek and Stary [2] determined chamazulene losses of 2.5%/month of storage.

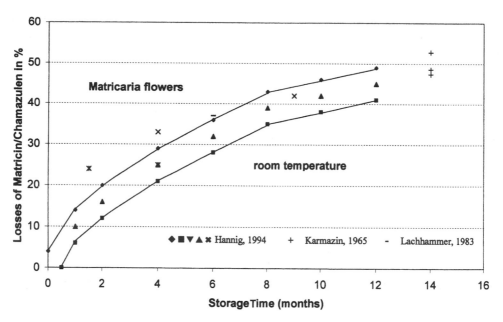

FIGURE 9.2 Losses of matricin and chamazulene in chamomile flower drugs during storage and the prediction interval for losses at room temperature and cold storage at 0 to +2°C [4, 7, 9, 10].

The reduction in the level of chamazulene in the essential oil during long-term storage of up to 25 months is, however, also strongly dependent on the prevailing storage temperature [5]: while, in conditions of 20 to 23°C the loss rate was 2.49%/month of storage, it was lowered at temperatures of 0 to +2°C to 1.06%/month of storage.

Chamomile flowers as "finely chopped plants in filter bags" also only showed a matricin loss of 3.33%/month of storage over 18 months.

Similarly, the level of **farnesene** decreases during storage. As the mean of 33 trade samples, Lachhammer [10] determined at room temperature a loss rate of 6.2%/month of storage. There are as yet no analytical values in the literature for changes in the important ingredient **bisabolol**, but there must clearly be losses, since both of its oxides A and B increase clearly in the essential oil of the chamomile flowers during storage as a function of temperature [5]. At room temperature, **bisabolol oxide A** showed an increase of 9.0%/month of storage; in cold storage it showed only an increase of 10.0%/month of storage. With **bisabolol oxide B** the extent of the increases is smaller: at room temperature 2.1%/month of storage, in cold storage 1.8%/month of storage [5].

The **cis-EN-IN-dicycloethers** are also not stable in storage. In cold storage conditions (0 to +2°C) in particular, the reduction rate of 1.85%/month of storage is much higher than the loss at room temperature of 0.58%/month of storage [5].

The **flavonoids**, which also have a therapeutic effect, are found to be far more stable in storage at higher temperatures than at lower temperatures [11, 12]. The **apigenins**, determined at the tongue blossoms of the chamomile drug, only increased at room temperature of 20°C by a mean of 5.5%/month of storage (i.e., by an absolute 28.4 mg/100 g drug a year), but at +2°C they increased by 101.6%/month of storage (i.e., by an absolute 526.8 mg/100 g drug in the course of a year). The apigenin-7-glycoside of the same flowers, in contrast, showed clear decreases: at 20°C, a decrease of 0.60%/month of storage (an absolute 292.0 mg/100 g drug a year) and at +2°C a decrease of 0.71%/month of storage (an absolute 346.7 mg/100 g drug a year). These shifts can be explained by a splitting of the sugar components of aglycon that occurred during storage, so that the value for apigenin increased.

The often very marked influence of temperature on the loss rates of some ingredients underlines the need to **cool freshly dried chamomile drugs immediately** and to **move** them **to colder storage conditions**.

9.5 PACKAGING

Dried chamomile flowers and herbs are very sensitive storage goods and so proper packaging during storage is absolutely essential, more so than for any other drug. The packaging must meet the following requirements:

- It must provide the necessary protection against the dried products getting moist again.
- It must provide sufficient protection against infestation with harmful insects of any type.
- It must minimize evaporation of the value-giving essential oils.
- It must prevent any mechanical stress on the dried products, which could encourage destruction of the flowers and herbs (formation of coarse powder).
- It must keep out light, since this encourages the oxidation of lipophilic constituents.
- It must keep out pollutants and prevent the content of the containers from being mixed up.
- It must be suitable for advertising on the outside surfaces.

The freshly produced raw material should be packed after all the parts of the plant have reached a uniformly stable moisture level and if necessary after a further subsequent treatment with cold outdoor air for cooling after about 2 weeks. The most suitable *packing materials* for whole flowers are wooden cases, firm cardboard, metal canisters, and other containers made from plywood, board, or metal. Mostly used are cardboards. Before filling, the dry matter must be *rattle-dry* right to the bottom of the flower. These containers can also be easily stacked in the storage-room. Jute sacks to enclose bulk deliveries, as are often used for industrial goods, are less suitable, since they do not meet all requirements and are also difficult to stack. Paper sacks made from three layers of paper are sometimes better, but can only be stacked ultimately using shelves or box pallets. Plastics should not be used for packaging or directly wrapping chamomile drugs, since various plastics may, under some circumstances, bind certain components of the essential oils from the dried drugs or may encourage sweating of the drugs in the inside of the package water vapor (condensation) in the event of fluctuations in temperature.

Basically, only drugs cooled to room temperature should be filled (to prevent condensation). All packaging materials must in principle be suitable for use with foods. Before they are used, they must be stored in clean, dry rooms.

Reused material must be cleaned and completely dried before being used again in order to prevent contamination.

Chamomile herbs may, under some circumstances, be kept in loose piles, if the storage rooms are very dry and largely wooden in structure.

Chamomile flowers should *not be pressed* or *compacted*. Herbs can be compacted, packed, and delivered in bales.

During a 12-month storage of chamomile flower drugs, Schilcher [15] detected a marked influence of the various packagings on the reduction of the quantities of essential oils, while the storage location and temperature were of lesser importance (Table 9.1): In sealed metal cans, only 10% of the essential oils were lost; in impregnated heat-sealed paper packs and plastic bottles (polyethylene), 12 to 18% of the essential oils were lost; while in untreated paper packs and plastic bags (type of plastic unknown), the losses over the same period reached 26 to 44%.

In tests by Karmazin and Zadinova [9], folding cardboard cartons impregnated with crystalline microwax also showed a slight superiority in terms of maintenance of quality, measured by the losses of essential oil (vol.%) and chamazulene (mg%) (mean value from three locations: essential oil 27.7%/chamazulene 47.4%). The wooden crates (28.2/48.6%) and the three-layer paper sack

TABLE 9.1

Influence of the Mode of Packaging on the Decline of the Essential Oils during Storage (According to Reference 14)

Mode of packaging	Temperature	Content vol% essential oils	Relative decline in %
Taken into storage:	—	0.50	0
Flores Chamomilleae			
Removed after 12 months:			
• **Metal cans**, sealed with adhesive strips	room ~	0.45	10
• **Bags** made from Diophan **coated paper** then heat-sealed	room ~	0.43	14
• **Plastic bottle** low pressure polythene with screw top	room ~	0.41	18
• Sealed tea packs with outside carton and **inner bags in flavor-seal, greaseproof parchment paper:**			
Storage place 1:	room ~	0.41	18
Storage place 2: dry cellar	10–15°C	0.44	12
• **Lacquered bags** made from uncoated paper (soda kraft paper), heat-sealed	room ~	0.37	26
• **Plastic pouch** sealed with metal clip	room ~	0.29	42
• **Plastic pouch** sealed with metal clip and additional outside carton	room ~	0.31	38
• **Standard paper bags**, loose product, folded over and adhered with Selotape:			
Storage place A	room ~	0.29	42
Storage place B	room ~	0.28	44

(28.6/52.8%) were almost the same, while the polyethylene bags (31.2/51.1%), RD glass (33.8/54.2%), and metal cases (34.7/57.6%) were considerably worse.

9.6 STORAGE MANAGEMENT

Drugs that have just been taken into storage should be housed initially in a separate room for **quarantine** until the presence of harmful agents and the quality classification have been clarified.

It is important that every individual pack unit is completely and precisely marked according to its origin with the following information:

- Drug type and form
- Origin
- Preparation state
- Classification
- Harvest year
- Weight unit
- Packaging date
- Active ingredient content

In the store, the packed, marked product must be kept in batches, clearly laid out and with no risk of confusion. It must be accessible at all times and it must be able to be removed according to the development of the quality and active ingredients.

Sensory and chemical constituent analysis should be carried out every 4 to 6 months.

Even packaged drugs should not be placed directly on the floor; they should preferably be kept on pallets or on shelves. A distance must be kept from the walls.

Basically, the *Guidelines for Good Agricultural Practice (GAP) for Medicinal and Aromatic Plants* (Chapter 5) [1] should also be followed for packing, storage, and transport.

Any disinfestation required may only be carried out by licensed personnel with approved gasification products in accordance with European and national regulations.

The possible storage period for drugs in powder is always considerably shorter.

In determining the sequence for removal from storage and utilization, the development of the active ingredients and the microbiological state must be taken into account.

REFERENCES

1. Anonymous (1998) *Guidelines for Good Agricultural Practice (GAP) of Medicinal and Aromatic Plants. Z. Arzn. Gew. pfl.*, **3**, 166–174.
2. Blazek, Z., Stary, F. (1961) Einfluß der Aufbewahrung auf die Qualität der Kamillenblüten. *Pharmazeutische Zentralhalle*, **100**, 366–371.
3. Blazek, D., Stary, F. (1962) Einige Bemerkungen zum Siebdurchlauf der Kamillenblüten. *Dtsch. Apoth. Ztg.*, **102**, 606–608.
4. Böttcher, H., Günther, I. (2000) Unpublished.
5. Dragland, S., Aslaksen, T.H. (1996) Storing of dried chamomile flowers at different temperatures and with different packaging. *Norsk Landbruksforsking*, **10**, 167–174.
6. Hannig, H.-J. (1993) Qualitatskontrolle von Drogen vom Anbau bis zum Fertigprodukt. *Herba Germanica*, **1**, 71–77.
7. Hannig, H.-J. (1994) Lagerhaltung von Arznei-und Gewürzpflanzen. *Herba Germanica*, **2**, 137–147.
8. Kabelitz, L. (1997) Korrekturmaßnahmen bei Qualitätsmängeln. *Z. Arzn. Gew. pfl.*, **2**, 120–126.
9. Karmazin, M., Zadinova, K. (1965) Veränderungen im Gehalt des ätherischen Öls und Chamazulens in der Droge *Flor. chamomillae* unter verschiedenen Lagerungsbedingungen. *Wiss. Zeitschrift der Karl-Marx-Universität Leipzig, Mathemat.-naturwiss. Reihe*, **14**, 459–461.
10. Lachhammer, B. (1983) *Qualität von Kamillendrogen aus dem Handel und Qualitätsveränderungen nach Lagerung*. Dipl. Thesis, Techn. University München, Dpt. Vegetables, Freising Weihenstephan.
11. Letchamo, W. (1991) Vergleichende Untersuchung über die nacherntetechnisch bedingten Einflüsse auf die Wirkstoffgehalte in der Droge bei Kamille-Genotypen. *Drogenreport*. Sonderausgabe zur Fachtagung in Erfurt, 129–134.
12. Letchamo, W. (1993) Effect of storage temperatures and duration on the essential oil and flavonoids of chamomile. *J. Herbs, Spices and Med. Plants*, **1** (3), 13–26.
13. Pank, F., Franz, Ch., Herbst, E. (1991) Richtlinien für den Integrierten Anbau von Arznei-.und Gewürzpflanzen. *Drogenreport* Sonderausgabe 1991, 45–64.
14. Schilcher, H. (1968) Lagerung von Kräutertees und Einzeldrogen. *Neuform-Echo*, **18** (6).
15. Schilcher, H. (1987) *Die Kamille - Handbuch für Arzte, Apotheker und andere Naturwissenschaftler*. Wissenschaftl. Verlagsgesellschaft, Stuttgart, Germany.

10 Chemical Analysis of the Active Principles of Chamomile

Heinz Schilcher, Peter Imming, and Susanne Goeters

CONTENTS

10.1 INTRODUCTION

Analytical tests of the active principles of chamomile have the following main objectives:

1. Qualitative proof of pharmacologically relevant active principles in *Matricariae flo*s
2. Impact of ecological and genetic factors on the qualitative and quantitative composition of the active principles
3. Elucidation of the biosynthesis of individual active principles of chamomile

The choice of the analytical method depends on the nature of the analysis. Under all circumstances, the method chosen should provide the necessary accuracy and reproducibility. For the development of a routine analysis, economic factors such as time and costs must be considered.

10.1.1 Test Regulations in Pharmacopoeias

Pharmacopoeias (Table 10.1) provide chemical, physical, and chromatographical test procedures to detect the identity and purity of the essential oil as well as the content of matricin.

TABLE 10.1
Test Regulations for Chamomile Flowers and Chamomile Preparations in Pharmacopoeias Since 1882

Year	Pharmacopoeia	Regulations
1882	*DAB (Deutsches Arzneibuch) (German Pharmacopoeia)*, 2nd edition	No test regulations
1893	*Pharmacopoea Helvetica*, 3rd edition	No test regulations
1894	*DAB (German Pharmacopoeia)*, 3rd edition	No test regulations
1897	Supplement to *DAB* 3, i.e., those drugs not being included in *DAB* 3, 2nd edition	No test regulations
1900	*DAB (German Pharmacopoeia)*, 4th edition	No test regulations
1901	*Svenska Farmakopen* 8	No test regulations
1905	*Pharmacopoeia of the USA*, 1900 edition	No test regulations
1905	*Pharmacopoea Nederlandica* 4	No test regulations
1905	*Farmacopea Española* 7	No test regulations
1906	3rd Supplement to *DAB* 4	No test regulations
1906	*Pharmacopoea Belgicae* 3	Density, solubility
1906	*Pharmacopoea Austria* 8	*100 mercis partis ne minus quam partes 15 extracti spirituosi praebeant*
1907	*Pharmacopoeia of Japan*, English edition	No test regulations
1907	*Pharmacopoea Helvetica* 4	Density, consistency on cooling
1908	*Pharmacopée Française*	No test regulations
1910	*DAB (German Pharmacopoeia)*, 5th edition	No test regulations
1916	4th Supplement to *DAB* 5	No test regulations
1926	*DAB (German Pharmacopoeia)*, 6th edition	Determination of essential oil content
1940	*Pharmacopoea Nederlandica* 5 (2nd printing)	No test regulations
1941	Supplement for *DAB* 6	No additional regulation to *DAB* 6
1941	*Pharmacopoea Helvetica* 6	Density, solubility, specific rotation
1948	*Pharmacopoea Danica*	No test regulations
1958	*Pharmacopoea Nederlandica* 6	Determination of essential oil content results in *blauwe druppeltjes*
1960	*Österreichisches Arzneibuch (Pharmacopoea Austria)* 9	Determination of essential oil content, identification by reaction with dimethylaminobenzaldehyde

TABLE 10.1

Test Regulations for Chamomile Flowers and Chamomile Preparations in Pharmacopoeias Since 1882 (continued)

Year	Pharmacopoeia	Regulations
1971	*Pharmacopoea Helvetica* 6	Determination of essential oil content, swelling factor, fluid extract, color reaction, pH-value, essential oil content
1968	*DAB (Pharmacopoeia of the Federal Republic of Germany)*, 7th edition	Determination of essential oil content and comparison of the blue color
1964/1975	*Pharmacopoeia of the German Democratic Republic*, 7th edition	Identification by reaction with dimethylaminobenzaldehyde, thin-layer chromatography, absorbance of xylene solution of steam distillate at 600 nm (standard: guaiazulene solution)
1975	*European Pharmacopoeia*, vol. III	Determination of essential oil content, test by reaction with dimethylaminobenzaldehyde, thin-layer chromatography
1976	*Pharmacopée Française* 9	Determination of essential oil content, which has to be of a dark blue color
1980	*British Pharmacopoeia*	Determination of essential oil content
1982	Standard Registration § 36 AMG 76 (Germany)	Determination of essential oil content, test by reaction with dimethylaminobenzaldehyde, thin-layer chromatography
1987	*DAB (Pharmacopoeia of the Federal Republic of Germany)*, 9th edition	a) Determination of essential oil content b) Identification by reaction with dimethylaminobenzaldehyde c) Test on purity by thin-layer chromatography
1987	*Pharmacopoea Helvetica* VII, supplement 1993, Flos and Extr. Liquid	a) Determination of essential oil content b) Identification by reaction with dimethylaminobenzaldehyde c) Identification by thin-layer chromatography, determination of essential oil content, gravimetric
1990	*Pharmacopoea Austria*, Flos	a) Determination of essential oil content b) Identification by reaction with dimethylaminobenzaldehyde c) Identification by thin-layer chromatography
1991	*DAB (German Pharmacopoeia)*, 10th edition, supplement 1993, Flos	a) Determination of essential oil content b) Identification by reaction with dimethylaminobenzaldehyde c) Identification by thin-layer chromatography
1997	*DAB (German Pharmacopoeia)* 1997, Extr. Fluid	a) Identification by thin-layer chromatography b) Determination of essential oil content, gravimetric
1997	*DAB 1997 (German Pharmacopoeia)* 1997, Matricariae aetheroleum	a) Identification by thin-layer chromatography b) Purity test by chromatographic profile

TABLE 10.1

Test Regulations for Chamomile Flowers and Chamomile Preparations in Pharmacopoeias Since 1882 (continued)

Year	Pharmacopoeia	Regulations
1997/ 2002	*European Pharmacopoeia*, Flos	a) Determination of essential oil content b) Identification by reaction with dimethylaminobenzaldehyde c) Identification by thin-layer chromatography
1997/ 2002	*European Pharmacopoeia*, Extr. Fluid	a) Identification by thin-layer chromatography b) Determination of essential oil content, gravimetric
2002	*European Pharmacopoeia*, Flos	a) Determination of essential oil content b) Determination of total apigenin-7-glucoside by liquid chromatography c) Identification by thin-layer chromatography
2002	*European Pharmacopoeia*, Extr. Fluid	a) Identification by thin-layer chromatography b) Determination of essential oil content, gravimetric
2002	*European Pharmacopoeia* Draft Monograph, Matricariae aetheroleum	a) Identification by thin-layer chromatography b) Purity test by gas chromatographic profile

Of course there are numerous other publications on the analysis of the active principles of chamomile [1–10, 17–24, 27–31, 33, 36, 39–49, 52–66, 68–81, 83–89, 92–119, 123, 127, 130, 132–139, 141–143, 148–157, 159, 160, 162, 163]. A summary and evaluation of the methods has been published and continually updated by Schilcher [116–121].

The most recent findings as well as established older test procedures are subject of the present summary.

10.2 ANALYSIS OF THE ESSENTIAL OIL

10.2.1 METHODS OF EXTRACTION

Two different methods — a steam distillation as described by Schilcher (Table 10.3) and an extraction using methylene chloride, n-hexane, and petroleum ether (at 40–60°C) — have been compared. The latter resulted in a higher yield of the essential oil and significant differences in the content of spiroethers and bisabolol oxide A as well as bisabolol oxide B, bisabolol, bisabolone oxide, and spathulenol.

Concomitant with the higher content of the essential oil, the amount of hydrophobic compounds such as fatty acids and carotinoids was increased. However, this will not influence the analysis by gas chromatography (GC) or (thin-layer chromatography (TLC). Instead of quantifying chamazulene — which is missing in the methylene chloride extract — the amount of matricin can be determined using TLC or GC. When extracting with methylene chloride, the amount of spiroethers was found to be increased by 68% on average compared to steam distillation, bisabolol oxide A by 42%, bisabolol oxide B by 20%, bisabolol by 12%, and spathulenol by 12%. Similar results

TABLE 10.2

Some Pharmacopoeias Containing Monographs on Roman Chamomile Flowers, Roman Chamomile Oil or Roman Chamomile Preparations (e.g., Extracts)

Year	Pharmacopoeia	Methods
1906	*Pharmacopée Belge* 3	No test regulations
1908	Pharmacopée Française	No test regulations
1934	*HAB* – Dr. Willmar Schwabe	No test regulations
1940	*Nederlandsche Pharmacopee,* 5th edition (Dutch Ph)	No test regulations
1941	*Pharmacopoeia Helvetica,* 5th edition	No test regulations
1954	*Farmacopea oficial Española* IX (Hisp IX)	No test regulations
1980	British Pharmacopoeia	No test regulations
1953	*Pharmacopée Belge* V	Flowers, volatile oil, chamomile water
1954	British Pharmaceutical Codex	ID: macroscop., microscop.
1959	Farmacopeia dos Estados Unidos do Brasil	No test regulations
1960	*Österreichisches Arzneibuch* (ÖAB 9), *Austrian Pharmacopoeia*	ID: macroscop. Assay: essential oil: steam distillation
1954	*Farmakopea Polska* III (Polish Ph.)	
1972	*Farmacopea ufficiale della Repubblica Italiana* VIII	
1976	British Herbal Pharmacopoeia	ID: a. macroscop., b. microscop., c. TLC
1976	*Pharmacopée Française* IX	ID: a. macroscop., b. microscop., c. TLC Assay: essential oil: steam distillation
1979	*Pharmacopoeia Helvetica* VI	
1980	British Pharmacopoeia	ID: TLC
1981	*Österreichisches Arzneibuch* 1981 (*Austrian Pharmacopoeia* 1981)	ID: a. macroscop., b. microscop., c. TLC
1983	British Herbal Pharmacopoeia	ID: macroscop., microscop., TLC
1987–1993	*Pharmacopée Française* X	ID: a. macroscop., b. microscop., c. TLC Assay: essential oil: steam distillation
1987	*Pharmacopea Helvetica* VII	
1988	British Pharmacopoeia	ID: TLC
1992	*DAB 10, German Pharmacopoeia* 10th edition	ID: a. macroscop., b. microscop., c. TLC Assay: essential oil: steam distillation
1993	*Pharmacopoeia Helvetica* VII	ID: a. macroscop., b. microscop., c. TLC Assay: essential oil: steam distillation
1993	British Pharmacopoeia	ID: a. macroscop., b. microscop., c. TLC Assay: essential oil: steam distillation
1997	*European Pharmacopoeia,* 3rd edition	ID: TLC Assay: essential oil: steam distillation
1999	British Pharmacopoeia	ID: a. macroscop., b. microscop., c. TLC Assay: essential oil: steam distillation
2003	*European Pharmacopoeia* 4.3 (German version)	ID: a. macroscop., b. microscop., c. TLC Assay: essential oil: steam distillation

have been published [31, 42, 91]. In order to determine the exact amount of certain components, an extraction time of 1 hour using methylene chloride is recommended.

Formerly published data still remain useful for comparisons. Most of them were obtained by steam distillation. For qualitative investigations, the TAS method, a thermic method developed by E. Stahl [138], can be applied. It is very sensitive, thus a small sample (e.g., three flower heads) is sufficient [119]. A further advantage of the TAS method is the reduction of time needed for the analysis, allowing the processing of a series of samples within short time. Labelling experiments are also possible using the TAS method [119]. However, the analysis is limited to the main components of the analysis.

TABLE 10.3
Quantitative Determination of the Essential Oil in Chamomile Flowers by Means of Steam Distillation [112,121]

Regulation	Test portion and degree of crushing	Distillation/ menstruum	Solvent in graduated tube	Speed of distillation	Time of distillation	Time of reading after distillation
Pharmacopoea Austria, 9th edition, 1960	20.0 g uncrushed	400 ml of water	Decalin	Allow to boil	3–4 h	5 min
Pharmacopoeia of the German Democratic Republic, 7th edition (DAB 7, DDR), 1964 and 1975	10.0 g uncrushed	135 ml of ethylene glycol, 15 ml of water, 0.2 g of silicone oil emulsion	—	Not specified; temperature at condenser not higher than 25°C	3 h	5 min
British Pharmacopoeia, 1968	No instruction	300 ml of water	Xylene	Allow to boil so that the lower part of the condenser stays cold	3–5 h	5 min
Pharmacopoeia of the Federal Republic of Germany, 7th edition, 1968 DAB 7, BRD, 1968	25.0 g of coarse powder	300 ml of water	Xylene	Allow to boil	2 h	15 min
Pharmacopoea Helvetica, 6th edition, 1971	5.0 g; degree of crushing not specified	500 ml of water, then extract with pentane; gravimetric determination similar to *DAB 6*				
European Pharmacopoeia - draft dated June 10, 1971	50.0 g uncrushed	500 ml of 0.5 *N* HCI	Xylene	3–4 ml/min.	4 h	10 min
Pharmacopoeia of CSSR, Ph.Bs. III, vol. I	10.0 g through sieve 3	300 ml of water	Decalin	Allow to boil	4 h + 30 min	5 min
Proposal H. Schilcher [112]	10.0 g uncrushed resp. passed through sieve Nr. 3 homogeneous mixture	300 ml of water	—	Cover of "Pilz'"" stage II, 220 volt, 300 watt = approx. 4–5 ml/min = approx. 40–45 drops/min	Condenser to be switched off after 3 h and distillation to be continued for approx. 5 min until oil is removed from condenser wall	15 min
European Pharmacopoeia, vol. III, 1975	50.0 g uncrushed	500 ml of 1% NaCI solution in a flask of 1000 ml	Xylene (1.0 ml)	3–4 ml/min	4 h	10 min

TABLE 10.3
Quantitative Determination of the Essential Oil in Chamomile Flowers by Means of Steam Distillation (continued)

Regulation	Test portion and degree of crushing	Distillation/ menstruum	Solvent in graduated tube	Speed of distillation	Time of distillation	Time of reading after distillation
Pharmacopoeia of the Federal Republic of Germany, 9th edition, 1987 DAB 9, FRG, 1987	30.0 g uncrushed	300 ml of water in a flask of 1000 ml	Xylene (0.5 ml)	3–4 ml/min	4 h	10 min
Pharmacopoea Helvetica, 7th edition, 1987	30.0 g uncrushed	300 ml of water in a 1-L flask	Xylene (0.5 ml)	3-4 ml/min	4 h	10 min
Pharmacopoea Austria, 1990	30.0 g uncrushed	300 ml of water in a 1-L flask	Xylene (0.5 ml)	3-4 ml/min	4 h	10 min
German Pharmac. – DAB 10	30.0 g uncrushed	300 ml of water in a 1-L flask	Xylene (0.5 ml)	3-4 ml/min	4 h	10 min
Ph. Eur., 1996, 3rd edition	30.0 g uncrushed	300 ml of water in a 1-L flask	Xylene (0.5 ml)	3-4 ml/min	4 h	10 min
Ph. Eur., 1997/2002	30.0 g uncrushed	300 ml of water in a 1-L flask	Xylene (0.5 ml)	3-4 ml/min	4 h	>10 min
Ph. Eur., 2002 Draft monograph	30.0 g uncrushed	300 ml of water in a 1-L flask	Xylene (0.5 ml)	3-4 ml/min	4 h	>10 min; cooling interrupted toward the end of the distillation

Note: Since 1987, the national European monographs on chamomile flowers were harmonized. In 1997, the monograph "Matricariae flos" of the *European Pharmacopoeia* 3rd Edition replaced the national monographs and is officially accepted in all member states of the European Pharmacopoeia Convention (31 states in 2003), including the 15 states of the European Union.

10.2.2 Volumetric and Gravimetric Determination of the Total Content

The gravimetric determination was shown to give better results than the volumetric determination [120, 121].

The volumetric quantification was carried out according to the guidelines of the DAB (*German Pharmacopoeia*) and gave the following results:

$x = 0.57\%$

$s = 7.572 \times 10^{-2}$

$vc\% = 13.28$ (vc% = relative standard deviation)

Using the same plant material, the gravimetric determination gave the following results:

x = 0.814%

s = 1.338 x 10^{-2}

vc = 1.64% (vc% = relative standard deviation)

With the t-value (Student test) calculated to be 10.05, a comparison of both methods is not possible. This explains the wide range of values found in the literature. The volumetric determination is the official method of the *Ph. Eur.*, but the gravimetric method can be applied to small samples.

10.2.3 THIN-LAYER CHROMATOGRAPHY

In a comparison of different thin-layer chromatography conditions, the use of silica gel plates (GF_{254}) as stationary phase and a mobile phase of benzene/ethyl acetate (95:5, V/V) was found to give the best results [109]. Benzene can be substituted by the less toxic toluene without loss in separation quality. The same solvent system is used for the analysis of chamomile oil and chamomile liquid extract according to the *German Pharmacopoeia* 1997. Alternatively, methylene chloride/ethyl acetate (98:2, V/V) [101] and toluene/ethyl acetate (93:7) [158] can be used with good results. The *European Pharmacopoeia* 1975, Volume III, recommended chloroform/benzene (75:25, V/V), which does not meet present standards of Maximum Working Concentration (MWC) and Technical Standard Concentration (TSC), but the choice of solvents of an up-to-date method should agree with these requirements (Table 10.4). A very simple mobile phase, pure chloroform, is specified by the *European Pharmacopoeia* 1997 for the identification of chamomile flowers by TLC [161].

The solvent system should be chosen according to the polarity of the compounds. For substances usually remaining close to the starting line (e.g., matricin), a mobile phase of toluene/ethyl acetate (80:20, V/V) gives useful retention times (e.g., matricin: RF 0.13). For the separation of the bisabolol oxides, silanized TLC plates can be used as the stationary phase in combination with 0.1% or better 0.2% glacial acetic acid in 50% toluene [36] as the mobile phase. There are several ways to detect the separated compounds. The best choice is a combination of an $SbCl_3$ solution and the EP reagent [109]; alternatively, anise aldehyde/sulfuric acid [149] or vanillin/sulfuric acid [158] as spray reagents work well.

TABLE 10.4
x R_F Values of Chamomile Constituents with Various Mobil Phases

Constituent	Benzene (or toluene) 95 Ethyl acetate 5 (v/v)	Dichloromethane 98 Ethyl acetate 2 (v/v)	Chloroform 75 Benzene or toluene 25 (v/v)
	x R_F after development over 12 cm		
trans-β-Farnesene	0.72	0.81	0.72
Chamazulene	0.68	0.78	0.69
cis-En-yne-dicycloether	0.46	0.71	0.46
trans-En-yne-dicycloether	0.42	0.68	0.42
Bisabolone oxide	0.38	0.64	0.39
Bisabolol	0.30	0.51	0.33
Spathulenol	0.21	0.42	0.24
Herniarin	0.19	0.32	0.21
Bisabolol oxide A	0.18	0.30	0.19
Bisabolol oxide B	0.13	0.27	0.16
Umbelliferon	0.02	0.04	0.02
Matricin	0.00	0.06	0.03

10.2.4 DENSITOMETRY

Chamazulene and bisabolol oxides were quantified by extracting their spots from the TLC plate followed by photometric determination [109]. Direct measurement of the main components on TLC has been developed using a Shimadzu CS-900 scanner system [120]. The spots were scanned in zig-zag mode at two different wavelengths. The peak height of the transformed signal was chosen for quantification (Figure 10.1).

After they had recognized chamazulene carboxylic acid (CCA) as a natural profen, the group of Imming extensively studied chemical and pharmacological aspects of this compound. Goeters [32] found the following chromatographic parameters to be suitable for both preparative and analytical purposes: silica gel, hexane/ethyl acetate/methanol 65:30:5, RF (CCA) 0.5. She quantified chamazulene carboxylic acid densitometrically when she looked for conditions that yielded highest CCA content in aqueous infusions; 8-min extraction at 100°C gave the best results (0.4–1.0 mg/ml using 10 g of flowers of the cultivar "Mabamille"). Chamazulene was not detected in the infusion; matricin was present, but was not quantified.

The content of en-yne-dicycloethers can be determined accurately by high-pressure liquid chromatography (HPLC) or by densitometry. The values of the densitometric determination were 20 to 35% higher than those obtained by gas chromatography. Published methods using photometry to determine bisabolol and en-yne-dicycloethers [44] and densitometry to quantify bisabolol [155, 156] have not been reproducible [44, 155, 156]. A colorimetric method to determine chamazulene [152] showed little improvement over a previously published method [109].

GC analysis showed advantages regarding the separation of bisabolol oxides A, B, C, chamazulene, and farnesene and is therefore recommended instead of TLC.

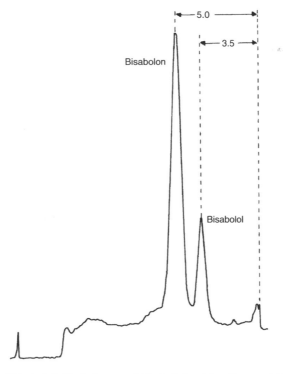

FIGURE 10.1 TLC scan of bisabolol, bisabolone, and bisabolol oxides. Scanner Model: Shimadzu CS-900 Chromato-Scanner. (a) Immerse TLC plate in saturated SbCl₃ solution for 5 sec; (b) dry at 110°C for 5 min; (c) immerse in EP reagent (acetic acid/phosphoric acid) for 5 sec; (d) dry in stream of cold air until odor of acetic acid is no longer detectable; (e) heat to 110°C for 10 min; (f) scan immediately. Measuring wavelength for bisabolol and bisabolone (violet spots): 530 nm; measuring wavelength for bisabolol oxides (red spots): 520 nm; reference wavelength for all: 700 nm.

10.2.5 Reaction Chromatography

The degradation of matricin to chamazulene can be detected while running TLC. Experiments have shown that matricin in extracts treated with steam for 5–10 min is gradually converted to chamazulene and three intermediates that gave three blue spots after TLC separation (see Figure 4.22 in Reference 121).

10.2.6 Gas Chromatography

The first gas chromatographic separation of the components in the essential oil was published in 1968 [49]. Since then, several GC methods have been developed [16, 29, 31, 43, 44, 47, 49, 72, 74, 112, 113, 155, 156]. Both polar and nonpolar columns were used; e.g., 5–10% Carbowax 20 M, 10% UCCW 982; 3% OV 17; 5% SE 30; 3% OV-1; 3% QF-1; 1,5% OV 101, and dexil 300. Based on a comparison of the methods mentioned above [90, 120, 122], optimum results were achieved using an OV 101 column of 50 min length and a capillary temperature of 120–170°C [118]. An OV-1 column gave a similarly good separation for bisabolol oxides A, B, C, bisabolol, and bisabolone oxide. The separation of the isomeric cis- and trans-en-yne-dicycloethers has been achieved by the following conditions depicted in Table 10.5, using bisabolol as the an internal standard. Its retention time was set to 100 and used as reference.

Besides calculations using an HP integrator (18850 A-GC-terminal), the absolute content of active principles in 100 g of plant material can be determined by the internal standard dilution method [16]. Several experiments were carried out using hexadecanol as the internal standard [31] Schilcher proved that guaiazulene has advantages over hexadecanol in several respects [121, 122].

TABLE 10.5
GC Retention Times of Chamomile Constituents

Compound	Retention time/sec	Relative retention time
Hexane	225	14
Farnesene	1047	66
Spathulenol	1340	85
Bisabolol oxide B	1580	100
Bisabolon	1647	104
Bisabolol	1676	106
Matricin/chamazulene	1827	116
Bisabolol oxide A	1936	123
Guaiazulene	2076	131
cis-en-yne-dicycloether	2546	161
trans-en-yne-dicycloether	2575	163

HP 5830 Chromatograph. Columns: OVI or OV 101, 50 m. Temperature program: 120–170°C, 5°C/min. Injection temp.: 225°C. FID temp.: 250°C. Gas flow: Helium at 1.2 ml/min (120°C) or 1.9 ml/min (30°C). Hydrogen at 1.1 bar and oxygen at 1.7 bar. Attenuation: 4 (up to 7). Paper feed: 1.5 cm/min. Integration with HP Integrator 18850 A.

10.2.7 HEADSPACE GAS CHROMATOGRAPHY

Headspace gas chromatography (HSGC) is an elegant method for the analysis of the components of the essential oil with high precision, provided a multiple headspace extraction (MHE) is performed. MHE prevents the matrix from interfering with the analysis. Hiltunen et al. have shown that HSGC is suitable for the analysis of the essential oil and individual volatile active principles of chamomile [39, 40, 41]. Using a DANI-HSS 3850 Automatic Head Space Sampler, very precise results were obtained. The analytes were separated by an OV-1-column with variation of the temperature from 140 to 200°C.

Stuppner and Bauer applied this method to the quantitative determination of the chamomile preparation components in water and ethanol, using an SF 52 column and a temperature range from 130–240°C [144, 145]. The results were influenced by the conditioning time and temperature, but mainly depended on the content of ethanol in the sample. Dilution to 5% ethanol in water guaranteed acceptable results. Compared to conventional GC, HSGC gave almost identical values for (−)-α-bisabolol and its derivatives, whereas values of en-yne-dicycloethers were quite low [145].

10.2.8 HIGH-PRESSURE LIQUID CHROMATOGRAPHY

The development of an HPLC separation protocol using a Bondapak C_{18} column under isocratic and gradient conditions (methanol-water mixtures) did not have advantages compared to published TLC and GC methods [122]. Other investigations also found HPLC a less reasonable "Strategy in Chromatography" [141] for the quantification of components of the essential oil. Nevertheless, HPLC is especially suitable for the separation of isomeric compounds. Azulenes separate very well on a Li-Chrosorb RP 8 column (reversed phase), eluting with methanol-water. This method allows the determination of the purity of isolated azulenes as well as the separation of isomeric azulenes (Figure 10.2) [121].

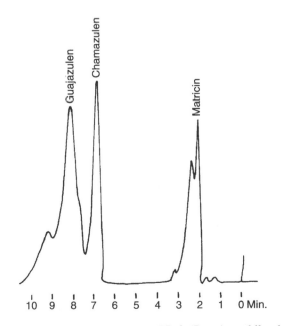

FIGURE 10.2 HPLC separation of azulenes. Column: RP 8 (7 μm); mobile phase: methanol/water 9/1; detection wavelength: 254 nm.

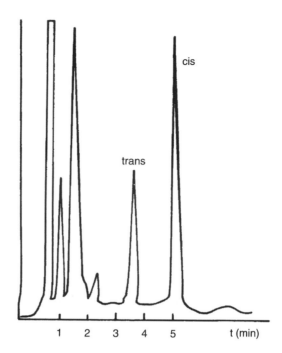

FIGURE 10.3 HPLC separation of isomeric en-yne-dicycloethers. Column: ODS sol-x; mobile phase: acetonitrile/acetic acid 2% 45/55.

For the quantification of matricin in chamomile flowers, Schmidt et al. [124, 126] recommended the direct HPLC analysis of a methylene chloride extract using a Nucleosil 100-5 C_{18} column and a gradient of acetonitrile/methanol/water (12:35:53) as solvent A and methanol as solvent B with detection by UV (244 nm).

Isomeric en-yne-dicycloethers can be separated on an ODS sil-x column (reversed phase) with 2% acetic acid in acetonitrile (45:55, V/V) (Figure 10.3). Under these conditions, neither artifacts nor losses are to be expected [121].

Similarly good separation results were reported by Schulz using water/methanol (70:30, V/V) [128]. The amounts found by HPLC analysis are about 16–39% higher than in GC. For the quantification of spiroethers, HPLC is best, whereas analysis by TLC is complicated and time-consuming.

Chamazulene carboxylic acid was quantitatively determined in human serum by a validated HPLC assay. EDTA plasma was acidified with phosphoric acid, extracted with diisopropyl ether, centrifugated, evaporated, and the residue dissolved in acetonitrile. The separation was done on an RP-18 column, eluting with acetonitrile-buffer pH 3 (4:6) and UV detection (286 nm) [164].

10.2.9 ENANTIOSELECTIVE HPLC

After several less satisfactory attempts [12, 15, 34], Günther et al. established an enantioselective HPLC method to separate the four stereoisomers of α-bisabolol [35]. Possible alterations of the genuine chamomile oil can be identified by quantification of the isomers. (–)-α-bisabolol has the strongest antiphlogistic effect of all stereoisomers [51]. It is the main component and indicates the quality of chamomile oil. Synthetic α-bisabolol is a mixture of four isomers. Its antiphlogistic activity is therefore weaker [51] (see also Chapter 4 on active constituents). Nevertheless, the synthetic compound advertised to be "identical to the natural one" is often used as a substitute or additive to natural chamomile oil. The preparative separation of the isomers of α-bisabolol has been achieved using tribenzoylcellulose as stationary phase (column: Superperformance, 10 mm × 150 mm, 10–20 μm). The eluent was a mixture of ethanol/isopropanol/water (400:300:400, V/V/V; flow rate 0.4 ml/min). The isomers were detected by a UV detector (type "Soma S-3702," 208 nm),

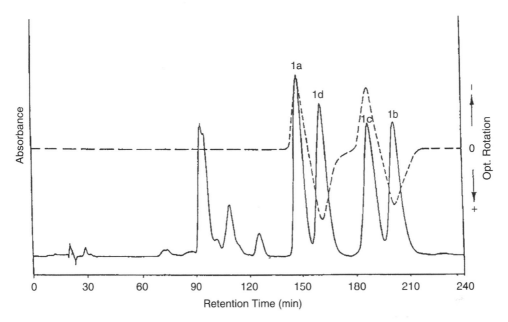

FIGURE 10.4 Analytical separation of the four α-bisabolol isomers on tribenzoylcellulose: 1a = (–)-α-bisabolol; 1d = (+)-epi-α-bisabolol; 1c = (–)-epi-α-bisabolol; 1b = (ı)-α-bisabolol. UV absorbance at 208 nm (full line), optical rotation at 546 nm (dotted line).

followed by a polarimeter (546 nm). Authentic α-bisabolol isomers obtained from other natural sources [34, 35] were used as reference substances (see Figure 10.4).

This was the first report on the separation of the four isomers without lengthy isolation and clean-up procedures. The samples can further be analyzed using isotopic mass spectrometry (MS) and ²H-NMR-spectroscopy [12, 15]. Columns with a wider diameter allow a semipreparative analysis of the isomers using low pressure (LP) LC conditions. The purity of isomers obtained is good (enantiomeric excess approximately 95–98%) [34, 35].

10.2.10 Droplet Countercurrent Chromatography

Droplet countercurrent chromatography (DCCC) was successfully applied by Becker et al. to separate both polar and apolar compounds [5]. The solvent was a mixture of hexane, ethyl acetate, nitromethane, and methanol (9:2:2:3). However, GC gave better results regarding the chromatographic separation compared to DCCC.

10.2.11 HPLC/MS and GC/MS

HPLC and GC combined with mass spectrometry allow the immediate determination of components in chamomile extract [16, 128]. Characteristic fragmentation of the compounds can be observed using different ionization methods: Electrospray Ionization (ESI) or Atmospheric Pressure Chemical Ionization (APCI). The MS data can be compared with data of reference substances in experiments or literature (see Figure 10.5).

LC/MS analysis provides quick results concerning the quality of a pure chamomile oil [16].

10.2.12 Review of the Analytical Possibilities

Schilcher's flow chart (Figure 4.26, page 89 in Reference 121) comprehensively shows further methods regarding the chemical analysis of chamomile oil.

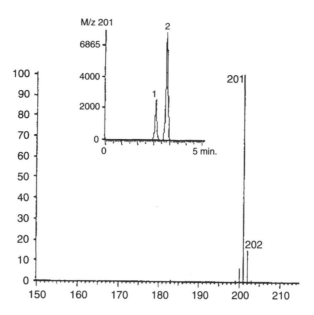

FIGURE 10.5 Thermospray LC-MS chromatogram (single ion mode m/z = 201) of an aqueous/ethanolic chamomile flower extract (concentrated) and mass spectra of the two detected peaks. 1 = cis en-yne-ether, 2 = trans en-yne-ether.

10.3 CHEMICAL ANALYSIS OF THE FLAVONOIDS

10.3.1 DETERMINATION OF THE FLAVONOIDS

Substances that interfere with the analysis of flavonoids (e.g., carotinoids) are usually removed by extraction. It has to be taken into account that the majority of apolar flavonoids will also be removed in this step, e.g., by the extraction with carbon tetrachloride [106]. Moreover, extraction by hot water results in 30–45% lower values for flavonoids compared to methanol extraction. It is therefore not possible to avoid the co-extraction of chlorophyll and related substances by using water.

The photometric determination of flavonoids has many advantages, although the absolute values are actually about 20–30% higher. Other methods as suggested by Reichling et al. showed no improvement [103]. The absolute content of flavonoids ranged between 1.0 and 2.5% in a study of 102 commercially available plants and determination according to References 106 and 17. Twelve samples of material of different origin cultivated by Schilcher showed values between 0.3 and 2.96% [121, 122].

10.3.2 PAPER CHROMATOGRAPHY (PC) AND THIN-LAYER CHROMATOGRAPHY

Flavonoids have musculotropic and neurotropic effects on spasmolysis. Therefore, there is a pharmacological interest in the content of flavonoids. It is known that flavonoid biosynthesis is influenced by environmental factors. The identification and quantification of flavonoids elucidates the chemotaxonomy of the plant material.

10.3.2.1 Paper Chromatography

Paper chromatography provides quick results using isopropanol/formic acid conc./water (2:5:5, V/V) as solvent [46]. Cyclic filter chromatography on Ederol round filters [111] is also possible.

10.3.2.2 Thin-Layer Chromatography

The development of TLC methods has been successful for the separation of about 30 flavonoids, among them the five main aglyca, apigenin, luteolin, quercetin, patuletin, and isorhamnetin, 20 flavonoid glycosides and so-called "lipophilic" methoxylated flavones [71]. This was followed by the densitometric determination of some main constituents such as apigenin-7-glucoside, quercetin-7-glucoside, quercetin-3-galactoside, and patuletin-7-glucoside.

Silica gel 60 F_{254} was used in combination with ethylacetate / formic acid conc. / water (10:2:3) [38], ethylacetate / formic acid / water (100:10:15) [71] or ethylacetate / formic acid / glacial acetic acid / water (100:11:11:26) [158] as eluent.

Aglyca were best separated on polyamide plates (Polygram DC11, MN) and a solvent system of methanol / methylethyl ketone / acetyl acetone (10 + 5 +1) [19] or toluene / methylethyl ketone / methanol (60 + 24 + 14) [103]. For the separation of aglyca on silica gel, toluene / ethylformiate/formic acid (50 + 40 + 10) is suitable [158]. Nineteen flavonoids were analyzed by this method [70], not including the methylated flavonoids (derivatives of the 6-methoxyquercetin and the 6,7-dimethoxykaempferol) [147]. The main components mentioned above can be quantified using a Shimadzu CS 900 scanner. The TLC plates were analyzed at two wavelengths (350 nm and reference wavelength of 710 or 460 nm). Pretreatment of the TLC plate was not necessary. The peak height of the signals was compared to standard substances. For the TLC separation, the solvent system of ethyl acetate/formic acid/water (10 + 2 + 3) can be used.

For qualitative analysis the TLC plate is sprayed with Natural Products (NP)-Polyethylene Glycol (PEG) reagent. NEU-reagent consists of 1% methanolic diphenylboric acid-β-ethylamino ester (NP) followed by 5% of ethanolic polyethylene glycol-4000 (PEG). The spots were detected by UV at 356 nm (see Figure 4.27 in Reference 121).

10.3.2.3 Two-Dimensional Thin-Layer Reaction Chromatography

Flavonoid glycosides as precursors of flavonoid aglyca can be detected using the two-dimensional thin-layer reaction chromatography [37, 67]. After the first development of the TLC plate (sorbens: Silica gel 60 F_{254}), the sugars were hydrolyzed by hydrochloric acid applied in a microwave vapor-blast process. Turning the TLC plate by 90° a second development was carried out (solvent system: ethyl acetate / formic acid / water 8 + 1 + 1 for both developments). The spots were detected by NP/PEG-reagent. Tschirsch and Hölzl adapted this method to acylated flavonoid glycosides, e.g., acylated apigenin-7-glucoside-derivatives from chamomile [146] (Figure 10.6). The acyl groups were removed by saponification, liberating the flavonoid glycoside. In this case concentrated ammonia was used after the first development (Figure 10.7). The duration and temperature of treatment varied from 1 min at room temperature to 1 hour at 70°C depending on the stability of the acyl group (sorbens: silica gel 60 F_{254}; solvent system: ethylacetate / formic acid water 100 + 10 + 5 for both developments, detection by NP/PEG-reagent). The separation of apigenin-7-β-D-(6"-O-acetyl)- and (4"-O-acetyl)glucoside was possible [146]. The method is particularly useful when no authentic reference substances are available.

10.3.3 HIGH-PRESSURE LIQUID CHROMATOGRAPHY

The first HPLC separation of quercetin, quercetin-7-glucoside, apigenin-7-glucoside, rutin, herniarin, umbelliferone, and two unidentified phenyl carboxylic acids was done by Schilcher [117, 118]. All compounds were separated on an RP 8 column running a gradient of methanol-water (15–80%). The method was further developed and perfected for the qualitative and quantitative determination of chamomile flavonoids by Redaelli et al. [99, 100] in 1981 and by Dölle et al. in 1985. A comparison of three methods is shown in Table 10.6.

The method Dölle published allows reproducible determinations of the chamomile flavonoids [18]. Method no. 3 uses the diode-array technique for detection, giving reliable results. Several later publications used this detection method [14, 82, 124, 125, 126, 128, 146].

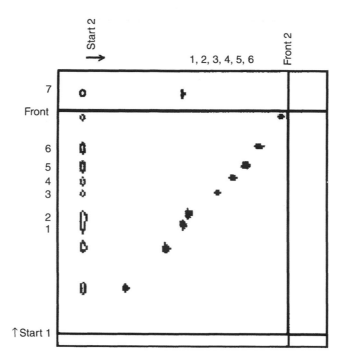

FIGURE 10.6 Two-dimensional thin-layer chromatogram without reaction of ammonia. 1 = Apigenin-7-glucosid; 2 = derivative 1; 3 = derivative 2; 4 = Apigenin-7-β-D-(6″-O-acetyl)glucoside; 5 = derivative 3; 6 = Apigenin-7-β-D-(4″-O-acetyl)glucoside; 7 = Apigenin-7-glucosid control.

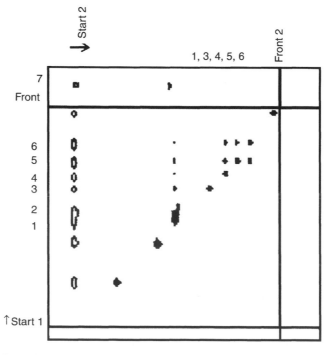

FIGURE 10.7 Two-dimensional thin-layer chromatogram after reaction with ammonia for 1 min. 1 = Apigenin-7-glucosid; 2 = derivative 1; 3 = derivative 2; 4 = Apigenin-7-β-D-(6″-O-acetyl)glucoside; 5 = derivative 3; 6 = Apigenin-7-β-D-(4″-O-acetyl)glucoside; 7 = Apigenin-7-glucosid control.

TABLE 10.6
Comparison of Three HPLC Separation Protocols (after Dölle et al. [18])

	Method		
	1ᵃ	**2**	**3**
Mobile Phase			
Eluent A	2000 ml water	1800 ml KH_2PO_4 (0.005 mol/l)	2000 ml KH_2PO_4 (0.005 mol/l)
	40 ml glacial acetic acid	175 ml methanol	14 ml dil. phosphoric acid
			(pH approx. 2.6)
	110 ml acetonitrile		
	16 ml dil. phosphoric acid		
	(pH approx. 2.55)		
Eluent B	acetonitrile	1750 ml methanol	1200 ml acetonitrile
		300 ml acetonitrile	600 ml methanol
Column			
Column material	RP 18	RP 8	RP 18
Particle size	5 μm	10 μm	5 μm
Column size	250 × 4.6 I.D. (steel)	250 × 4.0 I.D. (steel)	125 × 4.0 I.D. (steel)
Manufacturer	Perkin Elmer	Merck	Merck
Parameters			
Injected volume	15 μl	15 μl	15 μl
Temperature	37°C	35°C	37°C
Flow rate	1.0 ml/min	0.75 ml/min	1.0 ml/min
Sensitivity	21 × 10⁻⁴ A.U./cm	16 × 10⁻⁴ A.U./cm	64 × 10⁻⁴ A.U./cm
Wavelength	335 nm	350 nm	335 nm
Gradient	27–85% B within 28 min	23–85% B within 40 min	27–85% B within 22 min

ᵃ Modified according to Redaelli et al. [97–100].

Carle et al. successfully separated the acetylated isomers of the apigenin-7-*O*-β-glucoside using thermospray liquid chromatography/mass spectrometry (TSP LC/MS) [13]. The analysis was performed on an HPLC chromatograph HP 1090 L interfaced with an HP 5988A thermospray MS. The mass range scanned was m/z 150–800 using positive and negative ionization mode.

10.4　CHEMICAL ANALYSIS OF THE COUMARINS

The solvent system that was applied for the components in essential oil can be used to separate umbelliferone and herniarin — in the dichloromethane extract — on silica gel 60/Hf 254. Table 10.4 summarizes possible mobile phase systems as well as the RF values of the two coumarins. Both compounds show a "soft" blue fluorescence in short-wave UV light. Under long-wave UV, umbelliferone has a very strong, light-blue fluorescence and herniarin, a darker blue one. The strong fluorescence of both compounds can be used for the direct quantitative determination. It is best to use HPTLC plates and a mobile phase system of ether/toluene (1:1, V/V) saturated with 10% acetic acid [110]. The compounds can be determined in fluorescent light at 365 nm at a wavelength of 460 nm [147].

HPLC with UV/Vis detection rapidly and reliably allows the separation and identification of the coumarins in purified aqueous chamomile extracts [126, 128, 129]. Good results were obtained on an HP 1090 liquid chromatograph with microbore column. The mobile phase was a gradient of phosphoric acid (pH 2.8) in acetonitrile. The compounds were identified using an HP 1040 A HPLC detection system. The diode-array technology enables the simultaneous measurement of the absorbance in a range of 190 to 600 nm. An HP 1046 A programmable fluorescence detector can also be used [129] (see Figure 10.8).

For the isolation of both coumarins, sublimation is a suitable method [120].

10.5 CHEMICAL ANALYSIS OF THE CHAMOMILE MUCILAGE

The extraction of chamomile flowers with 96% ethanol precipitates large amounts of (natural) mucilage. The mucilage has to be demineralized in order to determine its viscosity and identify individual polysaccharide components. This can be done using amberlite IR-120, an acidic cation exchanger, and amberlite IR-45, a basic anion exchanger. Comparative hydrolytic tests [120] showed that hydrolysis with trifluoroacetic (TFE) acid (1 ml of 1% solution of chamomile mucilage + 1 ml 4N TFE, boiled for 30 min) is the most suitable method. Table 10.7 lists several solvent systems for the TLC separation of monosaccharides and urone acids.

Good separations of the monosaccharides and urone acids are shown in Figures 4.28 and 4.29 in Reference 121, page 92.

A selection of spray reagents is summarized in Table 10.8. On heating the plates, color reactions occur. However, for densitometric quantification, only immersed plates yield reproducible results.

Franz et al. [25, 26] extracted the polysaccharides with cold water for 7 hours, using chamomile flowers that were pre-extracted with petrol ether and methanol, followed by precipitation with ethanol (final ethanol concentration 80% G/G). For the fractionation of the polysaccharides, both ion exchange and gel permeation chromatography (GPC) are recommended. Ion exchange chromatography is performed on DEAE-Sephacel columns in phosphated form by successive elution with water/phosphate buffer (0.25/0.5/1.0 M) and water/NaOH (0.2 M). The polysaccharide fractions are detected via an anthrone test, dialyzed (MWCO 3500 D), and lyophilized. The latter is performed by medium

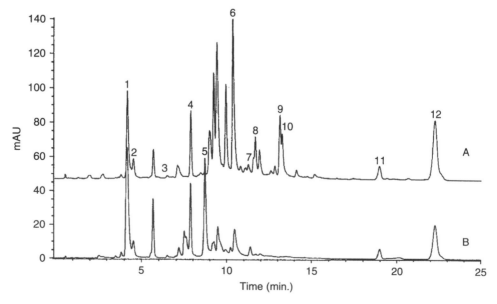

FIGURE 10.8 Reversed phase HPLC and UV/Vis spectra of an aqueous extract of fresh *Matricaria chamomilla*. A = flowers, B = leaves (eluent: diluted phosphoric acid, pH = 2,8/-acetonitrile; detection: 337 nm). 1 = chlorogenic acid, 2 = caffeic acid, 3 = umbelliferone, 4 = hydroxy-cinnamic acid derivate, 5 = luteolin-7-glucoside, 6 = apigenin-7-glucoside, 7 = herniarin, 8 = apigeninglucoside (not specified), 9 = apigenin-7-(6"-O-acetyl)-glucoside, 10 = apigenin, 11 = cis En-In-Ether, 12 = trans En-In-Ether.

pressure-GPC on HiLOAD 16/50 Superdex 75 or 200 columns. The fractions of polysaccharides were detected with an RI detector (e.g., ERC 7512 Benthron Scientific), dialyzed, and lyophilized. The determination of molecular weights was performed in the same GPC system [25].

10.5.1 QUANTITATIVE SPECTRAL DENSITOMETRIC DETERMINATION OF THE MONOSACCHARIDES AND URONE ACIDS

Spots of samples (concentration approx. 5 µg) and the sugar test solutions were applied on the TLC plate (silica gel 60 plates Merck) by microcaps. The test substances were dissolved in 10% of isopropanol and diluted to the final concentrations of 0.25, 0.5, 0.75, 1.0, 1.25, and 1.5 µg/µl. Galacturonic and glucuronic acid were used in concentrations ranging from 0.5 to 1.25 µg/µl [121].

The TLC plates were developed twice over a maximum length of 10 cm each. The solvent system was ethyl acetate-isopropanol-glacial acetic acid-water (60 + 30 + 5 + 5). After thorough drying of the TLC plates, the spots were detected by immersion in Scheffer-Kickuth reagent for 5 sec. The plates were dried and heated for 8 min at a temperature of 120°C, immediately followed by *in situ* measurement. Instrument parameters were as follows:

Apparatus: Zeiss chromatogram spectral photometer KM3
Wavelength: 385 nm
Gap width: 0.5
F-stop: 6
Gap measuring head plate: 2.5
Table speed: 200 mm/min.
Recorder 120 mm/min.
Evaluation: $F = h \times b_{h/2}$ (h = peak height, $b_{h/2}$ = peak width at half level) or by evaluation of the peak height

10.6 ANALYSIS OF CHAMOMILE: SUMMARY

- The gravimetric method gives exact quantitative determinations of the total essential oil. In case of problems with the exact quantitative determination of individual constituents, a 1-hour extraction with methylene chloride is recommended.
- For the analysis of the individual components in the essential oil, both TLC (e.g., on silica gel plates GF_{254}, mobile phase methylene chloride-ethyl acetate 98 + 2) and GC (OV 101 capillary column of 50 m length) are suitable. Very exact values of the spiroethers can be obtained by HPLC (Figure 10.3).
- In the quantification of the total flavonoids, lipophilic flavonoids should also be included. For the analysis of individual flavonoids, TLC is suitable (see Figure 4.27 in Reference 121, page 90). Quantification can be done with TLC scanners, applying the two-wavelength technique (e.g., on a Shimadzu CS 900).
- For the determination of coumarins, TLC or HPLC and UV/Vis detection are best when using the same extract as for the analysis of the essential oil. The coumarins are preferentially isolated by sublimation.
- For the analysis of components of mucilage, hydrolysis with trifluoroacetic acid is particularly suitable. The chromatographic separation of saccharides and urone acids can be performed equally well either by thin layer, ion exchange, or gel permeation chromatography. Quantification works well after immersion of the TLC plates in Scheffer-Kickuth reagent (see Figure 4.28 in Reference 121, page 90).

TABLE 10.7

RF Values of Monosaccharides and Uronic Acids with Various Mobile Phases on Commercial Silica Gel 60 TLC Plates (Merck)

Carbohydrate	RF value										
	1	2	3	4	5	6	7	8	9	10	11
Galactose	0.16	0.45	0.07	0.43	0.45	0.5	0.44	0.22	0.21	0.18	0.16
Glucose	0.21	0.52	0.1	0.49	0.57	0.51	a	0.24	0.24	0.24	0.23
Arabinose	0.23	0.64	0.13	0.57	0.64	0.65	0.48	0.28	0.27	0.31	0.30
Fucose	a	0.72	0.17	0.61	0.74	0.74	a	a	a	0.43	0.,7
Xylose	0.31	0.76	0.18	0.3	0.76	0.76	0.56	0.36	0.36	0.46	0.44
Rhamnose	0.39	0.88	0.28	0.69	0.85	0.86	0.63	0.2	0.2	0.62	0.57
Galacturonic acid	0	0	0	0.21	0	0	0.29	0.1	0.5	0	0
Glucuronic acid	0	0	0	0.23	0	0	0.0	0.14	0	0	0

1 : n-Propanol-Ethylacetate-Water   (70:20:10)   (V/V)   [3]
2 : Acetone-Water   (9010)   (V/V)   [95]
3 : Ethylacetate-aq. 2-Propanol (65%)   (65:35)   (V/V)   [142]
4 : Acetonitrile-Water   (85:15)   (V/V)   [27]
5 : Acetone-1-Butanol-Water   (50:40:10)   (V/V)   [4]
6 : Acetone-1-Butanol-Acetic acid-Water   (50:40:10:10)   (V/V)   [28]
7 : 1-Butanol-Acetic acid-Water   (80:30:30)   (V/V)   [163]
8 : Ethylacetate-Methanol-Acetic acid-Water   (60:15:15:10)   (V/V)   [36]
9 : 2-Propanol-Ethylacetate-Water   (50:40:10)   (V/V)   [94]
10 : Ethylacetate-2-Propanol-Water   (60:30:10)   (V/V)   [28]
11 : Ethylacetate-2-Propanol-Acetic acid-Water   (60:30 5:5)   (V/V)

a No separation of fucose and xylose and of glucose and arabinose.

TABLE 10.8

Color Reactions of Monosaccharides and Galacturonic Acid after Immersion of the TLC Plates and Development at 100°C

	Anisaldehyde-Sulfuric acid	Anilin-Diphenylamin	α-Naphthol-Sulfuric acid	Carbazol-Sulfuric acid	Scheffer-Kickuth Reagent
Galactose	Dark green	Blue	Red	Blue-grey	Yellow
Glucose	Blue	Blue-grey	Red	Blue-grey	Yellow
Arabinose	Light green	Green	Dark red	Blue-green	Yellow
Fucose	Green	Green	Red	Blue-green	Yellow
Xylose	Light green	Grey-green	Dark red	Blue-green	Yellow
Rhamnose	Green	Pale green	Orange	Greyish pink	Yellow
Galacturonic acid	Brown	Red-brown	Red-brown	Blue-green	Yellow

REFERENCES

1. (1971) Pharmacopoea Helv. Edit. VI.
2. (1978) *Ph. Eur. III*, 269–271.
3. Adachi, S. (1965) *J. Chromatog.* **17**, 295.
4. Baltes, W., Liesk, J., Domesle, A. (1973) *Chem. Mikrobiol. Technol. Lebensm.* **2**, 92.

5. Becker, H., Reichling, J., Wei-Chung, H. (1982) *J. Chromatogr.* **237**, 307.

6. Bohlmann, E., Herbst, Ch., Arndt, H., Schönowski, H., Gleinig, H. (1969) *Chem. Ber.* **94**, 3193.

7. Bournot, K. (1953) *Pharmazie* **8**, 174.

8. Breinlich, J. (1967) *Dtsch. Apoth. Ztg.* **107**, 1795.

9. Brieskom, C. H. (1965) *Arch. Pharm.* **298**, 505.

10. Brieskorn, C. H. (1954) *Arch. Pharm.* **287**, 503.

11. Carle, R. (1996) *Dtsch. Apoth. Ztg.* **136**, 2165–2176.

12. Carle, R., Beyer, J., Cheminat, A., Krempp, E. (1992) *Phytochemistry* **31**, 171–174.

13. Carle, R., Dölle, B., Müller, W., Baumeister, U. (1993) *Pharmazie* **48**, 304-306.

14. Carle, R., Dölle, B., Reinhard, E. (1985) *Planta Med.* **55**, 540–543.

15. Carle, R., Fleischhauer, I., Beyer, J., Reinhard, E. (1990) *Planta Med.* **56**, 456–460.

16. Carle, R., Fleischhauer, I., Fehr, D. (1987) *Dtsch. Apoth. Ztg.* **127**, 2451–2457.

17. Christ, B., Müller, K. H. (1960) *Arch. Pharm., Ber. Dtsch. Pharmaz. Ges.* **293**, 1033–1042.

18. Dölle, B., Carle, R., Müller, W. (1985) *Dtsch. Apoth. Ztg.* **125**, Nr. 43/Supplement 1, 14.

19. Egger, K. (1964) *Planta Med.* **12**, 265.

20. Evdokomoff, V., Tucci Bucci, B., Cavazutti, G. (1971) *Il Farmaco* **27**, 163.

21. Exner, J., Reichling, J., Cole, T. H. C., Becker, H. (1981) *Planta Med.* **41**, 198.

22. Franz, Ch. (1982) *Dtsch. Apoth. Ztg.* **122**, 1413.

23. Franz, Ch., Kirsch, C., Isaac, O. (1985) *Dtsch. Apoth. Ztg.* **125**, 43/Supplement 1, 20.

24. Frühwirth, H.,Krempler, F. (1979) *Ernährung/Nutrition* **3**, Nr. 1, 26.

25. Füller, E. (1992) Dissertation, Regensburg.

26. Füller, E., Franz, G. (1993) *Dtsch. Apoth. Ztg.* **133**, 4224–4227.

27. Gauch, R., Leuenberger, U., Baumgartner, E. (1979) *J. Chromatog.* **174**, 195.

28. Ghebregzabher, M., Rufini, S., Monaldi, B., Lato. M. (1976) *J. Chromatog.* **127**, 133.

29. Glasl, H. (1972) *Pharm. Ind.* **34**, 122.

30. Glasl, H. (1975) *J. Chromatogr.* **114**, 215.

31. Glasl, H., Wagner, H. (1976) *Dtsch. Apoth. Ztg.* **116**, 45.

32. Goeters, S. (2001) *Chemische und pharmakologische Charakterisierung eines anti-inflammatorisch wirksamen Pflanzeninhaltsstoffs, isoliert aus Asteraceen,* Dissertation, Marburg (Lahn).

33. Grahle, A., Höltzel, Chr. (1963) *Dtsch. Apoth. Ztg.* **103**, 1401.

34. Günther, K., Carle, R., Fleischhauer, I., Merget, S. (1993) *Arch. Pharm.* **326**, 617–618.

35. Günther, K., Carle, R., Fleischhauer, I., Merget, S. (1993) *Fresenius J. Anal. Chem.* **345**, 787–790.

36. Hagenström, U., Schmersahl, K.-J. (1954) *Planta Med.* **2**, 51.

37. Heisig, W., Wichtl, M. (1989) *Dtsch. Apoth. Ztg.* **129**, 2178.

38. Herout, V, Zaoral, M., Sorm, F. (1953) *Collect. Czechoslov. Chem. Commun.* **18**, 122.

39. Hiltunen, R., Vuorela, H., Laakso, I. (1985) in Baerheim-Svendsen, A., Scheffer, J.J.C. (Eds.), *Essential Oils and Aromatic Plants*, Martinus Nijhoff/Dr. W. Junk Publishers, Dordrecht, The Netherlands, 23–41.

40. Hiltunen, R., Vuorela, H., Laakso, I., von Schantz, M. (1984) *Farm. Tijdschr. Belg.* **61**, (3) 354.

41. Hiltunen, R., Vuorela, H., Laakso, J., von Schantz, M. (1984) *32. Vortragstagung der Gesellschaft für Arzneipflanzenf.,* Antwerp, *Abstracts* P 52.

42. Hölzl, J., Demuth, G. (1973) *Dtsch. Apoth. Ztg.* **113**, 671.

43. Hölzl, J., Franz, Ch., Fritz, D., Vömel, A. (1975) *Z. Naturforsch.* **30 c**, 853.

44. Hölzl, J.,Demuth, G. (1975) *Planta Med.* **27**, 37.

45. Hörhammer, L. (1961) *XXI. Int. Congr. Pharm. Sci.*, 4–8. Sept. 1961, Pisa, Italy.

46. Hörhammer, L., Wagner, H., Salfner, B. (1963) *Arzneim.-Forsch.* **13**, 33.

47. Isaac, O. (1969) *Präparative Pharmaz.* **5**, 189.

48. Isaac, O., Schimpke, H. (1965) *Arch. Pharm. Mitt. Dtsch. Pharm. Ges.* **35**, 133, 157 (review). C43.

49. Isaac, O., Schneider, H., Eggenschwiller, H. (1968) *Dtsch. Apoth. Ztg.* **108**, 293.

50. Isaac, O., Schneider, H., Eggenschwiller, H. (1968) *Dtsch. Apoth. Ztg.* **108**, 293.

51. Jakolev, V., Isaac, O., Thiemer, K., Kunde, R. (1979) *Planta Med.* **35**, 125–140.

52. Janecke, H., Kehr, W. (1962) *Arch. Pharm.* **295**, 182.

53. Janecke, H., Kehr, W. (1962) *Pharm. Ztg.* **107**, 40.

54. Janecke, H., Kehr, W. (1962) *Planta Med.* **10**, 60.

55. Janecke, H., Kehr, W., Weisser, W. (1962) *Dtsch. Apoth. Ztg.* **102**, 398.

56. Janecke, H., Weisser, W. (1964) *Planta Med.* **12**, 528.
57. Jones, J. K. N., Pridham, J. B. (1953) *Nature* **172**, 161.
58. Kaiser, H. (1950) *Pharm. Ztg.* **86**, 125.
59. Kaiser, H., Frey, H. (1938) *Dtsch. Apoth. Ztg.* **53**, 1385.
60. Kaiser, H., Frey, H. (1938) *Dtsch. Apoth. Ztg.* **53**, 1402.
61. Kaiser, H., Frey, H. (1939) *Dtsch. Apoth. Ztg.* **54**, 882.
62. Kaiser, H., Frey, H. (1942) *Dtsch. Apoth. Ztg.* **57**, 155.
63. Kaiser, H., Hasenmaier, G. (1956) *Arch. Pharm.* **289**, 681.
64. Kaiser, H., Hasenmaier, G. (1956) *Sci. Pharm.* **24**, 155.
65. Kaiser, H., Lang, W. (1951) *Dtsch. Apoth. Ztg.* **91**, 163.
66. Koedam, A., Scheffer, J. J. C., Baerheim Svendsen, A. (1979) *Chem. Mikrobiol. Technol. Lebensm.* **6**, 1.
67. Krüger, D., Wichtl, M. (1985) *Dtsch. Apoth. Ztg.* **125**, 45.
68. Kubeczca, K.-H. (1973) *Chromatographia* **6**, 106.
69. Kubeczca, K.-H. (1979) *Planta Med.* **35**, 291.
70. Kunde, R., Isaac, 0. (1979) *Planta Med.* **35**, 71.
71. Kunde, R., Isaac, 0. (1979) *Planta Med.* **37**, 124.
72. Lembercovics, E. (1979) *Sci. Pharm.* **47**, 330.
73. Lewe, W. A. (1964) *Pharm. Ztg.* **109**, 256.
74. Linde, H., Cramer, G. (1972) *Arzneim. Forsch.* **22**, 583.
75. Liptak, J., Verzar-Petri, G., Boldvai, J. (1980) *Pharmazie* **35**, 545.
76. Luckner, M. (1966) *Pharmazie* **21,** 620.
77. Luckner, M. (1966) *Prüfung von Drogen*, VEB Gustav Fischer Verlag Jena.
78. Luckner, M. (1969) *Der Sekundärstoffwechsel in Pflanze und Tier*, Gustav Fischer Verlag, Stuttgart, Germany.
79. Luckner, M., Bessler, 0., Luckner, R. (1966) *Pharmazie* **21**, 620.
80. Mechler, E., Kovar, K.-A. (1977) *Dtsch. Apoth. Ztg.* **117**, 1097.
81. Meyer, F. (1952) *Z. Naturforsch.* **7 b**, 61.
82. Miething, H., Holz, W. (1989) *Pharmazie* **44**, 784–785.
83. Mincso, E., Tyihak, E., Nagy, J., Kalasz, H. (1979) *Planta Med.* **36**, 296.
84. Moritz, 0. (1938) *Arch. Pharm.* **276**, 3760.
85. Mothes, K., Schütte, H. R. (1969) *Biologie der Alkaloide*, VEB Deutscher Verlag der Wissenschaften, Berlin.
86. Motl, 0., Felklova, M. Lukes, V., Jasicova, M. (1977) *Arch. Pharm. (Weinheim)* **310**, 210.
87. Motl, 0., Repcák, M. (1979) *Planta Med.* **36**, 272.
88. Motl, 0., Repcák, M., Sedmera, E. (1978) *Arch. Pharm.* **311**, 75.
89. Müller, K. H., Honerlagen, H. (1960) *Dtsch. Apoth. Ztg.* **100**, 309.
90. Naves, Y. R. (1947) *Helv. Chim. Acta* **30**, 278 (1947), ref. in Isaac, 0., Schimpke, H. (1965) *Arch. Pharm. Mitt. Dtsch. Pharm. Ges.* **35**, 133, 157.
91. Ness, A., Schmidt, P. C. (1995) *Dtsch. Apoth. Ztg.* **135**, 3598–3610.
92. Oelschläger, H., Volke, J., Lim, G. T. (1965) *Arch. Pharm.* **298**, 213.
93. Padula, L. Z., Rondina, R. V. D., Coussio, J. D. (1976) *Planta Med.* **30**, 274.
94. Papin, J. P., Udiman, M. (1979) *J. Chromatog.* **170**, 490.
95. Pifferi, P G. (1965) *Anal. Chem.* **37**, 925.
96. Poethke, W., Bulin, R. (1969) *Pharm. Zentralh.* **108**, 733.
97. Redaelli, C., Formentini, L., Santaniello, E. (1980) Poster *Intern. Kongress für Heilpflanzenforsch.*, Strasburg.
98. Redaelli, C., Formentini, L., Santaniello, E. (1981) *J. Chromatogr.* **209**, 110.
99. Redaelli, C., Formentini, L., Santaniello, E. (1981) *Planta Med.* **42**, 288.
100. Redaelli, C., Formentini, L., Santaniello, E. (1981) *Planta Med.* **43**, 412.
101. Reichling, J., Becker, H. (1977) *Dtsch. Apoth. Ztg.* **117**, 275.
102. Reichling, J., Becker, H. (1977) *Dtsch. Apoth. Ztg.* **117**, 275.
103. Reichling, J., Becker, H., Exner, J., Dräger, P.-D. (1979) *Pharm. Ztg.* **124**, 1998.
104. Reichling, J., Becker, H., Vömel, A. (1977) *Planta Med.* **32**, 235.
105. Ritschei, W A. (1960) *Sci. Pharm.* **28**, 280.
106. Roemisch, H. (1960) *Pharmazie* **15**, 33.

107. Salfner, B. (1963) Diss., Munich, Germany.
108. Schäfer, J. (1965) *Wiss. Zeitschr. Karl Marx Univ.,* Leipzig **14**, 435.
109. Schilcher, H. (1964) *Dtsch. Apoth. Ztg.* **104**, 1019.
110. Schilcher, H. (1965) *Dtsch. Apoth. Ztg.* **105**, 1069.
111. Schilcher, H. (1966) *Präp. Pharmazie* **3**, 1.
112. Schilcher, H. (1972) *Dtsch. Apoth. Ztg.* **112**, 1497.
113. Schilcher, H. (1973) *Planta Med.* **23**, 132.
114. Schilcher, H. (1974) *Dtsch. Apoth. Ztg.* **114**, 181.
115. Schilcher, H. (1977) *Dtsch. Apoth. Ztg.* **117**, 89.
116. Schilcher, H. (1977) *Planta Med.* **34**, 29–30, Heft 32 a.
117. Schilcher, H. (1980) Arbeitsmappe zur Arbeitstagung "*Ätherische Öle*" in Groningen 27–30 May 1980.
118. Schilcher, H. (1982) in *Vorkommen und Analytik ätherischer Öle, Ergebnisse der internationalen Arbeitsgruppe Ätherische Öle,* Georg Thieme Verlag, Stuttgart, Germany, 104–115.
119. Schilcher, H. (1984) *Dtsch. Apoth. Ztg.* **124**, 1433.
120. Schilcher, H. (1985) *Zur Biologie von Matricaria chamomilla, syn. "Chamomilla recutita (L.) Rauschert,"* Research report 1968–1981, Inst. Pharmakognosie and Phytochemie of the FU Berlin.
121. Schilcher, H. (1987) *Die Kamille - Handbuch für Ärzte, Apotheker und andere Wissenschaftler,* Wissenschaftliche Verlagsgesellschaft, Stuttgart.
122. Schilcher, H., Novotny, L., Ubik, K., Motl, 0., Herout, V. (1976) *Arch. Pharm.* **309**, 189.
123. Schmidt, F. (1969) *Dtsch. Apoth. Ztg.* **109**, 137.
124. Schmidt, P. C., Ness, A. (1993) *Pharmazie* **48**, 146–147.
125. Schmidt, P. C., Vogel, K. (1992) *Dtsch. Apoth. Ztg.* **132**, 462–468.
126. Schmidt, P. C., Weibler, K., Soyke, B. (1991) *Dtsch. Apoth. Ztg.* **131**, 175–181.
127. Schmitz, R. (1959) *Mitt. Dtsch. Pharm. Ges.* **28**, 105.
128. Schulz, H. (1992) *Dragoco Report* **2**, 59–70.
129. Schulz, H., Albroscheit, G. (1988) *J. Chromatogr.* **442**, 353–361.
130. Schwerdtfeger, G. (1963) *Dtsch. Apoth. Ztg.* **103**, 874.
131. Spegg, H. (1966) *Dtsch. Apoth. Ztg.* **106**, 959.
132. Stahl, E. (1970) *Chromatographische und Mikroskopische Analyse von Drogen,* G. Fischer Verlag, Stuttgart, Germany, 180.
133. Stahl, E. (1953) *Dtsch. Apoth. Ztg.* **93**, 197.
134. Stahl, E. (1954) *Chem. Ber.* **87**, 505, 1626.
135. Stahl, E. (1956) *Pharmazie* **11**, 633.
136. Stahl, E. (1957) *Pharm. Ztg.* **102**, 287.
137. Stahl, E. (1967) *Dünnschichtchromatographie,* 2nd Ed., Springer Verlag, Berlin, 50, 208, 224, 245, 321, 581.
138. Stahl, E. (1968) *J. Chromatogr.* **37**, 99.
139. Stahl, E. (1969) *Arzneim. Forsch. (Drug Res.)* **19**, 1892.
140. Stahl, E. (1969) *Arzneimittelforsch.* **19**, 1892.
141. Stahl, E. (1977) *Dtsch. Apoth. Ztg.* **117**, 1612.
142. Stahl, E., Kaltenbach, U. (1961) *J. Chromatog.* **5**, 351.
143. Stahl, E., Schütz, E. (1978) *Arch. Pharm.* **311**, 992.
144. Stuppner, H., Huber, M., Bauer, R. (1971) *Sci. Pharm.* **59**, 71.
145. Stuppner, H., Huber, M., Bauer, R. (1993) *Pharm. Ztg., Wiss.* **138/6** (2) 46–49.
146. Tschirsch, K., Hölzl, J. (1992) *Pharm. Ztg., Wiss.* **137**, (5) 208.
147. Tyihak, E., Sarkany-Kiss, J., Verzar-Petri, G. (1962) *Pharmazie* **17**, 301.
148. Tyihak, E., Sarkany-Kiss, J., Verzar-Petri, G. (1962) *Pharmazie* **17**, 301.
149. Tyihak, J., Sarkany-Kiss, J., Mathe, J. (1963) *Pharmaz. Zentralh.* **102**, 128.
150. Verzar-Petri, G., Bakos, G. (1979) *Mérés es Automatika* **XXVII**, 104
151. Verzár-Petri, G., Bhan, N. C. (1976) *Sci. Pharm.* **45**, 25.
152. Verzar-Petri, G., Cuong, B. N. (1973) *Sci. Pharm.* **45**, 220.
153. Verzár-Petri, G., Lembercovicz, E. (1976) *Acta Pharm. Hung.* **46**, 129.
154. Verzár-Petri, G., Marczal, G. (1976) *Acta Pharm. Hung.* **46**, 282.
155. Verzár-Petri, G., Marczal, G., Lembercovics, E. (1976) *Herba Hung.* **15**, 69.
156. Verzár-Petri, G., Marczal, G., Lembercovics, E. (1976) *Pharmazie* **31**, 256.

157. Vollmann, H. (1967) Dissertation, Univ. des Saarlandes, 8.
158. Wagner, H., Bladt, S. (1986) *Plant Drug Analysis, A thin Layer Chromatography Atlas*, 2nd Ed., Springer, Heidelberg, 186, 212.
159. Wichtl, M. (1954) *Sci. Pharm.* 22.
160. Wichtl, M. (1971) *Die Pharmakognostisch-chemische Analyse*, Akadem. Verlagsgesellschaft, Frankfurt, Germany.
161. Wichtl, M. (1993) DAB 10 Kommentar, 2. Nachtrag.
162. Willuhn, G., Röttger, P.-M. (1980) *Dtsch. Apoth. Ztg.* **120**, 1039.
163. Wolfram, M. L., Platin, D. L., De Lederkremer, R. M. (1965) *J. Chromatog.* **17**, 488.
164 Ramadan, M. (2005) Dissertation, Marburg/Lahn, Germany.

11 Pharmacology and Toxicology

Heinz Schilcher, Peter Imming, and Susanne Goeters

CONTENTS

11.1 PHARMACOLOGY — PHARMACOLOGICAL PROPERTIES

11.1.1 INTRODUCTION

Chamomile preparations are mainly used because of their antiphlogistic, spasmolytic, and carminative activity. However, their bacteriostatic and fungistatic properties should not be underestimated.

Application fields include dermatology, stomatology, otolaryngology, internal medicine, in particular gastroenterology, pulmology, pediatry, and radiotherapy [11, 56, 100]. The therapeutic effectiveness is in total due to the combined pharmacological and biochemical effects of several chamomile constituents. For therapeutic success, it is important to use standardized total extracts or the essential oil. The full spectrum of activity will not be reached by applying individual chamomile substances only [100].

11.1.2 Antiphlogistic Effect (Subdivided According to Active Principles)

11.1.2.1 Chamazulene, Matricin, Chamazulene Carboxylic Acid, and Guaiazulene

Even though its constitution was not known at the time, chamazulene was identified as the antiphlogistic principle of chamomile oil in a test system of chemosis caused by mustard oil in rabbit eye as early as 1933 [47], and again in 1942 in cavy eye [92]. Because these test systems were questioned later [17, 89], the activity of the "blue azulene" was confirmed in other test models. There was a positive effect on UV erythema of the cavy [50] and on heat-damaged mouse tail [22]. In the rat paw test according to Selye [131, 132], an antiphlogistic activity of the azulene was proven, although the effect of the total extract was much stronger. Guaiazulene was very helpful in the therapy of dermatitis caused by radiation [8, 30, 74, 88, 96, 100, 131]. It displayed weak antipyretic, analgetic, local anesthetic, and anti-histaminic activity, which contributed to the anti-inflammatory effect clinically observed [127].

The activation of the ACTH production and the corresponding functional increase of the adrenal cortex is one of the general effects of azulenes [86]. After applying guaiazulene to sensibilized cavies, a prevention of the antigen-antibody reaction, even the prevention of the anaphylactic shock [127, 131] was observed. Guaiazulene-1-sulfonic acid inhibited a serotonine-triggered shock [36]. Based on these results Stern et al. assumed that azulene prevents the liberation of histamine at the time of antigen and antibody binding, probably via an anti-serotonine effect [104]. A comparison of differently provoked edemas showed that the dextrane edema was inhibited strongly, whereas the activity in hyaluronidase-, formaldehyde-, and histamine-induced edema was only moderate. The authors assumed that the inhibition of histamine liberation, an anti-histaminic effect, an inhibition of 5-hydroxytryptamine liberation, as well as an anti-hyaluronidase effect and a decrease of the capillary activity [115] contributed to the total activity of the azulenes.

In the carrageenine-induced rat paw edema, chamazulene showed about double the antiphlogistic activity of guaiazulene [55]. Further investigations proved a significantly weaker antiphlogistic activity of guaiazulene compared with matricin and $(-)$-α-bisabolol [60]. Guaiazulene can be synthesized easily, but shows only half the antiphlogistic activity of the respective constituents of chamomile (matricin, chamazulene) (see Table 11.1 [60]).

In the carrageenine-induced rat paw edema, matricin was equally active to $(-)$-α-bisabolol 2 and 3 hours after peroral administration. After 4 hours, no significant activity differences were observed between the two substances. Chamazulene and guaiazulene were significantly less effective than bisabolol and matricin. While chamazulene remained antiphlogistically active for more than 4 hours, the effect of guaiazulene decreased. Guaiazulene thus showed a short-term effect [60].

A comprehensive investigation by Ammon with 5- and 12-lipoxygenase, cyclooxygenase, and peroxidation assays showed that chamazulene, but not matricin, inhibited 5-lipoxygenase [3]. Both compounds were equally ineffective regarding the formation of 12-HHT (hydroxyheptadecatriene acid) by cyclooxygenase. Only chamazulene showed a significant anti-oxidative effect.

Chamazulene carboxylic acid (CCA) is a degradation product of sesquiterpene lactones from *Asteraceae*, e.g., matricin. It was discovered not only to have a striking constitutional similarity with profens (e.g., ibuprofen, naproxen), but also to have S-configuration, the eutomeric configuration of profens [39, 38]. It was shown to be a potent anti-inflammatory agent and the second example of a cyclooxygenase-2 (COX-2) selective compound derived from medicinal plants [37, 95]. The inhibition of COX-2 results in anti-rheumatic therapy with greatly reduced side effects and holds the promise of anti-neoplastic therapy. COX-2 was discovered in 1993; it therefore came as a surprise that COX-2 inhibition seems to have been in use for thousands of years through a chamomile and yarrow constituent.

The antiphlogistic activity of CCA was assessed in two standard animal models [37]. Locally applied, it was approximately equipotent to S-naproxen. After oral application, its prodrug pivaloyl

TABLE 11.1
Comparison of the Anti-Inflammatory Effect in the Carrageenan-Induced Rat Paw Edema

	After 1 h			After 2 h			After 3 h		
	Molar titer (confidence interval)	Significance	ED_{50} (mMol/kg)	Molar titer (confidence interval)	Significance	ED_{50} (mMol/kg)	Molar titer (confidence interval)	Significance	ED_{50} (mMol/kg)
(−)-α-Bisabolol, M_R 222	1	—	2.69	1	—	2.95	1	—	2.43
Chamazulene, M_R 184	0.61 (0.38–0.94)	a	4.48	0.70 (0.46–1.08)	ns	4.27	0.49 (0.21–1.15)	ns	5.00
Guaiazulene, M_R 198	0.60 (0.39–0.89)	a	4.59	0.66 (0.46–0.95)	a	4.60	0.33 (0.13–0.80)	a	>7.07
Matricin, M_R 306	1.03 (0.62–1.72)	ns	2.60	1.27 (0.91–1.75)	ns	2.29	0.92 (0.61–1.41)	ns	2.68
Salicylamide, M_R 137	1.77 (1.25–2.51)	a	1.53	2.04 (1.45–2.86)	a	1.44	1.99 (1.28–3.10)	a	1.28

Anti-inflammatory effect in terms of molar titer with confidence intervals and of ED_{50} (mMol/kg p.o.) in the carrageenan-induced rat paw edema 1, 2, and 3 h after edema induction, oral application 1 h before edema induction. [60]

Calculation of molar titer:

$$\frac{\text{titer} \times \text{rel. molecular mass}}{222}$$

$$(222: M_R \text{ of bisabolol})$$

Calculation of molar ED_{50}:

$$\frac{ED_{50} \text{ (mg/kg)}}{\text{rel. molecular mass}}$$

Significance of the titer:

a = significant ($p = 0.05$)

ns = not significant ($p > 0.05$)

ester had approximately 75% aspirin activity. As predicted by docking studies with x-ray structural data of cyclooxygenases [28], CCA did not inhibit cyclooxygenase-1, but cyclooxygenase-2, its selectivity and activity (43.5% inhibition at 50 μM) being comparable to that of a standard COX-2 inhibitor, nimesulide (Aulin, Switzerland) [37].

Because of its structural resemblance with the auxin, indolylacetic acid, it was tested for auxin-binding [54]. The positive result suggests that sesquiterpene lactones may not only function as antifeedants, but also serve for the plant as reservoirs of auxins if they are convertible to azulenes like CCA. Especially the systemic antiphlogistic activities of camomile (and yarrow) thus are partly due to the conversion of guaianolides to "natural naproxen" CCA *in vivo*, especially under acidic conditions (stomach, inflamed tissue).

In a human pharmacokinetic study with volunteers, approved by the ethics commission, three volunteers took 500 mg matricin each orally. After 75 to 90 min, this gave rise to plasma peak levels of 1.3 to 2.2 μg of chamazulene carboxylic acid. The AUC was calculated as 14.7 μg*h/ml. So matricin indeed is a prodrug of the natural profen in man [134].

11.1.2.2 (–)-α-Bisabolol and Bisabolol Oxides

Certain types of chamomile contain up to 50% (–)-α-bisabolol in the essential oil; however, in the majority of types the oxides are more abundant [55].

Three years after its isolation, the antiphlogistic activity of (–)-α-bisabolol was determined in 1954. Its constitution was elucidated in 1968 [57].

After exposing cavies to UV light, (–)-α-bisabolol decreased the skin temperature. Farnesene and a bisabolol monoxide (not defined more specifically) showed a weaker effect [65]. (–)-α-bisabolol reduced the time to heal burns just like chamazulene. Further, it showed an improved blood circulation compared to chamazulene [133]. Histological experiments proved a stimulation of epithelization and granulation by (–)-α-bisabolol and farnesene [45]. A significant anti-inflammatory effect of (–)-α-bisabolol occurred in the carrageenine-induced rat paw edema and cotton-pellet granuloma. Guaiazulene and guaiazulene-1-sulfonic acid (azulene SN) had about the same activity that bisabolol had [62]. A later study found only half the activity for (–)-α-bisabolol compared to chamazulene [63]. However, this was again revised later [60].

There is no doubt that the antiphlogistic effect of (–)-α-bisabolol was proved in the carrageenine-induced rat paw edema and against UV erythema of the cavy, adjuvant arthritis, yeast-induced pyrexia of the rat, and in 5-lipoxygenase and cyclooxygenase tests [3, 61].

Regarding the effect on adjuvant arthritis of the rat, a dose of 500 mg/kg (–)-α-bisabolol was equivalent to 1.5 mg/kg of prednisolone. Even a dose of 2000 mg/kg bisabolol is far below the toxic dose and shows strong antiphlogistic activity. Protective properties of (–)-α-bisabolol on skin were tested in erythemae of cavies, comparing with salicylamide. After peroral administration, (–)-α-bisabolol weakly inhibited the development of an erythema. An increase of the dosage was limited because 2000 mg/kg was toxic for cavies. So the toxic dose of bisabolol was smaller for cavies than for rats.

Since percutaneous application of (–)-α-bisabolol showed inhibition of erythemae, the substance penetrated skin well [61].

Ammon proved that (–)-α-bisabolol was capable of inhibiting both 5-lipoxygenase and cyclooxygenase. Anti-oxidative activity was not observed, in contrast to chamazulene and ethanolic aqueous extracts [110]. In the peroxidation assay, even a propyleneglycol extract did not have any anti-oxidative effect [110].

(–)-α-bisabolol had antipyretic activity against yeast-induced pyrexia of the rat. The fever was reduced by 1.5°C for 2 hours after applying 2000 mg/kg (–)-α-bisabolol perorally [18]. The maximum effect was delayed compared with phenacetine [61].

(–)-α-bisabolol also decreased the production of mucopolysaccharides in cell cultures. The cells were obtained from embryonic spine tissue of mice and from fibroblast cultures.

A report published in 1979 mentioned that "α-bisabolol" (isomer not specified) inhibited the incorporation of radioactive sulfate into the proteochondroitine and protokeratane sulfates in calf cornea *in vitro*. At a concentration of $10^{-4} M$, α-bisabolol inhibited the incorporation by 15%. A higher dose of 10^{-3} M resulted in 58% inhibition, which is comparable with well-known anti-inflammatory agents [29].

Several investigations were undertaken to study if (−)-α-bisabolol could be substituted by the (+)-isomer found in *Populus balsamifera* or by the synthetic racemate [55, 60, 61]. The levogyrate form from chamomile oil proved to be twice as antiphlogistic as the racemate and the dextrogyrate form isolated from the oil of poplar buds. Interestingly, the racemate is weaker than expected, taking both (−)-α-bisabolol and (+)-α-bisabolol into account. The authors ascribe this "to the racemization" [61]. This does not make much sense except if they meant to say that the synthetic bisabolol they used actually was a mixture of all four stereoisomers; apart from that, they state that it did contain 23% of the nonnatural isopropenyl isomer with unknown therapeutic efficiency.

Farnesol and bisabolol oxides A and B were also tested, as well as the commercial Dragosantol, a mixture of racemic synthetic α-bisabolol isopropylidene and isopropenyl isomers.

The test results, summarized in Table 11.2 [61], show that bisabolol oxide B (the main constituent of "Argentinian" chamomile) and bisabolone oxide (the main constituent of "Turkish" chamomile) had only half the activity that (−)-α-bisabolol had. Bisabolol oxide A (the main constituent of "Egyptian" chamomile) had about a third and Dragosantol a quarter of the activity of (−)-α-bisabolol.

11.1.2.2.1 Protective and Curative Effect on Ulcers of (−)-α-Bisabolol and Protective Effect against Acetylsalicylic Acid

The traditional use of chamomile for gastro-intestinal diseases led to tests of (−)-α-bisabolol in different animal ulcer models. The substance showed activity [101] especially in the case of ulcus ventriculi [27, 71, 102, 120].

TABLE 11.2
Comparison of the Anti-Inflammatory Effect in the Carrageenan-Induced Rat Paw Edema, Dose-Effect Relationship [61]

Compound	ED_{50} (mg/kg p.o.)	Titer	Confidence interval of the titer[a]	S	P	L	No. of rats
(−)-α-Bisabolol	1465	1	—	—	—	—	231
(+)-α-Bisabolol	>2000[b]	0.595	0.395–0.898	00	00	00	51
(±)-α-Bisabolol, natural	>2000[b]	0.419	0.197–0.979	000	00	00	48
(±)-α-Bisabolol, synthetic	>2000[b]	0.493	0.277–0.877	000	00	00	51
Dragosantol	>3000[b]	0.260	0.125–0.539	000	00	00	30
Bisabolol oxide A	3164	0.337	0.179–0.635	000	00	00	36
Bisabolol oxide B	>2000[b]	0.591	0.364–0.958	00	00	00	36
Bisabolone oxide	>2000s)	0.431	0.223–0.832	00	00	00	18
Olive oil	>2000s)	0.516	0.271–0.982	00	00	00	18

00   significant with $p = 0.01$

000   significant with $p = 0.001$

P: parallelism

L: linearity of the dose-effect lines ($p = 0.01$)

S: significance of the titers compared to 1

[a] The confidence interval is given for the *p*-value the titer is significant compared to 1. Column "S" shows the *p*-value of the confidence interval.

[b] Largest tested dose; 50% effect (= inhibition of edema formation by 50%) not achieved.

Preparations of chamomile flowers or individual components were not effective in inhibiting acid secretion. Other experiments showed that (–)-α-bisabolol was capable of inhibiting the formation of ulcers induced by indometacin, stress, or alcohol [110]. It was equally effective to metiamide in indometacin-induced ulcus, and in the stress ulcus model it was even better. Just like metiamide, (–)-α-bisabolol accelerated the healing of ulcers induced mechanically (by heat coagulation) or chemically (by acetic acid). The curative and antiulcerogenous activity of a total chamomile extract was obviously not only due to (–)-α-bisabolol since the therapeutic effect was stronger than expected for the amount of bisabolol present. Actually, an ED of 0.25 mg/kg for the chamomile total extract was equivalent to 3.4 mg/kg of bisabolol. More than that, standardized total chamomile extracts protected against ethanol-induced ulcer. This is probably because individual compounds promote the synthesis of prostaglandins, which cause an increase in the mucose barrier that protects against ulcerogenous effects [94, 110]. In the pathogenesis of gastric and duodenal ulcers, the balance of aggressive (e.g., gastric acid) and defensive or protective factors (e.g., mucus of the gastric mucosa) is important, the latter being elicited by chamomile. The finding of an antiseptic effect, first observed by Isaac and Thiemer [58], was pointed out again by Appelt as an important reason for the traditional use of "manzanilla" (the Spanish name for chamomile) [4]. The antiseptic activity of (–)-α-bisabolol is not influenced by changes in pH [58].

Torrado et al. [113] studied the gastrotoxic influence of acetylsalicylic acid of different particle size in rats with concomitant application of (–)-α-bisabolol. At a dose of 200 mg/kg ASA, 0.8–80 mg/kg (–)-α-bisabolol showed a significant ($p < 0.05$) protective effect on gastric mucosa [113].

11.1.2.3 Further Active Principles of the Essential Oil of Chamomile Flowers and in Total Extracts

Cis- and *trans*-spiroether, two spirocyclic polyynes in the essential oil of chamomile flowers, have less antiphlogistic activity as compared to matricin, chamazulene, and (–)-α-bisabolol [16, 117]. The cis-enynedicycloether inhibited dextrane-induced edema in rats and prevented the decrease of plasminogen caused by dextrane. However, it did not inhibit paw edema induced by injections of serotonine, histamine, and bradykinine [16]. Verzár-Petri et al. did not find any antiphlogistic activity of spiroethers in generalized dextrane edema of rats [117]. Chemically these compounds are unstable; therefore, their contribution to the total activity may be low, but their participation is obvious in certain test models.

Choline was found in concentrations of 0.3% in flowers [76] and 0.6% in the herb [10]. Mucus, 10% in flowers [114], and flavonoids [23, 25, 26] contribute to the total antiphlogistic activity of alcoholic chamomile extracts.

In inflammation models induced by croton oil, Della Loggia proved that polysaccharides have anti-inflammatory activity [35].

A study focusing on mitochondria in rat liver showed an increase of the oxidative phosphorylation in the presence of chamomile extract in certain degrees of dilution [112]. ATP biosynthesis was activated simultaneously. The formation of active phosphates is only inhibited by doses that are not achieved *in vivo* [112]. With mitochondria from cavy liver it was shown that the creatine phosphate content increased time dependently and the amount of glucose-6-phosphate decreased [112]. A positive influence of chamomile components was observed on metabolic processes, providing energy and improving the regeneration of inflamed tissue. The use of chamomile extracts in dermatology and cosmetics gets some scientific justification through this (see also Chapter 2).

The antiphlogistic activity of chamomile oil obtained by distillation was studied right from the beginning. Krüger-Nilsen ascribed the antiphlogistic effect to the primary development of subliminal irritations [75]. A specific activity in the leucocytic defense mechanism could indeed be proven in 1952 [7]. The activation of the reticuloendothelial system (RES) was observed in tuberculotic mice [68]. The effect on arthritis caused by formaldehyde was likewise assumed to be due to the stimulation of the RES [73]. Grochulski and Borkowski demonstrated a significant decrease of high urea levels after applying chamomile oil to rabbits suffering from experimental glomerulone-phritis [40]. Within 12 days after the commencement of the therapy, the rabbits had completely recovered [40].

11.1.2.4 Flavonoids

While the spasmolytic effect of the chamomile flavonoids had been known for many years [42, 55], the antiphlogistic activity was proved much later [9, 20, 23, 25, 26, 126].

Della Loggia tested the flavonoids in the inflammation model induced by croton oil. He examined apigenin, luteolin, quercetin, and also rutin and myricetin. Antiphlogistic activity decreased in the following order: Apigenin > luteolin > quercetin > myricetin > apigenin-7-glucoside > rutin. Apigenin even exceeded the activity of indometacin and phenylbutazone. The experiments further showed that apigenin had both a positive influence on the vascular phase of the inflammation (e.g., edema) and on the cellular phase (e.g., the migration of leucocytes). The processes initiated were similar to those caused by nonsteroidal synthetic anti-inflammatory agents.

These results were confirmed by other experiments using different test systems. Apigenin and luteolin were effective and even influenced the metabolism of arachadonic acid.

Nevertheless, the exact mechanism of action has not been understood completely. Several possibilities are under discussion.

In summary, the traditional use of chamomile baths and infusions has been explained pharma-cologically by now. The topical antiphlogistic effect of aqueous chamomile preparations — which had been questioned before — seems to be true since alcoholic extracts contain both the essential oil and flavonoids. A reasonable content of flavonoids in preparations guarantees a high antiphlo-gistic activity.

11.1.3 ANTISPASMODIC EFFECT

Several components found in chamomile flowers are spasmolytic. Some flavonoids and certain components of the essential oil as well as the coumarins herniarin and umbelliferone are responsible for this activity, apart from their fungistatic activity [82].

11.1.3.1 Flavonoids

In 1914 Power and Browning were the first to find apigenin and an apigenin glycoside in chamomile flowers [93]. Their work was cited in numerous publications [11, 31, 53, 60, 77, 79, 98, 118] and followed by several studies concerning the antispasmic activity of aqueous chamomile extracts as well as of individual chamomile flavonoids [1, 45, 46, 48, 51, 52, 53, 60, 66, 67, 98].

According to Hava and Janku, apigenin inhibits the contractions of smooth muscle. The tests were performed using rat or rabbit duodenum. The contractions were induced by barium chloride, acetyl choline, and histamine [45, 66]. Apigenin intensified the adrenaline effect during the test. The contractions of seminal vesicle of cavy and of rabbit uterus were inhibited by apigenin after administering adrenaline. The spasmolytic effect of apigenin was regarded as nonspecific.

Later Hörhammer studied the effects of aqueous and methanolic chamomile extracts as well as of pure flavonoid substances on the convulsion of rabbit ileum induced by barium chloride and acetyl choline [51, 52, 53, 98].

The following results were observed:

1. The musculotropic effect was stronger than the neurotropic one. The chamomile fla-vonoids tested were mainly active on smooth muscle.
2. Alcoholic chamomile extracts seemed to have a stronger spasmolytic activity than aque-ous extracts.
3. Extracts from ligulate flowers were more active than those from tubular florets.

4. The antispasmodic effect of alcoholic extracts was not equivalent to the total effect of flavonoids estimated; therefore, further components were assumed to contribute to the spasmolytic effect. This was confirmed in later studies [1, 117].

5. The individual chamomile flavonoids vary in their antispasmodic activity. In general, the flavonoid aglyca were more active compared with the flavonoid glycosides. The compounds tested can be classified in descending activity as follows: apigenin, quercetin, luteolin, kaempferol, luteolin-7-glucoside, and apigenin-7-glucoside. 10 mg of apigenin were equieffective to about 1 mg of papaverine as for musculotropic effect. Patuletin-7-glucoside and a polyhydroxy-flavone [98] were even more effective than apigenin-7-glucoside [53].

6. The flavonoid aglyca of patuletin, apigenin, quercetin, and luteolin only showed about one fifth to one fourth the activity of papaverine in barium chloride-induced spasms. Their glycosides were about 100 times less active [52].

About 20 years later, chamomile extracts and their components, among them also the chamomile flavonoids already studied, were tested again for musculotropic-spasmolytic activity [1]. In gastro-intestinal diseases, symptoms like pain seem to better correlate with impaired motility than with the size of the lesion [109]. Therefore, a new test system was of great scientific interest in order to review previously published results. The results obtained not only confirmed the spasmolytic activity of chamomile flavonoids described by Hörhammer, they even proved a stronger spasmolytic activity of apigenin compared with papaverine. Apigenin mono-glycosides were about equieffective with the aglyca of luteolin, patuletin, and quercetin. Flavonoid diglycosides were about ten times less active. The results are summarized in Tables 11.3 and 11.4 [20].

11.1.3.2 Essential Oil

Junkmann and Wiechowski [67] discussed that the carminative effect of *Matricariae flos* was part of the intestinal spasmolysis. The essential oil did not show any effect on the autonomous inner-

---

TABLE 11.3

Antispasmodic Action of Various Fractions of the Hydrophilic Phase of the Complete Extract of *M. chamomilla* on Musculotropic Spasms of Isolated Guinea Pig Ileum (Barium Chloride 1×10^4 g/ml) (Data taken from Reference 20)

Substance	ED_{50} (g/ml) ($p \leq 0.05$)	ED_{50} Papaverine (g/ml) ($p \leq 0.05$)	Relative activity (Papaverine = 1) Titer	Relative activity (Papaverine = 1) Confidence intervals of titer ($p \leq 0.05$)
Fraction A	2.37×10^{-4} (1.30–4.31)	1.64×10^{-6} (1.43–1.88)	0.07	(0.006–0.008)
Fraction B	3.36×10^{-6} (2.86–3.95)	1.81×10^{-6} (1.61–2.03)	0.59	(0.48–0.73)
Fraction C	1.82×10^{-3} (1.48–2.24)	2.05×10^{-6} (1.70–2.26)	0.0011	(0.0008–0.0016)
Fraction D	7.22×10^{-3} (5.87–11.2)	$1,42 \times 10^{-6}$ (1.32–1.52)	No titer (approx. 0.0002)	
Fraction E	6.24×10^{-3} (4.84–8.04)	1.33×10^{-6} (1.21–1.46)	No titer (approx. 0.0002)	
Kamillosan[R]	1.09×10^{-3} (0.93–1.28)	1.37×10^{-6} (1.25–1.50)	0.0013	(0.0011–0.0015)

TABLE 11.4

Actions of Various Flavone Aglyca and Flavone Glycosides on Musculotropic Spasms of Isolated Guinea Pig Ileum (Barium Chloride 1×10^{-4} g/ml) (Data Taken from Reference 20)

Substance	ED_{50} (g/ml) ($p \leq 0.05$	ED_{50} Papaverine (g/ml) ($p \leq 0.05$)	Relative activity (Papaverine = 1) Titer	Confidence intervals of titer ($p \leq 0.05$)
Flavone aglyca				
Apigenin	8.02×10^{-7} (6.08–10.6)	2.10×10^{-6} (1.75–2.52)	3.29	(2.34–4.62)
Luteolin	4.56×10^{-6} (3.54–5.87)	1.70×10^{-6} (1.41–2.04)	0.44	(0.30–0.64)
Patuletin	2.63×10^{-6} (2.39 3.02)	1.78×10^{-6} (1.48–2.14)	0.68	(0.56–0.81)
Quercetin	2.22×10^{-6} (1.76–2.79)	1.70×10^{-6} (1.41–2.04)	0.71	(0.46–1.09)
Flavone glycosides				
Apigenin-7-(6″-O-acetyl)glucoside	$1 35 \times 10^{-5}$ (1.23–1.48)	3.67×10^{-6} (3.50–3.84)	0.36	(0.27–0.55)
Apigenin-7-glucoside	8.18×10^{-6} (6.20×10.8)	3.77×10^{-6} (3.14–4.53)	0.46	(0.39–0.55)
Apiin	3.80×10^{-5} (3.16 4.57)		0.08	(0.06–0.11)

vation of rabbit intestine nor on muscarine-induced contracture. Several years later, in 1955 [48] and 1957 [46], a spasmolytic activity of (–)-α-bisabolol and (+)-α-bisabolol on rat ileum was observed. Breinlich and Scharnagel [16] and Vezár-Petri et al. [117] reported a papaverine-like activity of the enyne dicycloethers.

Achterrath and Tuckermann [1] reviewed the data published so far and summarized the results as follows:

1. (–)-α-Bisabolol, the bisabolol oxides A and B as well as the chamomile oil itself have a papaverine-like musculotropic spasmolytic activity.
2. (–)-α-Bisabolol with a titre of 0.91 has the same strong musculotropic antispasmodic activity as papaverine. It is twice as active as bisabolol oxides A and B.
3. Cis-enyne dicycloethers are spasmolytic. There was no linear relationship between effect and dose; therefore, a comparison with the other substances was not possible.
4. The essential oil showed the lowest antispasmodic activity.
5. An alcoholic extract (Kamillosan®) had good spasmolytic activity.
6. The coumarin derivates umbelliferone and herniarin are antispasmodically active. It was not possible to relate their dose to the strength of activity of papaverine, so no titers were calculated.
7. Both hydrophilic (flavonoids) and lipophilic components (essential oil) contribute to the musculotropic antispasmodic effect. In order to standardize chamomile preparations, it is reasonable to include both groups of active principles. Carle and Gomaa summarized the most important experimental data in two tables (see Tables 11.5 and 11.6 [20]).

TABLE 11.5
**Effects of Kamillosan on Various Experimental Spasms of Isolated Guinea Pig Ileum;
Papaverine Was Used as Standard, Unless Otherwise Indicated (Data Taken from
Reference 20)**

Spasmogen (g/ml)	ED_{50} Kamillosan^R (g/ml) ($p \leq 0.05$)	ED_{50} Standard (g/ml) ($p \leq 0.05$)	Relative activity (Standard = 1)	
			Titer	Confidence intervals of titer ($p \leq 0.05$)
Barium chloride (1×10^{-4})	1.22×10^{-3} (0.925–1.61)	1.25×10^{-6} (1.14–1.37)	0.0011	(0.0008–0.0015)
Histamine dihydrochloride (5×10^{-7})	1.15×10^{-3} (1.00–1.32)	2.15×10^{-6} (1.75–2.64)	0.0019	(0.0014–0.0023)
Acetylcholine iodide (5×10^{-8})	2.47×10^{-3} (2.15–2.84)	Atropine: 2.87×10^{-9} (2.39–3.45)	0.00000116	(0.00000089–0.00000155)
Seratonine (5×10^{-7})	2.54×10^{-3} (2.37–2.72)	1.57×10^{-6} (1.06–2.32)	(approx. 0.00062)	
Bradykinine 5×10^{-8}	2.24×10^{-3} (1.78–2.82)	1.65×10^{-6} (1.40–1.94)	0.00071	(0.00052–0.00097)

TABLE 11.6
**Antispasmodic Action of Chamomile Oil, (–)-α-Bisabolol Oxides A and B on
Musculotropic Spasms of Isolated Guinea Pig Ileum (Spasmogen: Barium Chloride 1
× 10⁻⁴ g/ml) (from Reference 20)**

Spasmogen (g/ml)	ED_{50} Kamillosan^R (g/ml) ($p \leq 0.05$)	ED_{50} Standard (g/ml) ($p \leq 0.05$)	Relative activity (Standard = 1)	
			Titer	Confidence interval of titer ($p \leq 0.05$)
Chamomile oil	3.84×10^{-5} (3.19–4.62)	1.64×10^{-6} (1.16–1.84)	0.04	0.03–0.05
(–)-α-Bisabolol	1.36×10^{-6} (1.13–1.63)	1.12×10^{-6} 0.975–1.28	0.91	(0.71–1.17)
Bisabolol oxide A	5.63×10^{-6} (5.13–6.17)	2.60×10^{-6} (2.26–2.98)	0.46	(0.39–0.55)
Bisabolol oxide B	5.65×10^{-6} (5.15–6.19)	2.80×10^{-6} (2.49–3.14)	0.50	(0.44–0.56)

11.1.4 ANTIBACTERIAL AND ANTIMYCOTIC EFFECT

Although alcoholic chamomile preparations were used for a long time because of their antibacterial and antimycotic activity, microbiological tests of chamomile preparations of the essential oil and of individual substances were undertaken comparatively late [2, 21, 32, 99, 100, 105–109, 129]. In 1972, the first report on bacteriostatic and bactericidal activity of the essential oil was published [2]. Both Gram-positive bacteria, such as *Staphylococcus aureus* and *Bacillus subtilis*, and Gram-negative bacteria, such as *Escherichia coli* and *Pseudomonas aeruginosa*, were

TABLE 11.7

Antibacterial Activity of a 42 vol.% Alcoholic Extract against Gram-Positive and Gram-Negative Standard Microorganisms Originated in Mouth and Vagina (Data Taken from Reference 21)

Strain	Controls	Concentrations (mg/ml)[a]						Inoculum (time 0)[b]
		.31	.63	1.25	2.5	5.0	10.0	
B. megatherium	7.08 ± .12		—	7.11 ± .10	7.22 ± .10	7.06 ± .06	<2[c]	5.03 ± .13
C. albicans	5.55 ± .15	—	—	5.84 ± .12	5.77 ± .03	5.29 ± .14	4.91 ± .03[c]	3.14 ± .13
E. coli	7.65 ± .12	—	—	7.81 ± .08	7.78 ± .10	7.80 ± .09	7.23 ± .19[c]	5.13 ± .04
Kl. pneumoniae	8.23 ± .06	—	—	8.21 ± .13	8.13 ± .14	7.92 ± .08[c]	7.38 ± .07[c]	5.28 ± .02
L. icterohaemorr.	7.31 ± .15	7.00 ± .28[c]	6.48 ± .14[c]	4.98 ± .12[c]	<2[c]	<2[c]	—	5.17 ± .11
Ps. aeruginosa	6.87 ± .11	—	—	7.35 ± .16[c]	7.38 ± .08[c]	7.45 ± .02[c]	7.14 ± .05[c]	4.50 ± .21
S. aureus	7.29 ± .04	—	—	7.23 ± .06	7.24 ± .05	6.94 ± .02[c]	4.96 ± .02[c]	4.77 ± .08
S. epidermis	7.46 ± .03	—	—	7.99 ± .13[c]	7.42 ± .15	7.29 ± .01	6.61 ± .02[c]	4.72 ± .02
St. group B	6.52 ± .07	—	—	6.75 ± .06	6.59 ± .08	5.44 ± .15[c]	3.78 ± .11[c]	4.48 ± .04
St. faecalis	6.78 ± .05	—	—	7.39 ± .03[c]	7.30 ± .03[c]	7.29 ± .03[c]	6.42 ± .09[c]	5.01 ± .05
St. mutans	6.56 ± .03	—	—	6.80 ± .07	5.24 ± .12[c]	4.81 ± .08[c]	4.02 ± .06[c]	4.37 ± .14
St. salivarius	7.32 ± .05	—	—	7.31 ± .08	7.22 ± .08	4.32 ± .07[c]	3.48 ± .14[c]	3.75 ± .09

[a] Concentrations are expressed as mg of dry product c.f.u./ml of broth.

[b] Values are logarithms of the number of the c.f.u./ml of broth recovered at the end of incubation time (means ± S.D. of four determinations).

[c] Statistically different from controls ($p < 0.05$).

included. A significant effect on fungi, such as *Candida albicans*, was also observed [2]. Later, alcoholic chamomile extracts proved to be very effective against *Bacillus subtilis*, but had only weak bacteriostatic activity against *Staphylococcus aureus*, *Escherichia coli*, and *Bacillus mesentericus* [129].

(–)-α-Bisabolol has the strongest antibacterial activity compared with the oxides and enyne dicycloethers. It is active in low concentrations against *Staphylococcus aureus*, *Bacillus subtilis*, *Escherichia coli*, *Streptococcus faecalis*, and *Pseudomonas aeruginosa* and inhibits the growth of strains of *Bacterium phlei* that were resistant against standard anti-infectives [105, 108, 109]. (–)-α-Bisabolol and the enyne dicycloethers were fungistatic against *Candida albicans*, *Trichophytone menthagrophytes*, and *Trichophytone rubrum* at a concentration of 100 μg/ml. After an incubation of 30 minutes, a concentration of 1000 μg/ml (–)-α-bisabolol showed fungicidal effectiveness, whereas at the same concentration, enyne dicycloethers were fungicidal after 48 hours. Chamazulene also had fungistatic activity, but at higher concentrations [106, 107].

In contrast to the aforementioned results, a study from 1968 did not find any noteworthy bacteriostatic and fungistatic properties for the enyne dicycloethers against *Staphylococcus aureus*, *Streptococcus haemolyticus*, *Escherichia coli*, or *Candida albicans* [16]. (+)-α-Bisabolol obtained from *Populus tacamahaca* proved to be antibiotically active *in vitro* against *Mycobacterium tuberculosis* and other microorganisms [32].

A chamomile extract with an alcohol concentration of 42% (V/V) was antibacterial and showed activity against trichomonads [21]. Table 11.7 summarizes the antibacterial activity against Gram-positive bacteria strains (*Staphylococcus aureus* ATCC 12600, *Staphylococcus mutans*, group B Streptococcus, and *Streptococcus salivarius*).

Among Gram-negative bacteria, the highest antibacterial effect was observed with *Klebsiella pneumoniae*. There was little effect on the growth of *Escherichia coli*, and no inhibition of *Pseudomonas aeruginosa* ATCC 27853. The extract showed a strong bacteriostatic effect on *Bacillus megatherium* ATCC 96 and *Leptospira icterohaemorrhagia* PB-3.

TABLE 11.8

Comparison of Antibacterial Activity of Four Isolated Components and Defined Alcoholic Extract [21]

Bacteria[a]	Azulene		α-Bisabolol		Bisabolol oxides		Dicycloethers	
	Pure[b]	HEC[c]	Pure[b]	HEC[c]	Pure[b]	HEC[c]	Pure[b]	HEC[c]
S. aureus	4000	2	100	5	300	45	1,000	80
St. faecalis	500	4	2,000	10	50,000	90	15,000	160
E. coli	6000	4	100	10	300	90	50,000	160

[a] Data are referred to the 25% inhibition of growth.

[b] Concentrations of the pure compounds (µg/ml) inhibiting the growth of the bacteria; data from the literature (Szabo-Szalontai, 1976), incubation time 24 hours.

[c] Concentration of the components in the HEC dose having the same antibacterial activity; our data, incubation time 8 hours.

HEC = alcoholic chamomile extract.

Apart from this, there was a significant effect on *Trichomonas vaginalis*. A concentration of 2.5 mg chamomile extract per ml killed trichomonads effectively [21].

Table 11.8 names four main components of the essential oil having a considerably weaker antibacterial effect on *Staphylococcus aureus, Streptococcus faecalis,* and *Escherichia coli* compared to the total alcoholic chamomile extract. Further investigations will have to test the assumption that there is a contribution of hydrophilic components to the cumulative effect.

The contribution of the lipophilic components to this has been proven. As indeed various constituents contribute to the activity, of course a complex composition of a phytopharmaceutical is recommended. For antibacterial therapy with chamomile, standardized preparations should be preferred to preparations containing chamazulene only. This applies accordingly to other indications.

In 1963, the influence of components of chamomile and horseradish on the activity of Streptococcus toxins was tested. It became clear that the healing effect of chamomile in inflammation was not only due to antibacterial activity. Very small quantities of essential oil eliminated the effect of *Streptococcus* and *Staphylococcus* toxins on the upper respiratory tract, especially of the paranasal sinus. A petroleum ether extract of chamomile flowers inactivated the toxins visibly by preventing the hemolysis of blood cells. Allyl mustard oil from horseradish was second in this activity; chamazulene weakest [69].

An extended screening study for mycotic and antibacterial activity included chamazulene and (–)-α-bisabolol [99, 100]. The tests were carried out using the following organisms:

1. *Arthrodermataceae* (dermatophytes), *Trichophytone rubrum, Trichophytone menthagrophytes, Trichophytone tonsurans,* and *Trichophytone quinckeanum*
2. *Candida albicans*
3. *Escherichia coli* ATCC 32 902
4. *Staphylococcus aureus* ATCC 25 924

The minimal inhibitory concentration (MIC, µg/ml) was determined directly (dilution test) and indirectly ("holed plate" assay: diffusion test). The direct MIC test was carried out using four different growing media [99]. Fulcin S (Griseofulvin) and Dermowas were chosen as reference antimycotic and antidermatosis agents. Antibacterial activity was compared with Actidione (cyclo-heximide) and chloramphenicol. Chamazulene had a MIC of 1800 µg/ml, (–)-α-bisabolol of 1000 µg/ml against the tested dermatophyte strains. In the direct MIC test, a concentration of 200 µg/ml both for chamazulene and (–)-α-bisabolol showed a slight fungistatic activity.

There was no satisfactory effect against *Candida albicans* compared with Dermowas and no antibacterial effect on enterobacterial species compared with cycloheximide and chloramphenicol. The antibacterial activity of chamazulene and bisabolol against *Staphylococcus* ATCC 25 294 was about 45% of cyclohexamide and chloramphenicol. This was satisfactory [99]. These results as well as a second study with chamomile coumarins [82] confirmed the antibacterial and fungistatic activity of lipophilic components in chamomile, complementing other reports on antibacterial and fungistatic activity [21, 24, 32, 100].

Infection with *Helicobacter pylori* is now recognized as the primary cause of peptic ulcers and their recurrence. Compelling evidence has also been found linking *H. pylori* infection to gastric cancer. Given the high rate of patient morbidity and mortality associated with gastric cancer, any method by which one can reduce the occurrence of the disease or increase its early detection is desirable. But antibiotics have serious side effects. Therefore, searching for new nontoxic phyto-preparations with anti-*Helicobacter pylori* activity is urgent.

In recent studies chamomile oil extract was prepared using rotopulsed extraction of *Matricaria recutita* flowers by olive oil by Shikov et al. [103]. Coumarin and flavonoid derivatives, polyynes, bisabolol oxides, chamazulene, derivatives of humilones, and chlorophylls were found in this extract. The propagation of *H. pylori* in the control experiment was typical. Chamomile oil extract inhibited the production of urease at *H. pylori*. It is known that the level of activity of urease is very important for the survival of this microorganism in the stomach. It is possible to suppose that the mechanism of the therapeutic action of chamomile oil extract is based on inhibition of colony activity of *H. pylori* and an inhibiting effect on adhesion of this microorganism of phospholipid — lecithin. Thus, chamomile oil extract is promising for application in complex therapy of stomach ulcer and duodenal intestine, especially in patients with an allergic response to antibacterial drugs, and also in case of a resistance of the inducer to antibiotics.

11.1.5 Further Pharmacological Effects

High doses of 1,4-dimethyl-7-isopropylazulene (guaiazulene) influenced carcinogenesis by inhib-iting the development of metastases in animal experiments. Chamomile extracts and synthetic azulene were also shown to inhibit the growth of vaccination tumors [72]. This raised the question of how they interfered with tumor cell metabolism. Barton found [6] that guaiazulene caused damage of the respiratory mechanism of the tumor cell, an increase of fermentation, and finally a complete cessation of metabolism. It acted on the respiratory chain by inhibiting succinic dehy-drogenase in both liver and tumor cells. Thus, one pathway of the citric acid cycle in which the cytochrome system serves as acceptor is eliminated. Whereas the liver cell is in a position to compensate through the other systems (di- and triphosphopyridine nucleotides), this pathway is excluded for the tumor cell on account of its low enzymic activity. According to Emmrich et al., guaiazulene, structurally similar to chamazulene, as well as prednisolone, significantly reduced the granulation tissue produced in rats [33].

Chamomile extract (Kamillosan®, 34%) and 2.5 mg azulene SN (Na 1,4-dimethyl-7-isopropy-lazulenesulfonate) caused 24 and 94% inhibition of pepsin *in vitro* [111].

Della Loggia reported an activity of a freeze-dried aqueous extract (infusion) obtained from tubular florets of *Chamomilla recutita* on the central nervous system [24]. The tests were carried out with mice. The extract was applied intraperitoneally. The basal motility was reduced depending on the dose. A dose of 360 mg/kg chamomile extract reduced the initial motility by 92% without causing any relaxation of the muscles. The voluntary active movements (motoricity) were also decreased significantly. In doses of 160 and 320 mg/kg chamomile extract, slight hypnotic effects were observed. Sleep induced by hexobarbital was extended significantly. The aqueous preparation of chamomile flowers tested seemed to influence the state of activity of the central nervous system. However, the effects were much weaker than those of benzodiazepines: 1 mg/kg benzodiazepine applied perorally was already anxiolytic.

Nevertheless, it seems to be justified to apply chamomile infusion as a mild tranquilizer, a common use in traditional medicine [24]. Wolfmann et al. even characterized apigenin as a ligand for the benzodiazepine binding domain, causing an anxiolytic effect [122]. They write that "the results reported here demonstrate that apigenin is a ligand for the benzodiazepine receptor and possesses anxiolytic effects without evidencing anticonvulsant or myorelaxant actions." A recent paper [130] reported on the behavioral effects of acute administration of apigenin and chrysin in rats. Both flavonoids were equally able to reduce locomotor activity when injected in rats at a minimal ED of 25 mg/kg. However, while chrysin exhibited a clear anxiolytic effect when injected at a dose of 1 mg/kg, apigenin failed to exert this activity. The sedative effect of these flavonoids could not be ascribed to an interaction with GABA-benzodiazepine receptors, since it was not counteracted by the benzodiazepine antagonist flumazenil. To the contrary, the anxiolytic effect of chrysin, which was blocked by the injection of flumazenil, could be linked to an activation of the $GABA_A$ receptor unit.

11.2   TOXICITY AND SIDE EFFECTS OF CHAMOMILE PREPARATIONS AND INDIVIDUAL CHAMOMILE ACTIVE PRINCIPLES

11.2.1   ACUTE AND SUBACUTE TOXICITY

In the United States, chamomile oil has the GRAS status ("generally recognized as safe") and is admitted for food and cosmetics by the FDA.

The LD_{50} of chamomile oil exceeds 5 g/kg weight for acute oral toxicity in rats and acute dermal toxicity in rabbits [87]. Jakovlev and Schlichtegroll published values summarized in Table 11.9 [62]. Both the total chamomile oil and (–)-α-bisabolol proved to have almost no acute toxicity, unlike the two synthetic azulenes.

The extremely low toxicity of (–)-α-bisabolol after oral administration was confirmed later [41]. In dogs and Rhesus monkeys, tolerance of (–)-α-bisabolol after oral application was good. Undesirable effects were observed only with high doses. Doses of 12.6–15.9 ml/kg caused retching and vomiting; therefore, a determination of LD_{50} with dogs was not possible.

A 4-week subacute test in rats and dogs resulted in a toxicity of 1.0 and 2.0 ml (–)-α-bisabolol/kg. The prenatal development of rats and white New Zealand rabbits was not influenced

TABLE 11.9
Acute Toxicity According to Jakovlev and Schlichtegroll [62]

Substance	LD_{50} mouse, mg/kg, oral administration	LD_{50} rat, mg/kg, oral administration
(–)-α-Bisabolol	11,350	14,850
Guaiazulene	1,540	6,380
Acidified sulphonic natrium of guaiazulene	1,300	1,550
Essential chamomile oil	—	10,000–20,000
Phenylbutazone	625	530
Indometacin	24	24

by oral doses up to 1 ml/kg. None of the tested dosages caused any deformations. Intolerance was only observed with doses that were toxic for the adult animal by itself (approx. 3.0 ml/kg) [41].

The LD_{50} of cis-enyne dicycloether was determined in mice to be 670 mg/kg [16]. A pharmacological test of an aqueous chamomile extract (freeze-dried extract from 5 g chamomile flowers with 100 ml boiling water) did not show acute toxicity even in a dose of 1440 mg of extract/kg applied intraperitoneally [24].

11.2.2 Skin Reactions (Cell-Mediated Allergy Type IV)

Literature reports about irritative and allergic skin changes (contact dermatitis) as well as reactions of the respiratory tract and the mucous membranes (rhinitis, conjunctivitis, anaphylactic shock) are very inconsistent. A report about anaphylactic shock [13], supposedly caused by a chamomile preparation, gave reason to include chamomile flowers in a handbook of toxic plants and plant poisons. Hausen, University of Hamburg, re-examined this incident and stated that *Chamomila recutita* plays quite an unimportant role among other environmental allergens [42, 44]. Even before his investigations, it had been shown that hairless mice tolerated undiluted chamomile oil without any primary skin irritations [116], whereas a moderate irritant effect on the intact and scarified skin of rabbits was observed 24 hours after application. The same authors [116] as well as Kligman [70] could not find any skin irritation 48 hours after applying a small cloth soaked with chamomile oil on persons. Further reactions such as sensibilization or phototoxic effects could not be confirmed either, whereas Beetz had provoked contact dermatitis by applying pharmaceutical and cosmetic preparations containing chamomile [12].

According to Hausen, 46 of 51 reports on contact dermatitis by chamomile do not stand a critical re-examination [44]. In only five of these studies, a botanical identification was carried out; however, a determination of the origin or chemical type was never made. In at least 21 reports it was quite obvious that the contact allergies were caused by *Anthemis* species (particularly *Anthemis cotula*, stinking mayweed, dog chamomile). In English-speaking countries, very often there is no linguistic differentiation between genuine chamomile (*Matricaria recutita*) and Roman chamomile (*Chamaemelum nobile* (L.) All.). Both species contain contact allergens. The most potent in this context is the linear sesquiterpene lactone anthecotulide from *A. cotula,* which has strong contact allergenic activity in sensibilization tests. Up to 1.8% anthecotulide was found in *A. cotula,* which often contaminates wild collections. Yamazaki et al. [128] claimed to have isolated 7.3% (!) anthecotulide from *Matricaria recutita* L. One of the authors of this chapter (P.I.) asked the curator of the Herbarium of the University of Texas, Austin, to redetermine the voucher specimen of Yamazaki et al. As suspected, it was actually *Anthemis cotula* L. [121].

Hausen showed in Freund's complete adjuvant (FCA) test that anthecotulide was either missing completely in cultivated chamomile or it was contained in a concentration of 0.003–0.01% only, viz. in chamomile of Argentine origin (bisabolol oxide B type). Such low concentrations are not sufficient for irritative skin reactions. A test of 12 chamomile constituents (e.g., matricin) using animals sensibilized with a total chamomile extract turned out to be negative [42].

Chamomile preparations as used in several studies very likely contained preservatives, ointment bases, etc. Some additives known to be allergenic could have given falsely positive results regarding the active principles of a chamomile preparation causing allergy after skin contact or inhalation. Hausen [42] stated: "The results [of his own experiments] are thus in good correlation to the rareness of cases observed [in the Department of Dermatology, University of Hamburg, Germany] of a specific hypersensitivity to genuine chamomile."

Nevertheless, pollen allergy (pollinosis) caused by chamomile pollen was found relatively often due to multiple response to *Compositae* pollen (pollen of *Asteraceae*). This pollen allergy cannot be excluded completely when applying chamomile pharmaceuticals, since soluble antigenes of the pollen exines could possibly be found in extracts. Steam inhalation of chamomile flowers is most likely to cause allergies. A publication in the *American Journal of Contact Dermatitis* by Bjoern

M. Hausen summarized the results of 10 years' experience and experimental testing concerning the "*Compositae* Allergy" [43]. According to the patch test result, 3.1% of 3871 allergic patients showed a positive reaction towards a *Compositae* mixture consisting of ether extracts from feverfew, *Arnica montana*, German chamomile, yarrow, and tansy. In order to provoke a positive patch test, the concentration of the German chamomile extract had to be five times higher than that of *Arnica*. In individual patch tests, 70.1% responded positively to feverfew and 56.5% to German chamomile. Hausen suggested including at least five constituents in allergy test mixtures. He considered not only the sesquiterpene lactones but also polyynes probably to be responsible for the allergic reactions.

Recently, Jablonski and Rudzki reported clinical observations on 540 eczema patients who were treated with chamomile concentrate in the Dermatological Clinic in Warsaw [59]. Patients with a positive reaction to various standard substances in the epicutaneous test responded negatively to Kamillosan® concentrate in all 540 cases. This positive result regarding Kamillosan was ascribed to the fact that anthecotulide was absent.

REFERENCES

1. Achterrath-Tuckermann, U., Kunde, R., Flaskamp, E., Isaac, O., Thiemer, K. (1980) *Planta Med.* **39**, 38–50.
2. Aggag, M. E., Yousef, R.T. (1972) *Planta Med.* **22**, 140.
3. Ammon, H. P. T., Sabieraj, J., Kaul, R. (1996) *Dtsch. Apoth. Ztg.* **136**, 1821.
4. Appelt, G. D. (1985) *J. Ethnopharmacol.* **13**, 51–55.
5. Baker, P. M., Fortes, C. C., Fortes, E. G., Gazzinelli, G., Gilbert, B., Lopes, J. N. C., Pellegrino, J., Tomassini, T. C. B., Vichnewski, W. (1972) *J. Pharm. Pharmac.* **24**, 853.
6. Barton, H. (1959) *Acta Biol. Med. Gem.* **2**, 555.
7. Barton, H., Wendler, M. (1952) *Arch. Exp. Pathol. Pharmakol.* **215**, 573.
8. Bartunková, Z. (1956) *Cs. Dermatol.* **31**, 334.
9. Baumann, J., Wurm, G., Bruchhausen, F. (1980) *Arch. Pharm.* **313**, 330.
10. Bayer, J., Katona, K., Tardos, L. (1958) *Acta Pharm.- Hung.* **28**, 164.
11. Becker, H., Reichling, J. (1981) *Dtsch. Apoth. Ztg.* **121**, 1285.
12. Beetz, D., Kramer, H. J., Mehlhorn, H. Ch. (1971) *Dermatol. Monatsschr.* **157**, 505.
13. Benner, M. H., Lee, H. J. (1973) *Allergy Clin. Immunol.* **52**, 307.
14. Blum, A. L., Siewert, R. (1978) *Peptische Läsion im Lichte von Aggression und Protektion,* Demling, L., Rösch, W. (Eds.) Verlag Witzstrock, G., Baden-Baden.
15. Bohlmann, F., Zdero, C., Grenz, M. (1969) *Tetrahedron Letters* Nr. 28, 2417.
16. Breinlich, J., Scharnagel, K. (1968) *Arzneim.-Forsch.* **18**, 429.
17. Brock, N., Kottmeier, J., Lorenz, D., Veigel, H. (1954) *Arch. Exp. Pathol. Pharmakol.* **223**, 450.
18. Büchi, O. (1959) *Arch. Int. Pharmacodyn.* **123**, 140.
19. Busse, W. W., Kopp, D. E., Middleton, E. (1984) *J. Allergy Clin. Immunol.* **73**, 801.
20. Carle, R., Gomaa, K. (1992) *Drugs of Today* **28**, 559.
21. Cinco, M., Baufi, E., Tubaro, A., Della Loggia, R. (1983) *Int. J. Crude Drug Res.* **21**, 145.
22. Deininger, R. (1956) *Arzneim.-Forsch.* **6**, 394.
23. Della Loggia, R. (1985) *Dtsch. Apoth. Ztg.* **125, Suppl. I**, 9.
24. Della Loggia, R., Traversa, U., Scarcia, V., Tubaro, A. (1982) *Pharmacol. Research Comm. of the Italian Pharmacol. Society* **14**, 153.
25. Della Loggia, R., Tubaro, A., Zilli, C. (1984) *32nd Annual Congress for Medicinal Plant Research,* Antwerp, Abstracts L.16.
26. Della Loggia, R.; Tubaro, A., Dri, P., Zilli, C., Del Negro, P. (1986) *Plant Flavonoids in Biology and Medicine — Biochemical, Pharmacological and Structure-Activity Relationships,* Alan R. Liss, Inc., pp. 481–484.
27. Demling, L. (1975) in Demling, L., Nasemann, T. (Eds.) *Erfahrungstherapie — späte Rechtfertigung,* Int. Symposium Vienna, May 30–31, 1975, Verlag G. Braun, Karlsruhe, Germany.

28. Dettmering, D., Imming, P. Poster at the International Congress of the Society for Medicinal Plant Research, Zürich, Switzerland, Sept. 3–7, 2000.
29. Deutsche Offenlegungsschrift 24 26393 (Ludwig Merkle KG), ref. in Isaac, O. (1979) *Planta Mmed.* **35**, 118.
30. Docekal, B. (1956) *Cs. dermatol.* **31**, 340.
31. Dölle, B., Carle, R., Müller, W. (1985) *Dtsch. Apoth. Ztg.* **125 Suppl. I**, 14.
32. Dull, G. G., Fairley, J. L., Gottshall, R. Y., Lucas, E. H. (1956–1957) *Antibiot. Ann.*, 682, ref. in Isaac, O., Schneider, H., Eggenschwiller, H. (1968) *Dtsch. Apoth. Ztg.* **108**, 293.
33. Emmrich, R., Schade, U. (1964) *Z. inn. Med.*, Leipzig, **19**, 429.
34. Fewtrell, C. M. S., Gomperts, B. D. (1977) *Nature* **265**, 635.
35. Füller, E., Sosa, S., Tubaro, A., Franz, G., Della Loggia, R. (1993) *Planta Med.* **59**, Supplement A 666.
36. Giertz, H., Hahn, F. (1959) *Arzneim.-Forsch.* **9**, 553–555.
37. Goeters, S. (2001) Dissertation, Marburg (Lahn).
38. Goeters, S., Imming, P., Dullweber, F., Hempel, B. (2000) *Arch. Pharm.* **333** (Suppl. 1), 12.
39. Goeters, S., Imming, P., Pawlitzki, G., Hempel, B. *Planta Medica* (2001) 67, 292–294.
40. Grochulski, A., Borkowski, A. (1972) *Planta Med.* **21**, 289.
41. Habersang, S., Leuschner, F., Isaac, O., Thiemer, K. (1979) *Planta Med.* **37**, 115.
42. Hausen, B. M. (1985) *Dtsch. Apoth. Ztg.* **125 Suppl. I**, Nr. 43/24.
43. Hausen, B. M. (1996) *Amer. J. Contact Dermatitis* **7**, No. 2, 94.
44. Hausen, B. M., Busker, E., Carle, R. (1984) *Planta Med.* **50**, 229.
45. Hava M., Janku J. (1957) *Rev. Czech. Med.* **3**, 130.
46. Hava, M., Janku, J. (1958) *Cs. Fysiol.* **7**, 464.
47. Heubner, W., Grabe, E, (1933) *Arch. Exp. Pathol. Pharmakol.* **171**, 329.
48. Holub, M., Herout, V., Sorm, F. (1955) *Ceskoslov. Farm* **3**, 129, ref. in Isaac, O., Schneider, H., Eggenschwiller, H. (1968) *Dtsch. Apoth. Ztg.* **108**, 293.
49. Hope, W. C., Welton, A. F., Fiedler-Nagy, C., Batula-Bernardo, C., Cofley, J. W. (1983) *Biochem. Pharmacol.* **32**, 367.
50. Horáková, Z. (1952) *Cs. Fisiolog.* **1**, 148.
51. Hörhammer, L. (1961) *Dtsch. Apoth. Ztg.* **101**, 1178.
52. Hörhammer, L., Wagner, H. (1962) *Dtsch. Apoth.* **14**, 1.
53. Hörhammer, L., Wagner, H., Salfner, B. (1963) *Arzneim.-Forsch.* **13**, 33.
54. Imming, P., Goeters, S., Ford, Y., Blake, P., Marshall, J., Napier, R. (2002) unpublished results.
55. Isaac, O. (1979) *Planta Med.* **35**, 118.
56. Isaac, O., Kristen, G. (1980) *Med. Welt* **31**, 1145.
57. Isaac, O., Schneider, H., Eggenschwiller, H. (1968) *Dtsch. Apoth. Ztg.* **108**, 293.
58. Isaac, O., Thiemer, K. (1975) *Arzneim.-Forsch.* **25**, 1352.
59. Jablonski, S., Rudzki, E. (1996) *HG-Ztschr.-für Hauterkr.* **71**, 542.
60. Jakovlev, V., Isaac, O., Flaskamp, E. (1983) *Planta Med.* **49**, 67.
61. Jakovlev, V., Isaac, O., Thiemer, K., Kunde, R. (1979) *Planta Med.* **35**, 125.
62. Jakovlev, V., Schlichtegroll, A. V. (1969) *Arzneim. Forsch.* **19**, 615.
63. Jakovlev. V. (1975) in Demling, L., Nasemann, T. (Eds.) *Erfahrungstherapie — späte Rechtfertigung*, Int. Symposium Vienna, May 30–31, 1975, Verlag G. Braun, Karlsruhe, Germany.
64. Janecke, H., Weisser, W. (1964) *Planta Med.* **12**, 528.
65. Janku, J., Zita, C. (1954) *Cs. Fam.* **3**, 93.
66. Janku, J. (1981) Paper at 2nd Physiolog. Conf. Königgrätz, ref. in Becker, H., Reichling, J. (1981) *Dtsch. Apoth. Ztg.* **121**, 1285.
67. Junkmann, K., Wiechowski, W. (1929) *Arch. Path. Pharmakol.* **144**, 1–7.
68. Kato, L., Gözsy, B. zit., Tur, W., Joss, B. (1959) *Azulen im Lichte der medizinischen Weltliteratur*, Flyer of the company Th. Geyer KG, Stuttgart, ref. in Thiemer, K., Stadtler, R., Isaac, O. (1973) *Arzneim.-Forsch.* **23**, 756.
69. Kienholz, M. (1963) *Arzneim.-Forsch.* **13**, 920.
70. Kligman, A. M. (1973) *Report to RIFM* 31.1.1973, ref. in Habersang, S., Leuschner, F., Isaac, O., Thiemer, K. (1979) *Planta Med.* **37**, 115–123.
71. Koch, H. (1975) in Demling, L., Nasemann, T. (Eds.) *Erfahrungstherapie — späte Rechtfertigung*, Int. Symposium Vienna, May 30–31, 1975, Verlag G. Braun, Karlsruhe, Germany.

72. Kraul, M. A., Schmidt, E. (1957) *Arch. Pharm.* **290**, 66.
73. Kraul, M. A., Schmidt, F. (1955) *Z. inn. Med.* **10**, 934.
74. Kristen, G., Schmidt, W. (1957) *Arch. Pharm. Mitt. Dtsch. Pharm. Ges.* **27**, 105.
75. Krüger-Nilsen, B. (1934) *Arch. Exp. Pathol. Pharmakol.* **174**, 197.
76. Kuehl et al. (1957) *J. Amer. Chem. Soc.* **79**, 5576, ref. in Schilcher, H. (1966) *Präp. Pharmazie* **3**, 1.
77. Kunde, R., Isaac, O. (1979) *Planta Med.* **376**, 124.
78. Landolfi, R., Mower, R. L., Steiner, M. (1984) *Biochem. Pharmacol.* **33**, 1525.
79. Lang, W., Schwandt, K. (1957) *Dtsch. Apoth. Ztg.* **97**, 149.
80. Lee, T. P., Matteliano, M. T., Middelton, E. (1982) *Life Sa.* **31**, 2765.
81. Lengfelder, E. (1984) *Agents Actions* **15**, 56.
82. Mares, D., Romagnoli, C., Bruni, A. (1993) *Plantes Médicinales et Phytothérapie* **26**, 91.
83. Meyer, F. (1952) *Z. Naturforsch.* **7 b**, 61.
84. Middleton, E., Drzewiecki, G. (1982) *Biochem. Pharmacol.* **31**, 1449.
85. Middleton, E., Drzewiecki, G. (1984) *Biochem. Pharmacol.* **33**, 3333.
86. Millin, R., Stern, P., Kossak, R. (1955) *Z.R. Assoc. Anatom.* **55**, 438.
87. Moreno, O. M. (1973) *Report to RIFM* 31.10.73, ref. in Habersang, S., Leuschner, F., Isaac, O., Thiemer, K. (1979) *Planta Med.* **37**, 115.
88. Nöcker, J., Schleusing, G. (1958) *Münch. Med. Wschr.* **100**, 495.
89. Oettel, H., Wilhelm-Kollmannsperger, G. (1955) *Arch. Exp. Pathol. Pharmakol.* **226**, 473.
90. Opdyke, D. L. J.(1974) In *Food and Cosmetic Toxicology*, Oxford, 851.
91. Pearce, F. L., Dean Befus, A., Bienenstrock, J. (1984) *J. Allergy Clin. Immunol.* **73**, 819.
92. Pommer, Ch. (1942) *Arch. Exp. Pathol. Pharmakol.* **199**, 74.
93. Power, F., Browning, H. Jr. (1914) *J. Chem. Soc. London* **105**, 2280.
94. Robert, A., Lancaster, C., Hauchar, A. J., Nezamis, J. E. (1978) *Gastroenterology* **74**, 1086.
95. Robugen GmbH, Ger. Offen. DE 10065683 A1 5 July 2001 (*Chem. Abstr.* 2001, 135, 76647).
96. Roeckerath, W. (1950) *Strahlentherapie* **82**, 253.
97. Roth, L., Daunderer, M., Kormann, K. (1984) *Giftpflanzen-Pflanzengifte. Vorkommen, Wirkungen, Therapie*, ecomed Verlagsgesellschaft mbH, Landsberg-Munich, Germany.
98. Salfner, B. (1963) Dissertation, Univ. Munich, Germany.
99. Schilcher, H. (1985) *Zur Biologie von Matricaria chamomilla, syn. "Chamomilla recutita (L.) Rauschert,"* Research report 1968-1981, Inst. Pharmakognosie and Phytochemie of the FU Berlin.
100. Schilcher, H. (1987) *Die Kamille — Handbuch für Ärzte, Apotheker und andere Wissenschaftler,* Wissenschaftliche Verlagsgesellschaft, Stuttgart, Germany.
101. Schilcher, H. (1990) *Dtsch. Apoth. Ztg.* **130**, 555.
102. Schmidt, E. (1975) in Demling, L., Nasemann, T. (Eds.) *Erfahrungstherapie — späte Rechtfertigung,* Int. Symposium Vienna, May 30–31, 1975, Verlag G. Braun, Karlsruhe, Germany.
103. Shikov, A. N., Pozharitskaya, O. N., Makarov, V. G. et al. (1999) *Method of allocation of biologically active substances from plant material.* Patent Ru 214 1336 from Nov. 2, 1999.
104. Stern, P., Millin, R. (1956) *Arzneim.-Forsch.* **6**, 445.
105. Szabo-Szalontai, M., Verzár-Petri, G. (1976) 24. Jahresversammlung d. Ges. f. Arzneipflanzenforsch., Munich, Germany.
106. Szalontai, M., Verzár-Petri, G., Florián, E. (1976) *Acta Pharm.-Hung.* **46**, 232.
107. Szalontai, M., Verzár-Petri, G., Florián, E. (1977) *Parfümerie und Kosmetik* **58**, 121.
108. Szalontai, M., Verzár-Petri, G., Florián, E., Gimpel, F. (1975) *Dtsch. Apoth. Ztg.* **115**, 912.
109. Szalontai, M., Verzár-Petri, G., Florián, E., Gimpel, F. (1975) *Pharmaz. Ztg.* **120**, 982.
110. Szelenyi, J., Isaac, O., Thiemer, K. (1979) *Planta Med.* **35**, 218.
111. Thiemer, K., Stadtler, R., Isaac, O. (1972) *Arzneim.-Forsch.* **22**, 1086.
112. Thiemer, K., Stadtler, R., Isaac, O. (1973) *Arzneim.-Forsch.* **23**, 756.
113. Torrado, S., Torrado, Susana, Agis, A., Jiménez, M. E., Cadorniga, R. (1995) *Pharmazie* **50**, 141.
114. Tubaro, A., Zilli, C., Redaelli, C., Della Loggia, R. (1984) *Planta Med.* **51**, 359.
115. Uda, T. (1960) *Nippon Yak. Zasshi* **56**, 1151; ref. in *Chem. Abstr.* **50**, 4058 (1962).
116. Urbach, F., Forbes, P. D. (1973) *Report to RIFM* 18.3.73, ref. in Habersang, S., Leuschner, F., Isaac, O., Thiemer, K. (1979) *Planta Med.* **37**, 115.
117. Verzár-Petri, G., Szegi, J., Marczal, G. (17979) *Acta Pharm. Hungarica* **49**, 13.
118. Wagner, H., Kimayer, W. (1957) *Naturwiss.* **44**, 307.

119. Weiss, R. E. (1982) *Zschr. Phytotherapie* **3**, 439.
120. Weiss, R. F. (1974) *Lehrbuch der Phytotherapie*, Hippokrates-Verlag, Stuttgart, Germany.
121. Wendt, T. (2003) Private communication (Plant Resources Center, Herbarium, University of Texas, Austin).
122. Wolfmann, C., Viola, H., Wasowski, C., Levi de Stein, M., Silveira, R., Dajas, F., Medina, J. H., Paladini, A. C. (1995) *Journ. Neurochem.* **65**, 167.
123. Wolters, B. (1969) *Planta Med.* **17**, 42.
124. Wolters, B. (1975) *Dtsch. Apoth. Ztg.* **115**, 213.
125. Wolters, B. (1976) *Dtsch. Apoth. Ztg.* **116**, 667.
126. Wurm, G., Baumann, J., Geres, V. (1982) *Dtsch. Apoth. Ztg.* **122**, 2062.
127. Yamasaki, H., Irino, S., Uda, A., Uchida, K., Onho, H., Saito, N., Kondo, K., Jinzenji, K., Yamamoto, T. (1958) *Nippon Yakurigaku Zasshi* **54**, 362; ref. in *Chem. Abstr.* **53**, 10525 (1959).
128. Yamazaki, H., Miyakado, M., Mabry, T. J. (1982) *J. Nat. Prod.* **45**, 508.
129. Zajz, K. A., Arkadjewna, H. W., Iljina, W. A. (1975) *Farmatsiya (Moskva)* **24**, 41, ref. in Isaac, O., Kristen, G. (1980) *Med. Welt* **31**, 1145.
130. Zanoli, P., Avallone, R., Baraldi, M. (2000) *Fitoterapia* **71** (Suppl. 1), S117-S123.
131. Zierz, P., Kiessling, W. (1953) *Dtsch. Med. Wschr.* **78**, 1166.
132. Zierz, P., Lehmann, A., Craemer, R. (1957) *Hautarzt* **8**, 552.
133. Zita, C. (1955) *Cas. Lék. ces.* **94**, 203, ref. in Isaac, O. (1979) *Planta Med.* **35**, 118.
134. Ramadan, M. (2005) Dissertation, Marburg/Lahn, Germany.

12 Traditional Use and Therapeutic Indications

Heinz Schilcher

CONTENTS

12.1 TRADITIONAL USE

Chamomile has been known for centuries and is well established in therapy. In traditional folk medicine it is found in the form of chamomile tea, which is drunk internally in cases of painful gastric and intestinal complaints connected with convulsions such as diarrhea and flatulence, but also with inflammatory gastric and intestinal diseases such as gastritis and enteritis.

Externally chamomile is applied in the form of hot compresses to badly healing wounds, such as for a hip bath with abscesses, furuncles, hemorrhoids, and female diseases; as a rinse of the mouth with inflammations of the oral cavity and the cavity of the pharynx; as chamomile steam inhalation for the treatment of acne vulgaris and for the inhalation with nasal catarrhs and bronchitis; and as an additive to baby baths. In Roman countries it is quite common to use chamomile tea even in restaurants or bars and finally even in the form of a concentrated espresso. This is also a good way of fighting against an upset stomach due to a sumptuous meal, plenty of alcohol, or nicotine. In this case it is not easy to draw a line and find out where the limit to luxury is.

12.2 CLINIC AND PRACTICE

12.2.1 PRELIMINARY REMARK

The suitability of the empirical application of chamomile flower has been confirmed by intense research work over the last years. As a matter of fact, however, experience has shown that a usual chamomile tea domestically produced with boiling hot water only contains a small portion (1–3%) of the essential oil to be found in the drug [30, 34]. The important constituents of the essential oil with an antiphlogistic effect are not water soluble and remain in the chamomile flowers. One part of the essential oil evaporates. But the tea does contain a number of water-soluble flavonoids with a spasmolytic and — according to latest tests — also an antiphlogistic effect if locally applied [14].

A systematic clinical research of chamomile preparations only started in the 1920s. In 1921 a chamomile extract under the name Kamillosan® was produced and introduced into trade. This made available a preparation containing all essential constituents of chamomile, viz., chamazulene in the form of the prestage of Matricin, (–)-α-bisabolol (INN: levomenol), bisabolol oxides, cis- and trans-spiroether, and flavonoids. Kamillosan® is produced from the chamomile varieties Degumille (DBP 24 02 802) and Manzana, being rich in azulene and bisabolol. One hundred milliliters of extract contains at least 150 mg of apigenin-7-glucoside, at least 150 mg of essential oil from chamomile flowers, at least 50 mg of levomenol, and at least 3 mg of prochamazulene/chamazulene (determined and calculated as chamazulene).

The following section reports the research results of this standardized alcoholic chamomile extract. Comparative tests with other chamomile preparations, with the exception of a few other tests, especially with alcoholic or aqueous chamomile extracts were not available before 1998. The most important components are chamazulene with an undisputed antiphlogistic effect [31] as well as (–)-α-bisabolol, a sesquiterpene alcohol with antiphlogistic, antibacterial, antimycotic, ulcus-protective, and musculotropical-spasmolytical properties. Both of them are main constituents of the essential oil. The chamomile flavones have a musculotropical, a spasmolytical, and an antiphlogistic effect if locally applied. Further important active principles are the bisabolol oxides A and B, cis- and trans-spiroether, coumarins, and mucilage.

The therapeutic value of chamomile preparations is not just based on one main active principle but on the combination of many individual ones (their effects were proved both clinically and by animal experiments and combine a total effect suitable for a wide range of indications). A chamomile extract containing all components in an optimum concentration should be given preference to a usual domestic infusion. When diluting such an extract with hot water the lipophile constituents are also partly incorporated in the tea, being most suitable both for internal as well as for external application. In addition the concentrated form of the chamomile extract makes use exceeding the usual chamomile therapy possible.

The proven antiphlogistic effect is therapeutically used by various specialized disciplines on a broad basis and in different ways. A particularly favorable judgment should be passed on the fact that the corresponding preparations can be applied both internally and externally.

12.2.2 DERMATOLOGY

Prof. Born (Dermatological Clinic in Freiburg) describes the effect as soothing and antiphlogistic. A chamomile extract is, for instance, applied for the irrigation of undermined margins of a wound, pouches, sinus tracts, and hip baths, correspondingly diluted or in concentrated form for swabbing inflammatory lesions of the mucosa [4]. It is duly emphasized that, as experience has shown, chamomile preparations are willingly accepted by the patient; medically, however, there is a well-founded understandable reserve, especially in view of allergies appearing occasionally. If an appropriate extract is applied there is no reason for such doubts. In their review article, "Evidence for the Efficacy and Safety of Topical Herbal Drugs in Dermatology: Part I: Anti-Inflammatory Agents," Hörmann and Korting [29] came to a similar conclusion in 1994.

This statement clearly shows that from the medical point of view preference is always given to a standardized finished medicament compared with the usual chamomile preparations, particularly with external treatment of infected tissues.

According to Weitgasser [60], in dermatology mainly acute suppurative dermatoses of different genesis needing an additional treatment with baths or fomentions/stupes are possible as indications. A chamomile extract leads to good therapeutic success in the case of acute suppurative dermatoses such as dermatitis caused by contact — allergic exanthemata, intertrigo, ulcus cruris, and eczemas — as well as with diseases of the mucosae such as stomatitis, pemphigus vulgaris, and others. After a short treatment time relief is noticed and complaints are minimized. In this connection the cooling effect causes a particularly pleasant feeling. The antiphlogistic and slightly anaesthetizing effect of chamomile extract makes the basic therapy considerably easier. In a controlled double-blind study the clinical effectiveness of the medicament Kamille-Spitzner® was examined [23]. The drying and epithelizing effect on suppurative dermatoses after an abrasive tattooing was taken as an objective study parameter. Statistically both the reduction of the discharging wound area and the drying tendency were marked more heavily than in the placebo group.

Within 2 days the application of fomentions led to a considerable improvement of the inflammatory symptoms of dermatitis statica and dermatitis caused by contact. In a unilateral comparative study the fomentions with the standardized chamomile medicament turned out to be superior to those applied with salt solution. The local tolerance was very good; side effects were not observed [15, 43]. Partial baths, rinses/irrigations, and fomentions with chamomile extracts as well as the use of ointments also led to a quick disinfection of infected wounds and ulcers such as ulcera cruris [7, 13, 21, 24, 28].

After a dermabrasion of the face, especially after dermashaving, a good smooth granulation and epithelization of the skin could be observed as a positive effect with chamomile extract [22]. Decubitus ulcers, frequently found with paraplegics, are successfully treated with appropriate chamomile bath additives. A special advantage of this form of application is that it does not cause any pain. When judging the therapeutic success, the rapid reduction of the bad smell — due to the necrotizing inflammation — also plays an important role. The chamomile bath has also proved to be a success with the local treatment of deep second-degree burns. Apart from an accelerated cleansing process of a wound a significant improvement of the granulation is also observed. Deep necroses are excised; superficial ones heal without proteolytic ferments [12].

Physicians report from observational studies and case studies concerning experiences with Kamillosan® Crème under practice conditions that the effectiveness and tolerance of the preparation are very good in more than 95% of cases. As with long-term topical application of corticosteroids, the majority of the interviewed people had already noticed side effects such as atrophy, teleangiectasiae, etc. The product was classified as a suitable supplementary therapy and as a possibility of eliminating corticosteroids. According to the indications the elimination ranges from 20–100%, depending on the group of patients.

A controlled clinical study with 161 patients suffering from inflammatory dermatoses on their hands, lower arms, and lower legs (irritative-cumulative dermatitis caused by contact, neurodermitis, allergic eczema caused by contact, lower leg eczema, dyshidrotic and seborrhoic eczema) had already been conducted before the interview action. In a unilateral comparative study it could be shown that an interval therapy with Kamillosan® ointment and 0.25% hydrocortisone ointment is superior to the single therapy with 0.75% fluorcortinbutylester and equieffective with 5% bufexamac ointment [54].

Another use for applying chamomile extract refers to diseases of the anal regions. Here partial baths have a particularly favorable influence on pruritus ani, the perianal eczema of different genesis, and external fistulae [20]. The same good results were achieved by hip baths after anal operations, above all after operations of fistulae. Chamomile extract is also recommended for partial hand baths, especially for the after-treatment of hand injuries and hand operations [28]. For the preoperative cleansing of the intestine, enemas are recommended, containing chamomile as an essential component and protecting against irritations of the mucosa [47]. In a comparative study an irritative dermitis caused by contact with sodium lauryl sulphate and treated with Kamillosan® Creme and 0.1% of hydrocortisone acetate, incorporated into the Kamillosan® ointment base, and the antiphlogistic effect of the three test substances was measured by means of profilometry in view of the reduction of the roughness of the skin [35].

12.2.3 STOMATOLOGY

In dental and oral medicine, as well as in orthodontics, therapy with chamomile extracts is an essential component of the medicamentous therapy for the treatment of gingivitis, stomatitis ulcerosa, and stomatitis aphtosa, i.e., for all inflammatory diseases of the gingiva and oral mucosa. Here the treatment with oral baths is most important, whereas chamomile steam inhalation is used after radical operations in the maxillary sinus, causing a very pleasant feeling [10, 36]. In a review publication about the most important herbal medicinal plant products with inflammation activities to the oral cavity and as an adjuvans for an improvement of the resistance of the oral mucosa, Schenk [50] comes to the conclusion that two of the chamomile preparations analyzed by him belong to the more suitable products that can be compared with sage tincture. At the same time, however, he also criticizes the lack of clinical tests in the region of the oral cavity.

Suitable extract preparations may successfully be applied with refractory diseases of the oral mucosa and the gingiva such as ulcers and aphtae. Decubitus ulcers caused by tartar, films (on the teeth), or badly fitted prostheses disappear quickly. Besides the alcoholic extract, a chamomile ointment can also be used in the region of the oral mucosa, for example, for the massage of gingiva, which proves to be favorable for the treatment of parodontosis [18, 60].

In the Centre for Dermatology and Venerology of the University of Frankfurt/Main, 78 out- and inpatients suffering from different skin diseases and diseases of the mucosa were treated with the product base and rinses, especially irrigations produced from a chamomile extract, over a period of 27 months.

All patients with diseases of the oral mucosa noticed a relief of their symptoms as well as a pleasant cooling effect quickly. The durable and intense influence on existing bad breath (halitosis) could be objectified. On top of that the pains of those patients suffering from habitual aphtae eased remarkably, especially after eating [43]. Inflammatory diseases of the oral mucosa and the throat region (pharynx) can be treated successfully by applying Kamillosan® oral spray [10].

12.2.4 DISCIPLINE OF MEDICINE DEALING WITH EAR, NOSE, AND THROAT (OTOLARYNGOLOGY)

Chamomile therapy is used in otolaryngology as much as in dermatology. First tonsillectomies have to be mentioned, where badly smelling coatings are left for about 10 days. A rinse with chamomile extract leads to an antiphlogistic effect and at the same time to a deodorant effect, so that a local antibiotic treatment finally turns out to be completely superfluous. According to Hinz [27], a standardized ethanolic-aqueous chamomile flower extract is suitable for the adjuvant therapy of *Angina lacunaris* and for the symptomatic treatment of herpangina often occurring in (early) childhood.

Also, after operations of the paranasal sinus [27] and in cases of inflammatory and painful esophageal diseases, chamomile extract has a pain-alleviating effect, and moreover it was ascertained esophagoscopically that the process of healing is quick. After operations and with inflammatory changes of the oral mucosa or those changes caused by radio therapy, chamomile rinses alleviate the complaints. With disturbances of salivary secretion due to x-ray dermatitis of mouth and throat, chamomile extract is alternatingly added to artificial saliva. Rinses of the maxillary sinus may also be carried out only with chamomile extract without application of antibiotics [39, 40]. In an observational report of a municipal hospital, treatment with chamomile extract of at least

10,000 patients suffering from the symptoms mentioned above did not lead to any unpleasant side effects [39].

With sinusitis as well as with children's occult sinusitis, chamomile steam inhalations are recommended [42, 53]. Rinses with chamomile extract proved to be successful with postoperative treatment and as an adjuvant therapy with radiation treatment of the oral, nasopharyngeal, and pharyngeal cavity [16, 19]. Saller [48] reported a significant effectiveness ($p < 0.01$) of chamomile flower steam inhalations, which were produced by means of an ethanol-aqueous chamomile flower extract [48].

Patients who have to undergo radiation treatment in the nose and throat region are frequently suffering from *pharyngitis sicca*, as with this method of treatment the mucosa and submucosa are strongly attacked. Also in this case rinses, especially irrigations with chamomile extract, mean not only a subjective improvement that is apparent by a pleasant cooling effect in the throat and by a reduction of bad breath, but also a pharyngoscopically and laryngoscopically noticeable decongestion of the mucosa and reduction of inflammation. This is also applicable to pharyngitis of other etiological causes. Dryness of the mouth frequently noticed as a side effect of the radiation therapy can be significantly reduced by chamomile extract [11].

In a more recent study at an ear, nose, and throat practice, the effectiveness of Kamillosan® inhalations, especially Kamillosan® oral spray, was tested with various diseases of the oral and pharyngeal cavity (among others, uncomplicated sinusitis, pharyngitis, tonsillitis, glossitis rhinitis, state after tonsillectomy, etc.). The duration of the treatment was between 6 and 9 days. After the treatment with Kamillosan® 96% of all patients noticed a subjective improvement of their complaints [58].

In cases of an inflammatory nasal mucous membrane, a rapid normalization with reduced or no crust formation can be achieved by applying chamomile ointment. The smell is not a disturbing one but is even felt to be pleasant, so that rinses, especially irrigations, can also be carried out. In this case a particular advantage is the fact that when swallowing no unpleasant or harmful side effects will occur [40].

Otitis externa can be treated successfully in children as well as adults by means of chamomile extract.

12.2.5 RADIATION THERAPY

Treatment of radiation damages in the region of the mouth, nose, and throat with chamomile extract — within the scope of an adjuvant therapy — was judged distinctly favorable by the patients [19, 40] as well as were the prophylactic effects [9] on the oral mucosa with patients irradiated or treated with chemotherapeutics ascertained by a U.S. working group. The reactions of mucosa of the rectum resulting from a highly dosed radiation therapy, frequently felt to be unendurable, can also be treated successfully with chamomile extract. For that purpose an enema is given three times a week; besides antiphlogistic properties, this also has a mild cleaning effect [3].

In gynecological radiation therapy, hip baths for the alleviation of painful skin reactions are successfully applied as well [54].

With radiation of large skin surfaces (e.g., mastocarcinoma) radiation erythemata as far as epidermolysis can be observed. Also in these cases application of chamomile extract leads to alleviation of pain during the time of treatment as well as to a quick regeneration of the skin after finishing the radiation [19].

In 1952 treatment of radiation dermatitis by means of ointment containing azulene was reported. As there was no pure chamazulene available from chamomile at that time and as the chamomile ointments did not have enough concentration of active principles, people used an ointment preparation with synthetic azulene (1-isopropyl-5-methylazulene[1]).

---

[1] Kamillen-Bad-Robugen, Producer: Robugen GmbH, Esslingen/N, Germany.

With the radiation of tumors of still-healthy skin, the preparation proved to be the best skin protection so that x-ray dermatitis, even with skin already damaged by radiation, could be avoided to a large extent.

This also referred particularly to the child's skin, which tolerated higher x-ray doses without showing any important formation of erythemata. Badly healing ulcerations disappeared rapidly after application of ointment containing azulene, without having to interrupt the x-ray or radium treatment [25].

At the university hospital of Helsinki the effects of Kamillosan® cream in comparison with an ointment consisting of almond oil were tested with acute reaction of radiation with patients suffering from a mastocarcinoma [38]. With additional treatment of chamomile cream a radiation erythema could not be avoided completely; however, no heavy reactions occurred with the chamomile cream — they could only be observed later, at the end of the radiation therapy.

12.2.6 Pulmology

In pulmology chamomile extract is appreciated for use as an inhalation treatment due to its established antiphlogistic effect. With patients suffering from chronic bronchitis with or without obstruction, after an inhalation treatment therapeutic effects in the tracheobronchial system could be proved in a test of longer duration. The healing process of the inflamed bronchial mucosa is improved, and a bronchial restriction of the lumen possibly existing at the same time goes down.

Inflammable swellings of the mucosa of a nonallergic and abacterial nature caused primarily by noxious agents (just think of the bronchial stimulus states due to chronic inhalation of tobacco smoke) respond well to inhalation therapy by means of chamomile extract [54].

12.2.7 Pediatrics

In pediatrics the protective effect of chamomile preparations on skin and mucosa for babies and infants as well as the antiphlogistic property with diseases of these tissues is most important and well known.

Chamomile extract is outstandingly suitable for delicate skin having the tendency of being dry and forming eczemata.

According to a pediatrician's open report, very good results were achieved with using chamomile ointment for the treatment of napkin dermatitis. In this practice the effect of Kamillosan® ointment on the treatment of various kinds of dermatitis was tested with 76 babies (between 1 and 10 months) and little children: 49 children had napkin dermatitis, especially an inflammation of the skin in the region where napkins are used; 9 of 22 cases healed up completely within a week; another 10 improved considerably. Special emphasis is placed on easing off all complaints such as pains and itching, which affect the general condition very much. Babies' eczemas and perioral dermatitis could also be influenced positively, although within the short time of treatment according to expectations only partial success was achieved [56].

In pediatrics the antiphlogistic effect with inflammations of the mucosa is mainly used for the treatment of sinusitis. Principally the exudative and festering sinusitis is, also in pediatrics, a range of indication for chamomile steam inhalation being one of the most effective remedies. With the inhalation of chamomile steam, produced by boiling chamomile flowers moderately in a lot of water, allergic symptoms can very occasionally be caused by evaporating pollen allergens. Normally, however, inhalation of chamomile steam produced from an alcoholic extract caused no allergic reaction.

With more comprehensive defects of substance and surfaces of inflammation in the region of skin and mucosae, chamomile baths or irrigations are not only useful for the regeneration of the injured integument, but were also subjectively found to give a pleasant feeling. The chamomile bath is an essential part of the treatment of sensitive skin in the anal and genital areas of young babies, but is also meant for cleaning wound and burn surfaces as well as for skin defects and those of the mucosa (Lyell syndrome).

In pediatrics chamomile extract therapy is advisable in the following cases [52, 53]:

- For sensitive skin care of babies and immobilized children as well as seriously ill, chronically ill, and disabled children, mainly suffering from immobilized cerebral pareses, and wetting and defecating the bed.
- For the treatment of an inflamed skin or skin defects. The main fields of indication are, for example, dermatitis ammoniacalis, scald and burn areas and exfoliative dermatitis.
- For the treatment of inflammations of the nose and the paranasal sinus by application of a chamomile bath and inhalation.

12.2.8 GYNAECOLOGY

According to reports of various gynaecological hospitals, chamomile extract proves to be a suitable remedy for the treatment of bartholinitis, vulvitis, and mastitis and in rare cases secondarily healing episiotomies [10].

Hip baths and irrigations are principally indicated for the postoperative treatment of vaginal operation wounds [32, 33] as well as for the therapy of inflammable diseases in the genital area. Reference 44 reports about the antiphlogistic effect of Kamillosan® ointment in comparison with a nonsteroidal ointment in case of episiotomies, with colpitis senilis, and about the improvement of the healing of wounds after surgical operations carried out by laser in gynecology after taking a chamomile (hip) bath.

12.2.9 GASTROENTEROLOGY

The spasmolytic and antiphlogistic effects of chamomile products are taken advantage of when treating gastrointestinal diseases of different kinds. Acute gastritis and enteritis, for example, are regarded as empirical fields of indication for chamomile. Colitis can also be treated successfully with chamomile, and irritations of the colon, for example, respond particularly well if such a state has developed from chronic constipation coming along with spasms. The effect of chamomile extract with diseases of the stomach and the duodenum was clearly proved by a number of tests. Thus, gastro-bioptic [6, 24, 41] and cytologic [8, 37, 61] tests as well as controls of the gastric juice were carried out [8, 41].

In a multcenter study of 104 outpatients with complex complaints of pressure on the stomach, sensation of repletion, eructation, heartburn, loss of appetite, nausea, and sickness without any corresponding organic findings, a 6-week therapy with Kamillosan® was carried out. Further specific medicaments were excluded. It was proved that the most frequent symptoms showed the best rate of success, always provided the symptom in question disappeared completely. As expected, there was the least influence on loss of appetite but 61% of the cases could still be influenced. Side effects and incompatibilities were found in no cases [57]. This means that with mostly nonspecific vegetatively overlapped gastric troubles, the therapy with Kamillosan® only without organic findings is most suitable.

According to Weiss [59], with different modes of gastric troubles that can be classed under the general term of "dyspepsia," the internal administration of chamomile tea or preparations from chamomile extracts is appropriate. With gastric erosions or gastric ulcers the so-called "Roll" method of treatment is recommended, in which the patient, after drinking the chamomile tea, rotates his prone position every few minutes. Apart from the spasmolytic and thereby mainly subjectively analgetic effect, the conducive effect of chamomile to the healing of wounds is much better, as experience has shown. In view of the side effects turning up quite often with the application of modern acid blockers of the cimetidine or ranitidine type (H_2 antagonists) and the relatively high rate of relapse, the chamomile rotation method, which can also be combined well with traditional antacida if necessary, is currently still quite justified [51, 59].

In pediatrics chamomile extract is successfully applied due to its carminative and spasmolytic effect with diseases of the gastrointestinal tract and the effect as such is said to set in immediately after taking the preparation [53]. A comprehensive summary of all therapeutic possibilities of chamomile preparations with diseases of the gastrointestinal tract was published by Schilcher [51].

12.3 PROOF OF EFFECTIVENESS BY MEANS OF FLUVOGRAPHY, REFLEX PHOTOMETRY, AND PROFILOMETRIC JUDGMENT

A publication issued in 1982 reported about successful tests to objectively prove the effectiveness by means of fluvography according to Hensel [26] and the transcutaneous measurement of oxygen according to Eberhardt and Mindt [17]. Chamomile extract containing the active principles of chamomile in a standardized high concentration (according to manufacturers' indications being regarded as effective[1]) was applied on 10 resp. 3 volunteers. Provided the work is done accurately, both testing methods are reliable and objective.

The fluvographical findings achieved after the chamomile extract had an effect showed a reduction of the (blood) circulation of the skin in all 10 cases, manifesting an antiphlogistic effect. Likewise, pO_2 of the skin of the 3 testees decreased as well under the influence of the tested chamomile extract. Although only the hemodynamic part of the antiphlogistic effect can be covered by the reduction of blood circulation of the skin, the findings can be harmonized well by the empirical application of chamomile [55].

The objective proof of the effectiveness of a chamomile cream in comparison with a hydrocortisone ointment was also measured precisely by means of reflex photometry [2]. The medium AUC values for the three test substances differed significantly: 56.5 for the neutral cream foundation, 70.3 for the Kamillosan® cream, and 101.4 for the hydrocortisone ointment. Thus, in this test system the anti-inflammatory effect of Kamillosan® cream reached 69% of the effectiveness of the hydrocortisone preparation [2].

As a third objective measuring system to judge the therapeutic effect of a dermatologic externum the reconstruction of the anatomical structure of the epidermis including the corneous cellular layer is — besides the fading of an erythema — also suitable to be covered objectively. For this the dermatological university clinic in Bonn used a surface measuring instrument as it is applied in metallurgy to judge the roughness of metal surfaces [45]. This so-called profilometric judgment was taken as a comparing study of Kamillosan® ointment, the galenic basis of the ointment, and an ointment with 0.1% hydrocortisone acetate. Of the three ointments only the Kamillosan® ointment had a significant smoothing effect compared with the other two test preparations [45].

12.4 GALENIC PREPARATIONS OF GERMAN CHAMOMILE

The following preparations are produced from the entire plant:

1. From fresh aerial (aboveground) parts of *Matricaria recutita* L. a fresh press juice is obtained, that is mainly sold in health food stores.
2. Dried, herbal parts, the so-called chamomile herb, are finely cut and packed in filter tea bags and sold as foodstuff. Usually the herb is cut after harvesting the flower heads two or three times. The proportion of flower heads in the mixture normally is 5 to 20%.
3. From dried or (more rarely) fresh chamomile flower heads a blue volatile oil is obtained by various steam distillation techniques. "Chamomile (essential) oil" is used for cham-

---

[1] Kamillen-Bad-Robugen, Producer: Robugen GmbH, Esslingen/N, Germany; new trade name is Kamillin® -bath.

omile ointments, creams, and spray preparations. Additionally, it is technically used for the standardization of ethanolic aqueous extracts (chamomile tincture).

4. Dried and pure chamomile flower heads (purified of stems) are used for infusions and herbal teas. Pharmacopoeia-grade chamomile flower heads have to be proved for a minimum content of volatile oil according to the pharmacopoeial monograph.

5. Sterile aqueous extracts of pharmacopoeial-quality flower heads are used for eyedrops, mostly in single-dose quantities.

6. Fluid extracts and tinctures with varying ethanol–water mixtures are prepared from dried or deep frozen flower heads, fluid extracts usually being in the ratio ethanol/water 1:1, tinctures in the ratio 1:5 or 1:10. Extracts of high quality should be standardized on constituents which contribute to efficacy.

   Examples for a well-standardized extract read as follows:

   "100 g of an ethanolic-aqueous extract contain 150–300 mg of blue essential chamomile oil with 50 mg (–)-α-bisabolol and 3 mg chamazulene together with 150–300 mg apigenin-7-glucoside" or "100 g of an ethanolic-aqueous extract contain 170 mg of blue essential chamomile oil with 50 mg (–)-α-bisabolol, along with 10–40 mg free apigenin," and as a third example: "100 g of an ethanolic-aqueous extract contains 200 mg of blue essential chamomile oil, and additionally 150 mg apigenin-7-glucoside."

7. For the preparation of chamomile gels, ointments, and creams the ethanol-aqueous extracts are concentrated to viscous extracts (Latin: *extractum spissum*) and incorporated into the respective dermatologic vehicles.

8. Through further and total evaporation of the liquid of a chamomile extract a dry extract is obtained that is used for the preparation of tablets, capsules, and coated pills.

9. Chamomile tablets are usually prepared of dried and purified powder of chamomile flower heads.

10. For the preparation of chamomile bath oils chamomile flower heads of pharmacopoeial quality are extracted with natural plant oils or neutral oils (e.g., Miglyol).

11. For anthroposophic chamomile preparations the roots of German chamomile are extracted with ethanolic aqueous solvents.

REFERENCES

1. Alban, S., Franz, G. (1990) *Sozialpädiatr. Prax. Klin.,* **12**, 863.
2. Albring, M., Albrecht, H., Alcorn, G., Lücker, P. W. (1983) *Meth. Findings Exp. Clin. Pharmacol.,* **5**, 575.
3. Blumenberg, E.-W., Hoefer-Janker, H. (1972) *Radiologie,* **12**, 209.
4. Born, W.: Personal communication to company Homburg (letter of August 6, 1980), ref. in T. Nasemann, R. Patzelt-Wenczler (Eds.), *Kamillosan im Spiegel der Literatur*, pmi-Verlag Frankfurt/Main (1991).
5. Breinlich, J. (1967) *Dtsch. Apoth. Ztg.,* **107**, 1795.
6. Broi, G. L. da (1960) *Minerva Gastroenter.,* **6**, 147.
7. Brugger, A. W. (1950) Inaug.-Diss., Ludwig-Maximilians University of Munich, Germany.
8. Brühl, W. (1952) *Dtsch. Med. Wschr.,* **77**, 11.
9. Carl, W., Emrich, L. S. (1991) *J. Prostethic Dentistry,* 66, 361.
10. Carle, R., Isaac, O. (1987) *Zschr.-f. Phytoth.,* **8**, 67.
11. Cauwenberge, P. (1979) Expert report, HNO clinic, Gent.
12. Contzen, H. (1975) in Demling, L., Nasemann, T. (Eds.), *Erfahrungstherapie — späte Rechtfertigung*; Verlag G. Braun, Karlsruhe, Germany.
13. Degreef, H. (1977) Expert report, Dermatol. Univ. Klin., Leuven/Belg.
14. Della Loggia, R. (1985) *Deutsche Apoth. Ztg.,* **125**, Suppl. 1, 9–11.
15. Demling, L., Nasemann, T. (Eds.) (1975) *Erfahrungstherapie — späte Rechtfertigung*; Verlag G. Braun, Karlsruhe, Germany.

16. Dewulf, L., de Thibault de Boesinghe, L. (1977) *Tijdschrift voor Geneeskunde, 33,* 169.
17. Eberhard, P., Mindt, W., Jann, F., Hammacher, K. (1975) *Medical and Biological Engineering, 5,* 436–442.
18. Eck, J. (1924) Inaug.-Diss., Heidelberg.
19. Eykenboom, W. (1976) Expert report, Radiotherap. Institut Rotterdam.
20. Frank, H. (1980) Expert report, Regional Hospital Pfarrkirchen/Ndby.
21. Friedrich, H. C. (1979) Expert report, Dermatol. Univ. Clinic, Marburg.
22. Friedrich, H. C. (1978) *Z. Hautkrankheit, 53,* 793.
23. Glowania, H. J., Raulin, Chr., Swoboda, M. (1986) *Z. Hautkr., 62,* 1266.
24. Hammerl, H., Henk, O., Pichler, O. (1962) *Wien. Med. Wschr., 112,* 583.
25. Hellriegel, W., Kreudel, W. (1952) *Strahlentherapie, 86,* 241–248.
26. Hensel, H., Bender, F. (1956) *Pflügers Arch. Ges. Physiol., 263,* 603.
27. Hinz, D. (1995) *Therapiewoche, 8,* 478.
28. Holle, F. (1979) Expert report, Chirurg. Poliklinik Univ. Munich, Germany.
29. Hörmann H. P., Korting, H. C. (1994) *Phytomedicine, 1,* 161.
30. Isaac, O. (1968) *APV Informationsdienst, 14,* 156.
31. Isaac, O., Schimpke, H. (1965) *Arch. Pharm. Mitt.-Dtsch. Pharm. Ges., 35,* 133, 157.
32. Kaltenbach, E.-J. (1980) Expert report, Univ.-Frauenklinik Freiburg.
33. Kepp, R. (1978) Expert report, Univ.-Frauenklinik Gießen.
34. Kohlstaedt, E., Staab, E., Kesper, W. (1946) *Pharmazie, 1,* 218.
35. Kreysel, H. W. (1991) in Nasemann, T., Patzelt-Wenczler, R. (Eds.), *Kamillosan im Spiegel der Literatur.* pmi-Verlag Frankfurt/Main.
36. Kristen, K. (1975) in L. Demling, T. Nasemann, T. (Eds.), *Erfahrungstherapie — späte Rechtfertigung,* Verlag G. Braun, Karlsruhe, Germany.
37. Lange, E. (1959) *Wien. Med. Wschr., 109,* 658.
38. Maiche, A. G, Gröhn, P., Mäki-Hokkonen, H. (1991) *Acta Oncol., 30,* 395.
39. Matzker, J. (1978) Expert report, Hospitals of Köln, Hospital Holweide.
40. Matzker, J. (1975) in Demling, L., Nasemann, T. (Eds.), *Erfahrungstherapie — späte Rechtfertigung,* Verlag G. Braun, Karlsruhe, Germany.
41. Mauro, G. (1958) *Rass.- Ital.- Gastroenter., 5,* 5.
42. Münzel, M. (1975) *Selecta, 24,* 2258.
43. Nasemann, T. (1975) *Z. allg: Medizin, 51,* 1105.
44. Nasemann, T., Patzelt-Wenczler, R. (1991) *Kamillosan im Spiegel der Literatur,* pmi-Verlag Frankfurt/Main.
45. Nissen, H. P., Biltz, H., Kreysel, H. W. (1988) *Z. Hautkrebs, 63,* 184.
46. Patzelt-Wenczler, R. (1985) *Deutsche Apoth. Ztg., 125,* Suppl. 1, 12–13.
47. Richter, R. (1975) *Schweiz. Rundschau med. Praxis, 64,* 689.
48. Saller, R., Beschorner, M., Hellenbrecht, D., Bühring, M. (1990) *Eur. J. Pharmacol., 183,* 728.
49. Sauer, H. (1990) *Jatros HNO, 5,* 11.
50. Schenk, R. (1988) *Zahnärztliche Praxis, 8,* 290.
51. Schilcher, H. (1990) *Dtsch. Apoth. Ztg., 130,* 555.
52. Schilcher, H. (1999) *Phytotherapie in der Kinderheilkunde,* 3rd ed., Wissenschaftliche Verlagsgesellschaft mbH, Stuttgart, Germany.
53. Schmid, F. (1975) in Demling, L., Nasemann, T. (Eds.), *Erfahrungstherapie — späte Rechtfertigung,* Verlag G. Braun, Karlsruhe, Germany.
54. Schmidt, O.-P. (1975) in Demling, L., Nasemann, T. (Eds.), *Erfahrungstherapie — späte Rechtfertigung,* Verlag G. Braun, Karlsruhe, Germany.
55. Sorkin, B. (1982) *Kosmetika, Aerosole, Riechstoffe, 55,* 9.
56. Stechele, U. (1979) Expert report from a pediatric practice. Ref. in Nasemann, T., Patzelt-Wenczler, R. (Eds.) *Kamillosan im Spiegel der Literatur,* pmi-Verlag Frankfurt/Main (1991).
57. Stiegelmeier, H. (1978) *Kassenarzt, 18,* 3605.
58. Troll, W. (1990) *Jatros HNO, 5,* 18. Ref. in Nasemann, T., Patzelt-Wenczler, R. (Eds.) *Kamillosan im Spiegel der Literatur,* pmi-Verlag Frankfurt/Main (1991).
59. Weiss, R. F. (1987) *Kneipp-Blätter, 1,* 4.
60. Weitgasser, H. (1979) *Z. Allg. Medizin, 55,* 340.
61. Zetzschwitz, E. J. (1957) *Münch. Med. Wschr., 99,* 117.

Index

A

Abrasion, mechanical, 174
Acetophenone, 2-hydroxy-4, 6-dimethoxy-, 64
Achillea millefolium, 57, 110
Actidione, 257
Action, medicinal, 28
Active principles, 13, 92, 93, 101, 221, 222, 224, 230, 231, 250, 253, 258, 259, 266, 269, 272
Adelphocoris lineolatus, 171
Administration, 23–25, 27, 33, 246, 248, 258, 271
Adonis aestivalis, 40, 110
Adverse reactions, 33
Afghanistan, 79
AFLP, 102
Agriotes spp., 170
Agropyron repens, 110
Albugo trapogonis, 167
Allergen, 24, 27, 63, 259, 270
Allergenic potential, 24
Allergenicity, 27
Allergic, type IV allergic patients, 27
Allergy, 24, 33, 259, 260
 Compositae, 27, 260
 contact, 24, 27
Allergy type IV, 259
Altitude, 91, 105
Anatolia, 43, 105
Angelica roots oil, 203
Angina lacunaris, 268
Anisic acid, 72, 73
Anthecotulid, 24, 27, 63, 259, 260
Anthemis, 24, 30, 41–45, 48–51, 104, 259
Anthemis arvensis, 42, 43, 45, 47, 48, 51
Anthemis austriaca, 34, 43, 48, 51
Anthemis cotula, 24, 30, 42, 43, 46–48, 51, 63, 259
Anthemis nobilis, 9, 12, 30, 40–45, 167
Anthemis tinctoria, 43, 46, 47, 52
Anthrenus verbasci, 172
Anti-dermatosis agents, 257
Anti-inflammatory effect, 24, 26, 32, 246– 249, 272
Antipeptic activity, 32
Apera spica-venti, 110
Aphid, 169, 171, 172
Aphis fabae, 171
Aphis gossypii, 158

Apigenin, 24, 25, 27, 31, 33, 64–69, 112, 216, 235, 239, 251–253, 258, 273
Apigenin-7-glucoside, 13, 15, 16, 23, 25–27, 33, 64–68, 111, 112, 180, 183, 216, 224, 235, 236, 239, 251–253, 266, 273
Apigenin-7-*O*-glycoside, 33, 64–68, 236, 237, 253
Apigenine-glycoside, 15, 64, 65, 67, 68, 236, 237
Apiin, 64, 68
Apion confluens, 171
Apis mellifera, 155
Appetite, loss of, 271
Application, 14, 17, 23, 27, 28, 32–35, 58, 93, 95, 116, 117, 144, 246–248, 250, 257–259, 266–272
 intradermal, 33
Arabia, 43
Arabino-galactane glycoproteins, 71
Arabinose, 71, 240
Arctium lappa, 110
Argentina, 42, 48, 63, 70, 79, 81, 82, 86, 91, 92, 97–100, 102, 104, 105, 141, 142, 144–146, 148, 150, 151, 154, 162, 249, 259
Arianta spp., 171
Arion spp., 171
Arnica, 260
Aroma / flavoring agents, 29, 127, 202, 203
Aromadendrene, 62
Artemisia, 24
Artemisia arborescens, 56
Artemisia monogyna, 110
Artemisia vulgaris, 27
Arthritis, 248
Arthrodermataceae, 256
Artificial drying, 117, 150
Asclepios, 1
Ash, 14, 30, 71, 127, 128
 insoluble, 30, 127, 128
Asia, 30, 43, 44, 105
Asia Minor, 43, 44, 79
Aspirin activity, 248
Assay, 14, 15, 22, 31, 225, 246, 248, 256
Asteraceae, 29, 33, 150, 156, 246, 259
Astylus gayi, 154
Athous niger, 170
Australia, 43, 79
Austria, 9, 10–12, 43, 101, 103, 105, 222, 223, 225–227